INDUSTRIAL RHEOLOGY

INDUSTRIAL RHEOLOGY

with particular reference to
foods, pharmaceuticals, and cosmetics

P. SHERMAN

Unilever Research Laboratory
Welwyn, Hertfordshire, England

1970

ACADEMIC PRESS · LONDON & NEW YORK

ACADEMIC PRESS INC. (LONDON) LTD
Berkeley Square House
Berkeley Square
London W1X 6BA

U.S. Edition published by
ACADEMIC PRESS INC.
111 Fifth Avenue
New York, New York 10003

Printed in Great Britain by
HARRISON AND SONS LIMITED
BY APPOINTMENT TO HER MAJESTY THE QUEEN,
PRINTERS, LONDON, HAYES (MIDDX.) AND HIGH WYCOMBE

To
Gina Lucy

Preface

Many years ago two admirable texts appeared which provided the practical scientist with a comprehensive introduction to theoretical and experimental rheology and to their application to a wide range of problems relating to materials of industrial importance. The books in question were R. Houwink's "Elasticity, Plasticity, and Structure of Matter" and G. W. Scott Blair's "A Survey of General and Applied Rheology". Since that time there have been many advances in both pure and applied rheology, but the many other rheology books which have appeared in the English language have dealt primarily with advanced theoretical rheology or with a restricted field of application. Consequently, these books have not offered a really suitable modern introduction to rheology for the uninitiated, with previous knowledge neither of the subject nor of higher mathematics. The aim of this book is to fill the void in view of the rapidly growing interest in the application of rheological principles to industrial problems.

The subject matter is arranged within six chapters which cover theoretical rheology, instrumentation for measuring rheological parameters, and some industrial applications of rheological concepts, in this order. Mathematical presentations have been limited to the treatment of those concepts for which it is considered impossible to obtain a true appreciation of their significance by any other means. Chapter 1 deals mainly with the definition of parameters and terms in regular use; also with the analysis of those general phenomena which may be encountered. Chapter 2 describes a wide range of instrumentation which can be used for quantitative or qualitative rheological evaluation of materials. It is not feasible to describe all the available instrumentation within a book of limited size, any more than it is possible to describe all the industrial applications of rheology, and it is doubtful whether this is desirable. Instead, to simplify the approach, and to make it easier for the reader to select a suitable instrument for his purpose, materials are divided into three arbitrary classes, viz. solid, semi-solid, and fluid, and a limited number of instruments is described for studying each group. The range of instrumentation described is sufficiently comprehensive to offer methods for studying all materials that may be encountered, with few exceptions. Sufficient details are provided about the construction and mode of operation of each instrument to enable the reader to evaluate its potential. Chapter 3 is devoted to a detailed survey of the rheological properties of dispersed systems and the factors which influence these properties. Many of the materials described subsequently fall within this category, and such information can be used in the case of the simpler structured materials to interpret their rheological behavior on a molecular level. With materials of more complex structure it is not possible yet to go much beyond a phenomenological analysis of the data. Chapters 4 and 5 aim, as far as it is possible, to illustrate the application of the principles laid down in the preceding chapters to certain food, pharmaceutical, and cosmetic products. Here again it was necessary to limit the range of examples which could be dealt with in some detail, but it is hoped that they will suffice to illustrate the diverse applications of rheological principles, and also the basic similarity between the problems encountered with many different materials.

Foods, pharmaceuticals and cosmetics were selected as the fields of application because the rheological phenomena covering their practical usage have much in common. For example, an analogy can be drawn between the type and magnitude of the shearing forces operating when butter is spread on bread and when a cosmetic cream is spread on human skin. Similarly, a consumer's sensory response when consuming foods or many types of pharmaceutical preparation will be governed by comparable factors. It is hoped that the similarity between many of the examples quoted in Chapters 4 and 5 will impress upon the reader the importance of not restricting his attention to the perusal of published work within his own sphere of interest. A cross fertilisation of ideas is highly desirable, and this can be achieved only by maintaining an awareness of rheological developments in other fields, determining how relevant these are to his own problems, and adapting them wherever possible.

The final chapter discusses the correlation of instrumental and sensory assessment by consumers of consistency and texture during product evaluation. The emphasis here is on the application of the "texture profile" concept, originally developed by scientists at the General Foods Research Center in New York, U.S.A., which analyses the relevant attributes which contribute to the all-embracing term "texture" or "consistency". As yet this approach has been applied only to mastication of foods, but there is no reason why in suitably modified form it should not be applied to the examination of other materials, and their non-masticatory usage. Instrumental evaluations should be made, strictly speaking, under such conditions that the samples are subjected to the same mechanical forces as those which operate during product evaluation by the consumer. Very little information is available at present about these conditions, and much more work is required on this very thorny problem.

The author is deeply indebted to Dr. G. W. Scott Blair for having sacrificed so much of his leisure to read the manuscript of this book. His many comments and suggestions have proved both relevant and beneficial, as have the many stimulating discussions which the author has enjoyed with Dr. Scott Blair over many years.

September 1969   *Philip Sherman*

Contents

Preface vii

CHAPTER 1

NOMENCLATURE AND GENERAL THEORY

1. Types of Mechanical Deformation 1
   A. General Terminology for Very Small Deformations of Solids . . 1
   B. General Terminology for Fluid Flow 8
   C. General Terminology for Linear Viscoelastic Behavior . . . 12
   D. Large Deformations 28
2. Nutting's General Law of Deformation. 30
Bibliography 31

CHAPTER 2

EXPERIMENTAL METHODS

1. Introductory Comments on Rheological Test Methods 33
2. Quantitative Rheological Test Methods. 35
   A. General Comments on Viscometers 35
   B. Limitations of the Four Principle Categories of Commercially Available Viscometers 38
   C. Comparison of Flow Data obtained with the Principle Types of Viscometer 54
   D. Vibrating Reed Viscometer 55
   E. Viscometric Studies at Very Low Shear Rates, and/or Very Low Shearing Stresses 55
   F. Parallel Plate Plastometer 59
   G. Viscometric Measurement of Dynamic Rheological Properties . . 60
   H. Viscometric Study of Stress Relaxation 66
   I. Modulus of Rigidity of Gels 67
   J. Creep Compliance-Time Studies on Solids at Low Shearing Stresses . 69
   K. Dynamic Test Methods for Solids 76
3. Qualitative Rheological Test Methods 81
   A. Viscometers for Fluids 81
   B. Penetrometers 83
   C. Hesion Meter 85
   D. Sectilometer 88
   E. FIRA-NIRD Extruder for Soft Solids 88
   F. Bread and Cereal T.N.O. Panimeter for Solids 90
   G. Rotary Cutter 93
   H. Indentation of Hard Solids by a Falling Sphere 94
Bibliography 95

CHAPTER 3

RHEOLOGY OF DISPERSED SYSTEMS

1. Introduction 97
2. Rheology of Flow 98
   A. Power Laws Relating Stress and Rate of Shear in non-Newtonian Flow   98
   B. Theories of non-Newtonian Flow Based on Structure Breakdown by
      Shear 101
   C. Rheological Behavior at Low Rates of Shear 125
3. Factors Arising From the Constituents of a Dispersed System Which
   Influence Its Viscosity 127
   A. Internal Phase 130
   B. Continuous Phase 157
   C. Emulsifying Agent 158
   D. Rheological Properties of the Adsorbed Film of Emulsifier . . 162
   E. Electroviscous Effect 164
   F. Hydrocolloids, Pigments, and Crystals 168
4. Rheological Properties of Dispersed Systems at Very Low Shearing Stresses 169
5. Rheological Changes in Dispersed Systems During Aging . . . 172
Bibliography 180

CHAPTER 4

RHEOLOGICAL PROPERTIES OF FOODSTUFFS

1. Fats, Margarine, and Butter 185
2. Ice Cream 198
3. Chocolate 207
4. Cake, Bread, and Biscuits 215
5. Cheese 229
6. Fruit and Vegetables 235
7. Dough 252
8. Meat 274
9. Hydrocolloids 288
10. Milk and Cream 305
Bibliography 316

CHAPTER 5

RHEOLOGICAL PROPERTIES OF PHARMACEUTICAL AND COSMETIC
PRODUCTS

1. Introduction 323
2. Rheological Studies on Ingredients of Pharmaceutical and Cosmetic
   Products 327
   A. Suspending Agents 327
   B. Ointment Bases 330
   C. Model Emulsions 335
3. Pharmaceutical and Cosmetic Products 353
   A. Lotions 353
   B. Penicillin Suspensions 358
   C. Vaccines 361
   D. Creams and Ointments 362
   E. Pressurised Foams 365
Bibliography 367

CHAPTER 6

THE CORRELATION OF RHEOLOGICAL AND SENSORY ASSESSMENTS OF CONSISTENCY

1. Introduction 371
2. Consistency Profiling 372
   A. Szczesniak's Classification of Textural Characteristics . . . 372
   B. Amended Classification of Textural Characteristics 377
   C. Basis for Classifying the Consistency Characteristics of Pharmaceutical and Cosmetic Products. 378
3. Correlation of Instrumental and Sensory Assessments of Consistency . 380
   A. Steady Shear Stress Tests 380
   B. Scoring Procedures and Sensory Assessment 381
   C. Instrumental Evaluation of Food Texture with a Modified M.I.T. Tenderometer 383
   D. Magnitude of Shear Forces Operating in Sensory Assessment as a Basis for Instrumental Assessment 385
   E. Usage of Statistical Procedures 390

Bibliography 391

Nomenclature 393

Author Index. 407

Subject Index 415

"Where there is no wisdom there is no reverence, and where there is no reverence there is no wisdom; where there is no understanding there is no knowledge, and where there is no knowledge there is no understanding"

"ETHICS OF THE FATHERS"
Chapter 3

CHAPTER 1

Nomenclature and General Theory

1. Types of Mechanical Deformation · · · · · · · 1
   A. General Terminology for Very Small Deformations of Solids · · 1
   B. General Terminology for Fluid Flow · · · · · · 8
   C. General Terminology for Linear Viscoelastic Behavior · · · 12
   D. Large Deformations · · · · · · · · · 28
2. Nutting's General Law of Deformation · · · · · · 30

Bibliography · · · · · · · · · · · 31

1. TYPES OF MECHANICAL DEFORMATION

A. General Terminology for Very Small Deformations of Solids

1. Stress

Consider a volume element in the shape of a small cube (Fig. 1.1) which has its edges parallel to the x, y, z axes of a rectangular coordinate system. The cube is assumed to be both structurally homogeneous and isotropic, i.e. it has identical properties in all three planes. A tangential force (F) is now applied so that the upper surface and all other parts of the cube, with the exception of the base, are slightly displaced from their original positions. Stress is the internal force acting upon unit area (F/A) of the cube. It acts in three directions on all six faces of the cube, so that there are altogether 18 components of stress. However, the stress components acting on opposite faces of the cube are identical, so that the state of stress is defined by nine components

$$p_{ij} = \begin{pmatrix} p_{xx} & p_{xy} & p_{xz} \\ p_{yx} & p_{yy} & p_{yz} \\ p_{zx} & p_{zy} & p_{zz} \end{pmatrix} \tag{1.1}$$

where p_{ij} is the stress tensor. The tensor concept can be explained as follows. In an isotropic medium, i.e. one in which the properties are identical in all

1

directions, stress and strain are related by a vector equation of the form

stress = constant × strain,

with stress and strain having the same direction. When the medium is not isotropic, however, stress and strain will not be in the same direction. It is necessary then to use a more general mathematical expression in which the constant in the vector equation is replaced by a factor which indicates not only a change in magnitude, but also a change in direction. This factor is known as a tensor.

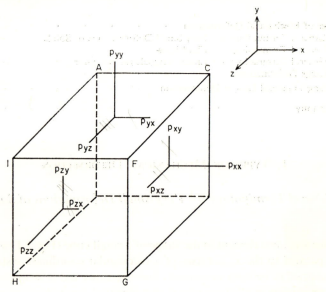

FIG. 1.1. Force components acting on cube shaped volume element

Since $p_{xy} = p_{yx}$, $p_{xz} = p_{zx}$, and $p_{yz} = p_{zy}$, only six independent stress components are required to define the state of stress. Any component which is parallel to the plane in which the stress operates is known as a tangential stress (shearing component), while any component which is perpendicular to this plane is known as a normal stress (tensile component or pressure, according to the direction in which it acts).

2. *Simple Shear*

Deformation of the cube along the x axis can be regarded as arising from the shear in parallel, along straight lines, of infinitely thin layers of the material of the cube.

Referring to Fig. 1.2 it will be seen that the displacement along the x axis is a linear function of the y coordinate, and that

$$\left.\begin{array}{l} \text{displacement } (\gamma) = AB = CD = FE = IJ \\ \text{shear} = AB/IH \;\; = \tan \alpha \\ \qquad \approx \alpha \text{ (angle of shear), when } \alpha \text{ is very small.} \end{array}\right\} \qquad (1.2)$$

Deformation of the cube does not result in any change in its volume.

3. Strain

The strain tensor (ε_{ij}) has six components, as had the stress tensor, and the deformation is defined in three-dimensional Cartesian coordinates by

$$\varepsilon_{ij} = \begin{pmatrix} \varepsilon_{xx} & \tfrac{1}{2}\varepsilon_{xy} & \tfrac{1}{2}\varepsilon_{xz} \\ & \varepsilon_{yy} & \tfrac{1}{2}\varepsilon_{yz} \\ & & \varepsilon_{zz} \end{pmatrix} \qquad (1.3)$$

where

$$\varepsilon_{xx} = \varepsilon_{yy} = \varepsilon_{zz} = 0, \quad \varepsilon_{xy} = \tfrac{1}{2}\gamma, \quad \varepsilon_{yz} = \varepsilon_{zx} = 0.$$

From Fig. 1.1 it is apparent that half the displacement takes the form of strain, and that half takes the form of rotation, therefore, the ε_{xy}, ε_{xz}, ε_{yz}, components of stress should be halved, so that definition of ε_{ij} is somewhat more difficult than for p_{ij}.

The significance of Eq. (1.3) will be more easily understood by considering a volume element with coordinates (x, y, z) in the undeformed cube. When a stress is applied and the cube is deformed, this point moves to new co-ordinates $(x+\xi, y+\delta, z+\zeta)$ with ξ, δ, and ζ defining the components of displacement. When the latter is infinitesimally small, then the strain components can be defined by partial differentials†

$$\varepsilon_{xx} = \frac{\partial \xi}{\partial x}, \quad \varepsilon_{yy} = \frac{\partial \delta}{\partial y}, \quad \varepsilon_{zz} = \frac{\partial \zeta}{\partial z}$$

$$\varepsilon_{xy} = \frac{\partial \xi}{\partial y} + \frac{\partial \delta}{\partial x}, \quad \varepsilon_{yz} = \frac{\partial \delta}{\partial z} + \frac{\partial \zeta}{\partial y}, \quad \varepsilon_{xz} = \frac{\partial \xi}{\partial z} + \frac{\partial \zeta}{\partial x} \qquad (1.4)$$

so that ε_{xx}, ε_{yy}, ε_{zz}, represent the relative elongation of the volume element in the x, y, and z planes respectively, while the strain components ε_{xy}, ε_{yz}, and ε_{zx}, are the elongations in the xy plane (about the z axis), yz plane (about the x axis), and zx plane (about the y axis) respectively.

In an ideal elastic solid the deformation (strain) occurs instantaneously,

---

† Partial differentiation is used when it is necessary to differentiate functions of two or more independent variables; since the variables are independent the effects arising from them can be treated separately.

and it disappears instantaneously when the stress is removed, so that the original geometry is wholly regained. When the applied stress exceeds the yield value the original shape is not wholly recovered when the stress is removed, i.e. a plastic deformation is exhibited, and this involves some structural breakdown. It will be shown elsewhere that both elastic and plastic deformation can be studied by means of mechanical stress–strain–time tests. When a viscous liquid is subjected to a shear stress the displacement which develops continues after the stress is removed, in contrast to the behavior of solids.

4. *Pure Shear*

In simple shear of a cube the two faces in one of the planes remain unchanged (Fig. 1.1), but the four faces in the other two planes rotate to an equal degree around the stationary faces. The deformation component of simple shear, i.e. without the superimposed rotation, is known as pure shear.

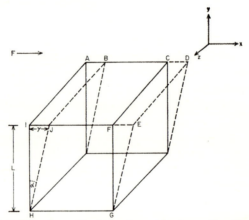

FIG. 1.2. Pure shear of volume element in shape of a cube

Assuming that the faces in the z plane remain unchanged, and extension occurs in the x plane from unity to $1+\varepsilon$, then a contraction will ensue in the y plane from unity to $(1+\varepsilon)^{-1}$. Thus, in pure shear (Fig. 1.2)

$$\varepsilon_{xx} = \varepsilon \qquad\qquad \varepsilon > 0$$
$$\varepsilon_{yy} = (1+\varepsilon)^{-1} - 1 \approx -\varepsilon \qquad\qquad (1.5)$$
$$\varepsilon_{zz} = \varepsilon_{xy} = \varepsilon_{yz} = \varepsilon_{xz} = 0$$

5. *Ideal Elastic Deformation*

An ideal elastic material is one which immediately deforms to some strained condition when a stress is applied, so that no time factor is involved.

The strain components are unique functions of the stress components, and vice versa. When the stress is removed the strain immediately disappears. Under these conditions Hooke's law applies, i.e. there is a linear relationship between stress and strain.

6. *Simple Tension*

When the cube shown in Fig. 1.1 is stretched in one direction only, e.g. along the x axis (Fig. 1.3), the force producing the extension acts only in

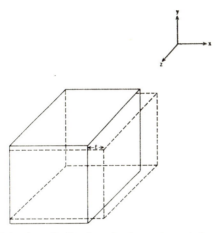

FIG. 1.3. Extension of cube shaped volume element along the x axis

this plane, and not in the y and z planes. The stress tensor p_{ij} is now given by

$$p_{ij} = \begin{pmatrix} p_{xx} & 0 & 0 \\ 0 & 0 & 0 \\ 0 & 0 & 0 \end{pmatrix} \tag{1.6}$$

Extension in the x plane is accompanied by lateral contraction in the y and z planes. If the contractions in the y and z planes are equal, the strain tensor is given by

$$\varepsilon_{ij} = \begin{pmatrix} \varepsilon_{xx} & 0 & 0 \\ 0 & -\mu\varepsilon_{xx} & 0 \\ 0 & 0 & -\mu\varepsilon_{xx} \end{pmatrix} \tag{1.7}$$

where μ is Poisson's ratio, and $\varepsilon_{yy} = \varepsilon_{zz} = -\mu\varepsilon_{xx}$.

7. *Poisson's Ratio*

Poisson's ratio is the ratio between the relative lateral contraction and the relative longitudinal elongation when extension occurs in one plane only, as under the conditions described in Section 5.

$$\mu = -\frac{\varepsilon_{yy}}{\varepsilon_{xx}} = \frac{-(\partial D/D)}{(\partial L/L)} \tag{1.8}$$

where ∂D is the infinitesimal lateral contraction, and ∂L is the infinitesimal longitudinal elongation (Fig. 1.1). When the volume remains unchanged $\mu = \frac{1}{2}$, but for many materials there is an increase in volume and $\mu < \frac{1}{2}$.

8. Shear Modulus

The shear, or rigidity, modulus of a material which undergoes ideal elastic (Hookean) deformation is the ratio between shear stress and shear strain in simple shear (Fig. 1.2)

$$G = \frac{p_{xy}}{2\varepsilon_{xy}} = \frac{(\partial F/A)}{(\partial \gamma/L)}. \tag{1.9}$$

9. Compression (Bulk) Modulus

In compression a uniform isotropic pressure produces a deformation which reduces the volume, but does not alter the shape. The compression modulus is the ratio of the change in hydrostatic pressure to the relative

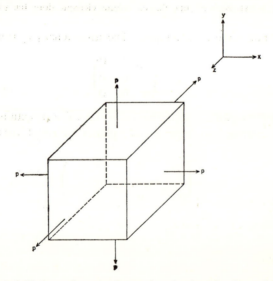

FIG. 1.4. Hydrostatic state of stress in cube shaped volume element

change in volume. The coefficient of compressibility is the reciprocal of the compression modulus.

A hydrostatic state of stress is one in which all normal stresses are equal and all shearing stresses are zero (Fig. 1.4). When the normal stresses are positive the condition is referred to as hydrostatic tension, and when they are negative the condition is referred to as hydrostatic pressure. In Cartesian coordinates

$$p_{ij} = \begin{pmatrix} p_{xx} & 0 & 0 \\ 0 & p_{yy} & 0 \\ 0 & 0 & p_{zz} \end{pmatrix} \tag{1.10}$$

with $p_{xx} = p_{yy} = p_{zz}$. In hydrostatic tension these three stress components are positive, whereas for hydrostatic pressure they are all negative.

$$K = \frac{(\partial F/A)}{(\partial V/V)} = \frac{\partial P^a}{(\partial V/V)} . \tag{1.11}$$

10. Young's Modulus

Young's modulus is the ratio of relative change in normal (tensile) stress to relative change in length (Fig. 1.2)

$$E = \frac{p_{xx}}{\varepsilon_{xx}} = \left(\frac{\partial F/A}{\partial L/L} \right) \tag{1.12}$$

11. Relationships Between Elastic Constants

A completely anisotropic homogeneous solid has 21 independent elastic moduli, although there are 36 in all. For a completely isotropic solid the elastic constants depend on only two independent parameters (λ and α_e), as follows

$$\left. \begin{aligned} k_{11} = k_{22} = k_{33} = \lambda + 2\alpha_e \\ k_{12} = k_{23} = k_{13} = \lambda \\ k_{44} = k_{55} = k_{66} = \alpha_e \end{aligned} \right\} \tag{1.13}$$

the subscripts 11, 22 and 33 refer to the reciprocals of the elongations in the x, y, and z planes which are produced by stress components in these planes. Similarly subscripts 44, 55, and 66 refer to the yz, xz, and xy planes respectively. The parameters λ and α_e are usually known as Lamé's constants.

All four elastic constants, discussed in Sections 7–10, can be defined in terms of λ and α_e

$$\mu = \frac{\lambda}{2(\lambda + \alpha_e)} \qquad (1.14)$$

$$G = \alpha_e \qquad (1.15)$$

$$K = \frac{3\lambda + 2\alpha_e}{3} \qquad (1.16)$$

$$E = \left(\frac{3\lambda + 2\alpha_e}{\lambda + \alpha_e} \right) \alpha_e. \qquad (1.17)$$

It is readily apparent from Eqs (1.14)–(1.17) that the four elastic constants are interrelated, and that provided two of them are known it is possible to calculate the other two. Some of these interrelationships are as follows

$$G = \frac{3EK}{9K - E} \qquad (1.18)$$

$$K = \frac{E}{3(1 - 2\mu)} = \frac{EG}{9G - 3E} = G\left[\frac{2(1 + \mu)}{3(1 - 2\mu)} \right] \qquad (1.19)$$

$$E = \frac{9GK}{3K + G} = 2G(1 + \mu) = 3K(1 - 2\mu) \qquad (1.20)$$

$$\mu = \frac{E - 2G}{2G} = \frac{1 - E/3K}{2} \qquad (1.21)$$

B. General Terminology for Fluid Flow

1. Newtonian Flow

Whereas an ideal elastic solid produces an elastic displacement when a shear stress is applied, a fluid produces viscous flow. Newtonian flow implies that the rate of viscous flow is proportional to the shear stress, but as we shall see later many fluid systems of interest do not fall within this category.

Newtonian flow can be conveniently explained by means of a simple model (Fig. 1.5) which assumes that the liquid consists of many layers in parallel, rather like the model adopted for an elastic solid in Section 1.A.2.

The lowermost layer (*b*) is held stationary, and a force *F* is applied to the uppermost layer (*a*) so that it moves with a constant velocity (*v*) in the direction of *F*. The liquid between *a* and *b* does not move with velocity *v* also, but instead it varies with the distance from *a*. In the layer adjacent to *a* it is still *v*,

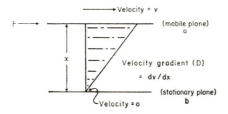

Fig. 1.5. Model to illustrate Newtonian flow

but it then falls away from layer to layer until at *b* it is zero. *F* is opposed by the viscous drag between the parallel layers which are moving with different velocities. The rate of change in fluid velocity is given by d*v*/d*x*, where *x* represents the distance between planes *a* and *b*. It is usually known as the velocity gradient (*D*), which is equal to the rate of shear, while the force per unit area (*F*/*A*) applied to *a* represents the shearing stress (*p*). Following the definition given for Newtonian flow

$$D = \frac{1}{\eta}(p) \tag{1.22}$$

while, on a micromolecular scale, in Cartesian coordinates,

$$p_{xy} = p_{yx} = \eta D$$

and

$$D = 2\frac{\partial \varepsilon_{xy}}{\partial t}.$$

The proportionality constant in Eq. (1.22) is the Newtonian viscosity of the fluid.

2. *Non-Newtonian Flow*

Most materials of practical interest show more complex behavior than Newtonian fluids, with a more complicated relationship between *D* and *p*. In fact, it is not feasible now to talk in terms of viscosity, since it varies with *D*. At any value of *D* the proportionality between *D* and *p* can be represented by an expression which resembles Eq. (1.22)

$$D = \frac{1}{\eta_{\text{app}}} (p) \tag{1.23}$$

but the proportionality factor is now an "apparent viscosity", and it is constant only for this one value of D.

There are four principal types of non-Newtonian flow (Fig. 1.6).

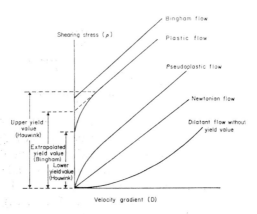

FIG. 1.6. Types of flow behavior

(a) Bingham (1922) pointed out that the D–p curves for many fluids approximated to a straight line with an intercept p_0 on the p axis. Thus

$$D = \frac{1}{\eta_{\text{app}}} (p - p_0) \tag{1.24}$$

where p_0 is the yield value or yield stress. Flow commences only when the shear stress exceeds p_0. Fluids which give this form of D–p plot are called ideal plastic materials or Bingham bodies.

(b) Most fluids show a pronounced curvature in their D–p curves when p_0 is exceeded. In some cases the increase in D grows progressively larger per unit increase in p, i.e. η_{app} decreases, until a limiting value of p is reached. Beyond this point there is a linear relationship between D and p, so that η_{app} is now constant. This type of flow, plastic flow, is exhibited by many dispersed systems which undergo structural alteration during shear.

(c) When there is a curvilinear relationship between D and p of the type described under (b), but flow begins as soon as a stress is applied, then pseudoplastic flow is exhibited.

(d) A D–p curve in which D decreases as p increases may indicate dilatant flow. Alternatively, it may be due to shear thickening. A yield value may, or may not, be present. When there is a yield value the flow is the reverse of plastic flow, and when there is no yield value it is the reverse of pseudoplastic flow. Many powders, and closely packed dispersions are dilatant. When they are sheared the packing geometry becomes looser due to an initial increase in volume before the individual particles can move past one another.

Use of the term "yield value" has been extended in recent years (Houwink, 1958) to define the inflexion point in a plastic flow curve. The intercept on the p axis which is obtained by extrapolating the linear portion of the curve is termed the extrapolated yield value. The value of p at which linear flow commences is called the upper yield value.

Since the viscosities of pseudoplastic, plastic, and dilatant systems vary with rate of shear, viscosity measurements which are made at only one rate of shear have little significance. Particularly when a comparison is being made between the flow properties of two, or more, non-Newtonian systems it is essential that their apparent viscosities should be measured over a wide range of D. Because one of the systems has the larger viscosity at one value of D one should not conclude that this will be the case at all other values of D.

3. *Thixotropy*

When a low D is applied to concentrated dispersed systems it is often found that a steady value of p is not recorded immediately. Instead, the latter increases over a period of time due to structure breakdown until eventually it reaches a steady value, when an equilibrium is established between the rate of breakdown and the rate of structure redevelopment. The time interval required for equilibrium to be established decreases as D increases. If D is now reduced to zero the structure redevelops over a period of time.

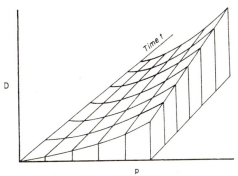

FIG. 1.7. Three dimensional D–p–t diagram (Houwink, 1958)

This phenomenon is often referred to in the literature as thixotropy, although in point of fact, this term was originally introduced to define an isothermal sol (fluid) $\rightleftharpoons$ gel (solid) transformation (Freundlich, 1935). In its present connotation thixotropy is used in a broader sense to describe isothermal reversible structural transitions which are associated with a steady increase in shear followed by a reduction to the original value. Because of the time factor involved in their structural breakdown and redevelopment the flow properties of thixotropic systems should be represented by a three dimensional D–p–t diagram (Fig. 1.7) rather than the two dimensional D–p curves. The viscosity (η_t) at any time t is given by

$$\eta_t = \eta_f + (\eta_i - \eta_f) \exp(-t/\lambda_t) \tag{1.25}$$

where η_i is the initial viscosity, η_f is the steady state viscosity after infinite time, and λ_t is the relaxation time (Umstätter, 1935).

If, instead of keeping D constant, it is first increased over some decades and then reduced, without waiting for the equilibrium structural state to be established at each rate of shear, the upward p curve does not coincide with the downward curve. Instead a hysteresis loop is obtained, which may, or may not, retain the original yield value (Fig. 1.8), depending on the shear history.

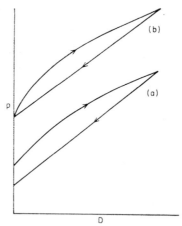

FIG. 1.8. Hysteresis effects in a thixotropic emulsion (a) yield value altered (b) yield value unaltered

C. General Terminology for Linear Viscoelastic Behavior

1. Creep Compliance

Many dispersed systems and solids show both solid (elasticity) and fluid (viscosity) behavior when they are subjected to a sudden, instantaneous, constant shear stress, provided sufficient time is allowed for the test, and the stress is large enough to prevent the sample showing pure elasticity. They behave like solids in the initial stages of shear, and subsequently they exhibit fluid behavior "in the sense that work of shearing deformation is not completely conserved, as in solids, nor is it completely dissipated as in fluids" (Frederickson, 1964). The increase in deformation (strain) as a function of time is called creep, and the creep compliance at any time t is the ratio of the strain to the constant shear stress. When the creep compliance–time curves for different shear stresses all coincide the behavior is linear viscoelastic. With many food materials, for example, the creep compliance at any given time t, where $t > 0$, increases as the shear stress increases. Such materials show non-linear viscoelastic behavior, and interpretation of the experimental data is much more difficult. However, in many cases, provided the strain is very small, the materials will be linear viscoelastic over a narrow range of shear stresses.

Fig. 1.9 shows a typical creep compliance–time curve. It can be subdivided into three principal regions.

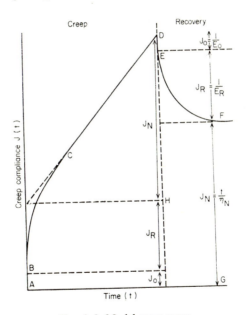

FIG. 1.9. Model creep curve

(1) A region of instantaneous compliance $J_0(A–B)$ in which bonds between the primary structural units are stretched elastically. If the stress was removed at B then the sample would completely recover its original structure.

$$J_0 = \frac{1}{E_0} = \frac{\varepsilon_0(t)}{p} \qquad (1.26)$$

where E_0 is the instantaneous elastic modulus, and $\varepsilon_0(t)$ is the instantaneous strain $(t \to 0)$.

(2) A time dependent retarded elastic region with a compliance $J_R(B-C)$. In this region bonds break and reform. All bonds do not break and reform at the same rate, however, the weaker bonds breaking at smaller values of t than the stronger ones.

In its simplest form, using mean values for the various parameters, the equation for this part of the curve is

$$J_R = J_m[1 - \exp(-t/\tau_m)] = \frac{\varepsilon_R(t)}{p} \qquad (1.27)$$

where J_m is the mean compliance of all the bonds involved, and τ_m is the mean retardation time $(= J_m \eta_m,$ or $\eta_m/E_m)$, where η_m is the mean viscosity associated with retarded elasticity. A more detailed relationship, based on a distribution of bond strengths in preference to a mean value, takes the form

$$J_R = \sum_i J_i[1 - \exp(-t/\tau_i)] = \sum_i J_i\{1 - \exp[-t/(J_i\eta_i)]\} \qquad (1.28)$$

so that $E_m (= 1/J_m)$ and η_m are now replaced by a spectrum of retarded elastic moduli $E_1, E_2, E_3, \ldots E_i$, and viscosities $\eta_1, \eta_2, \eta_3, \ldots \eta_i$, respectively.

(3) A linear region of Newtonian compliance, $J_N(C-D)$. Following rupture of some of the bonds, i.e. the time required for them to reform is longer than the test period, the particles, or units, thus released will flow past one another, and

$$J_N = \frac{t}{\eta_N} = \frac{\varepsilon_N(t)}{p} \qquad (1.29)$$

where ε_N is the strain in this region.

When the stress is removed at D the recovery usually follows a similar pattern to the creep compliance. An instantaneous elastic recovery $(D-E)$ is followed by a retarded elastic recovery $(E-F)$. Since bonds between structural units were broken in the $C-D$ region of the creep compliance–time curve a part of the structure is not recovered. This is represented by $F-G$, which is equivalent to $D-H$.

The whole creep compliance–time plot of Fig. 1.8 is defined in detail by the sum of Eqs (1.26), (1.28) and (1.29)

$$J(t) = J_0 + \sum_i J_i[1 - \exp(-t/\tau_i)] + \frac{t}{\eta_N}. \qquad (1.30)$$

When $t \to \infty$

$$J(t) = J_0 + \sum_i J_i + \frac{t}{\eta_N} \qquad (1.31)$$

When $t \to 0$

$$J(t) = J_0 \qquad (1.32)$$

A detailed analysis of the retarded elastic region ($B-C$) can be made using Inokuchi's (1955) graphical procedure.

In Eq. (1.28) let

$$\sum_i J_i - \frac{\varepsilon_R(t)}{p} = Q \qquad (1.33)$$

then, also

$$\sum_i J_i \exp\left(-t/\tau_i\right) = Q \qquad (1.34)$$

so that Q represents the distance, at any time t, between the extrapolated linear part of the curve and the curved portion.

When $\ln Q$ is plotted against t, for several values of Q, a straight line should be obtained at large values of t, from which a single retardation time (τ_1)

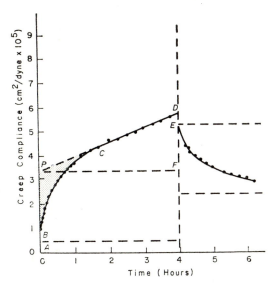

FIG. 1.10. Typical creep curve for ice cream containing 10% fat (Shama and Sherman, 1966)

and creep compliance (J_1) can be derived by extrapolating the line back to zero time. The intercept on the ln Q axis gives J_1, and τ_1 is obtained from the slope of the line. These two values are inserted in Eq. (1.28), and if it does not adequately define the shape of the B–C region of the experimental curve, a second plot is required of ln $[Q - J_1 \exp(-t/\tau_1)]$ against t to determine the magnitude of the second retardation time (τ_2) and J_2. This plot should also be linear at large values of t provided $\tau_1 > \tau_2$. If Eq. (1.28) still does not define the retarded elastic region of the curve when the values

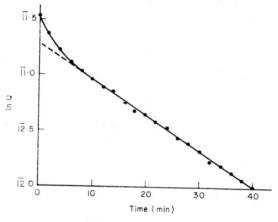

(a) Plot of ln Q v. time

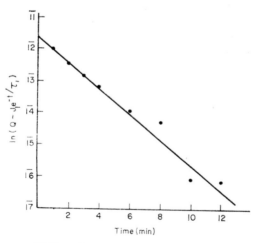

(b) Plot of ln $[Q - J_1 \exp(-t/\tau_1)]$ v. time

FIG.1.11 Analysis of retarded elastic compliance region of Fig. 10 (Shama and Sherman, 1966)

of J_1, J_2, τ_1, and τ_2 are introduced, it is necessary to make a further plot of $\ln [Q - J_1 \exp(-t/\tau_1) - J_2 \exp(-t/\tau_2)]$ against t. This procedure is repeated until such time that sufficient retardation times and compliances have been calculated for Eq. (1.28) to represent adequately the form of B–C in Fig. 1.8.

To illustrate this method for analysing retarded elastic compliance let us examine the creep compliance curve for frozen ice cream (Fig. 1.10). Two plots of the form described were required, i.e. $i = 2$ in Eq. (1.28), for it to give a theoretical curve which closely resembled the retarded elastic region of the experimental curve (Fig. 1.11).

If Eq. (1.27) replaces Eq. (1.28) in Eq. (1.30), then by rearrangement

$$\exp(-t/\tau_m) = \exp[-t/(J_m \eta_m)] = \left[\frac{J(t) - (t/\eta_N) - J_0 - J_m}{J_m} \right] \qquad (1.35)$$

so that if the Naperian (natural) logarithm of the right hand side of the equation is plotted against t, the slope of the line is $1/\tau_m [= 1/(J_m \eta_m)]$. J_m can now be calculated since $J(t)$, J_0, and t/η_N can all be obtained from the creep compliance–time curve, and this leads to η_m.

2. Retardation Spectrum

The mathematical treatment described in the preceding section leads to a finite number of retardation times. When these are introduced into Eq. (1.30) it gives a theoretical creep compliance–time curve which approximates to the experimental curve. In practice, because the basic structural units differ in size a material shows a wide range, or spectrum, of retardation times, but the differences between adjacent times in the spectrum are quite small. The complete spectrum may extend over several orders of magnitude in the time scale.

Equation (1.30) can be rewritten now in integral form

$$J(t) = J_0 + \int_{-\infty}^{\infty} L(\tau) [1 - \exp(-t/\tau)] \, d \ln \tau + \frac{t}{\eta_N} \qquad (1.36)$$

where $L(t)$ is the retardation spectrum. A reasonable approximation for the variation of $L(t)$ with $\ln t$ can be derived by Schwarzl's method (Leadermann, 1958), in which the elastic component of creep compliance $[J(t) - (t/\eta_N)]$ is plotted against $\ln t$ and the gradient of the curve is measured at selected points

$$L(t) \approx \frac{d}{d \ln t} \left[J(t) - \frac{t}{\eta_N} \right]. \qquad (1.37)$$

The spectra obtained in this way take the form of distribution functions which are dimensionless, but actually they have the dimensions of a compliance.

3. *Mechanical Analogues for Describing Creep Behavior*

When a material exhibits linear viscoelastic behavior in creep its shear history can be represented by a mechanical model which, at least qualitatively, behaves in an analogous manner. Electrical models have been used also (see Van Wazer *et al.*, 1963), but as mechanical models are the more common form of representation present discussion is confined to them. Models such as these help one to visualize the fundamental processes which are involved in viscoelasticity.

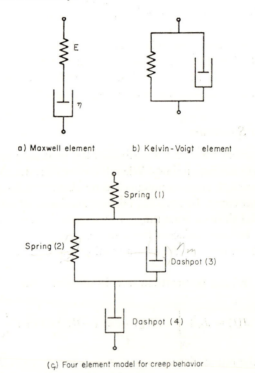

a) Maxwell element   b) Kelvin-Voigt element

(c) Four element model for creep behavior

Fig. 1.12.

The basic elements of any mechanical model are a helical spring which obeys Hooke's law, and a dashpot which consists of a cylindrical container filled with a Newtonian liquid in which a loosely fitting plunger can move upwards or downwards. Shear is represented by an extension of the spring, which has an elastic modulus E_0. The extension of the spring is proportional to the applied force, or shearing stress, while the friction constant of the dashpot represents Newtonian viscosity.

A spring in series with a dashpot forms a Maxwell body (Fig. 1.12a). When a stress (p) is applied the strain (ε) in the dashpot is identical with that on the spring and the strains are additive. The spring stretches instantaneously by p/ε_0 while the dashpot elongates steadily at the rate of p/η_N ($= d\varepsilon/dt$). A spring and dashpot in parallel (Fig. 1.12b) form a Kelvin–Voigt body in which the spring and dashpot undergo equal strains. The dashpot offers a damping resistance to elastic equilibrium, so that the spring now has an elastic modulus E_R and the dashpot has a viscosity η_m (see Eq. (1.27)). Accordingly, the retarded elastic response, as depicted by the Kelvin–Voigt body, is given by

$$\eta_m \frac{d\varepsilon}{dt} + \frac{\varepsilon}{J_m} = p \qquad (1.38)$$

and

$$\varepsilon = pJ_m \{1 - \exp[-t/(\eta_m J_m)]\} = pJ_m [1 - \exp(-t/\tau_m)] \qquad (1.39)$$

Creep compliance with time is represented basically by the four elements (Fig. 1.12c) of a Maxwell body and a Kelvin–Voigt body in series. The total strain when stress is applied is given by

$$\varepsilon(t) = pJ_0 + pJ_m [1 - \exp(-t/\tau_m)] + (pt/\eta_N). \qquad (1.40)$$

According to Fig. 1.12c, if a shear stress is suddenly imposed spring (1) stretches instantaneously. This is followed by the retarded elastic stretching

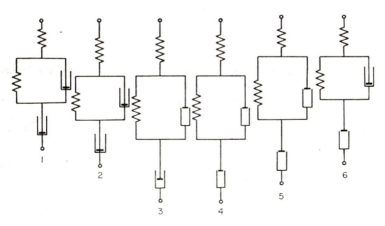

FIG. 1.13. Simple mechanical model showing strain–time behavior

of spring (2), and the gradual extension of dashpot (4) (Fig. 1.13). When the shear stress is removed spring (1) contracts instantaneously to its original length, but spring (2) does so more slowly owing to the damping effect exerted by dashpot (3). Any extension of dashpot (4) is unaffected by removal of the shear stress.

In view of what has already been said about the concept of a spectrum of retardation times, the complete mechanical analogue for linear viscoelastic behavior would contain an infinite number of Kelvin–Voigt bodies in series (Fig. 1.14) to depict the retarded elastic compliance.

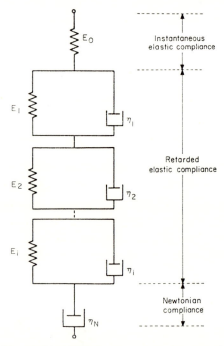

FIG. 1.14. Detailed mechanical model for linear viscoelastic behavior of a W/O emulsion

Non-linear viscoelastic behavior presents a more complex problem of representation by mechanical models because the creep compliance is related to the stress. Reiner (1963) proposed the incorporation of "transmutation" units consisting of a spring in series with a shear pin, which is bridged loosely by a dashpot (Fig. 1.15). When the applied stress exceeds a critical value the shear pin ruptures, and this releases the elastic energy stored in the spring and the deformation is transferred to the dashpot. In practice, of course, there will be a spectrum of shear pin strengths, so that the pins will rupture over a wide range of shear stresses.

4. *Stress Relaxation*

In a creep compliance study the shear stress is kept constant while the strain increases with time, but in a stress relaxation study the strain is kept constant and the stress decreases with time.

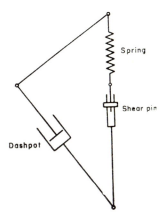

Fig. 1.15. "Transmutation" unit (Reiner, 1963)

If a strain is suddenly imposed on a Maxwell body (Fig. 1.12a), and it is maintained at this degree of deformation throughout the test, then the strain decreases exponentially in accordance with

$$p(t) = p_0 \exp\left(-t/\lambda_t\right) = p_0 \exp\left(-\frac{t}{\eta/G}\right) \tag{1.41}$$

where λ_t is the mean relaxation time, and G is the modulus, or stress/strain ratio. A Kelvin–Voigt body does not show stress relaxation under these conditions, but a combination of this body with a Maxwell body shows stress relaxation according to the complex equation (Alfrey, 1948)

$$p(t) = A_c \exp\left(-\alpha_c t\right) + B_c \exp\left(-\beta_c t\right) \tag{1.42}$$

where A_c and B_c are constants whose values depend on the shear history of the sample, and

$$2\alpha_c = \left[\frac{\eta_m^2}{E_0\,\eta_N} + \frac{E_m}{\eta_m} + \frac{E_0}{\eta_m}\right] + \sqrt{\left\{\left[\frac{\eta_m^2}{E_0\,\eta_N} + \frac{E_m}{\eta_m} + \frac{E_0}{\eta_m}\right]^2 - \frac{4E_m E_0}{\eta_m\,\eta_N}\right\}}$$

and

$$2\beta_c = \left[\frac{\eta_m{}^2}{E_0\,\eta_N} + \frac{E_m}{\eta_m} + \frac{E_0}{\eta_m}\right] - \sqrt{\left\{\left[\frac{\eta_m{}^2}{E_0\,\eta_N} + \frac{E_m}{\eta_N} + \frac{E_0}{\eta_m}\right]^2 - \frac{4E_m\,E_0}{\eta_m\,\eta_N}\right\}}$$

where $E_m = 1/J_m$.

If the sample is deformed instantaneously at $t = 0$ to a deformation d_1, and maintained in this condition, then

$$A_c = d_1\,\frac{E_0{}^2\,(1/\eta_m + 1/\eta_N) - \beta_c\,E_0}{\alpha_c - \beta_c}$$

$$B_c = E_0\,d_1 - A_c$$

Since the relaxation process can be regarded as involving a continuous spectrum of Maxwell elements (Fig. 1.12c) in parallel, the stress relaxation following the deformation d_1 at time $t = 0$ is

$$p(t) = d_1 \int_0^\infty G\,(\lambda_t)\exp\,(-t/\lambda_t)\,\mathrm{d}\lambda_t \qquad (1.43)$$

Following Eq. (1.36) the relaxation modulus is given by

$$G(t) = G_0 + \int_{-\infty}^\infty H\,(\lambda_t)\exp\,(-t/\tau_R)\,\mathrm{d}\ln\lambda_t \qquad (1.44)$$

where $H(t)$ is the relaxation spectrum.

An approximate estimate of $H(t)$ is obtained (Leadermann, 1958) from a relationship which is analogous to Eq. (1.37)

$$H(t) \approx -\frac{\mathrm{d}}{\mathrm{d}\ln t}\,G(t) \qquad (1.45)$$

Both H and L have their distinctive advantages, the former providing greater detail about short time processes, and the latter about longer time processes.

5. *Dynamic Modulus and Viscosity*

In creep tests a constant stress is imposed on the sample at time $t = 0$, while in a stress relaxation test a constant strain is imposed at time $t = 0$. The stress, or strain, can also be made to vary sinusoidally with time, and this technique provides data which are especially valuable for small values of t.

When a stress of frequency v Hz, or ω ($=2\pi v$) radians/sec is applied the amplitude of the strain will be proportional to the amplitude of the stress if the sample is linear viscoelastic. The time t for which the stress is applied is $1/\omega$. Although the strain responds sinusoidally to the stress it will not be in phase with the latter (Fig. 1.16). The strain (ε^*) can be subdivided vectorially

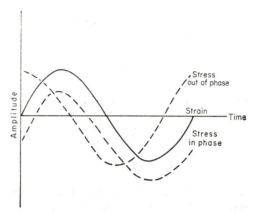

FIG. 1.16. Sinusoidal variation in stress and strain

into two components one of which is in phase with the stress while the other lags 90° behind the stress. Similarly the stress (p^*) can be subdivided into two components one of which is in phase with the strain, while the other is 90° ahead of the strain.

The complex strain is given by

$$\varepsilon^* = \varepsilon_0 \exp(i\omega t) \tag{1.46}$$

and the complex stress by

$$p^* = p_0 \exp[i(\omega t + \delta)] \tag{1.47}$$

where δ is the phase lag between the stress and strain amplitudes (Fig. 1.16). It follows that the complex shear modulus (E^*) at a given frequency, in terms of the stress/strain ratio, is

$$E^* = E' + iE'' \tag{1.48}$$

where E', the ratio of the stress component in phase with the strain/strain, is the storage or real shear modulus, and E'', the ratio of the stress component out of phase with the strain/strain, is the loss or imaginary shear modulus. The moduli for other deformation processes, e.g. bulk or tension moduli, can be defined in a similar way. Alternatively to Eq. (1.48), the complex shear compliance (J^*) is given by

$$J^* = J' - iJ'' \tag{1.49}$$

where J' and J'' are the storage and loss compliances respectively. J' is a measure of the energy stored and recovered per cycle when different systems are compared at the same stress amplitude, and J'' is a measure of the energy lost as heat per cycle (Ferry, 1961).

The ratio of the optimum stress/optimum strain in the amplitude–time curve is given by

$$|E^*| = \sqrt{(E'^2 + E''^2)}.\tag{1.50}$$

Furthermore

$$\frac{E''}{E'} = \tan \delta \tag{1.51}$$

where $\tan \delta$ is the loss tangent in shear. From Eqs (1.48) and (1.51) it follows that

$$E' = |E^*| \cos \delta \tag{1.52}$$

and

$$E'' = |E^*| \sin \delta.\tag{1.53}$$

In addition to a complex shear modulus viscoelastic materials also possess a complex viscosity (η^*). This is defined in a similar way to Eq. (1.49).

$$\eta^* = \eta' - i\eta'' \tag{1.54}$$

where η', the ratio of the stress in phase with the *rate* of strain/strain, is the dynamic viscosity, and η'' is the ratio of the stress out of phase with the rate of strain/strain.

The rate of strain in sinusoidal deformations is

$$\frac{d\varepsilon}{dt} = i\omega\varepsilon_0 \exp(i\omega t) \tag{1.55}$$

so that it follows from Eq. (1.46) that

$$\eta^* = \frac{G^*}{i\omega} \tag{1.56}$$

and that the two components of the complex viscosity are related to the two components of the complex shear modulus in the following way

$$\eta' = \frac{G''}{\omega} \tag{1.57}$$

and

$$\eta'' = \frac{G'}{\omega} \tag{1 58}$$

The mechanical analogues discussed in Section C.3 can now be interpreted in terms of the dynamic shear parameters. A Maxwell element's behavior (Fig. 1.12a) can be defined by

$$E' = \frac{E_m \omega^2 \tau_m}{1 + \omega^2 \tau_m^2} \tag{1.59}$$

$$E'' = \frac{E_m \omega \tau_m}{1 + \omega^2 \tau_m^2} \tag{1.60}$$

$$\eta' = \frac{\eta_m}{1 + \omega^2 \tau_m^2} \tag{1.61}$$

where E_m is the shear rigidity of the spring, and η_m is the dashpot viscosity. For a Kelvin–Voigt body the following relationships apply

$$E' = E_m \tag{1.62}$$

$$E'' = E_m \omega \tau_m = \omega \eta_m \tag{1.63}$$

$$\eta' = \eta_m \tag{1.64}$$

6. Interrelationships Between the Retardation and Relaxation Spectra, and the Dynamic Moduli

When one type of spectrum has been calculated in detail over the complete time scale, the dynamic shear moduli can be derived from the following relationships.

$$E' = G_0 + \int_{-\infty}^{\infty} \left(\frac{H \omega^2 \tau_m^2}{1 + \omega^2 \tau_m^2} \right) d \ln \tau_m, \tag{1.65}$$

$$E'' = \int_{-\infty}^{\infty} \left(\frac{H \omega \tau_m}{1 + \omega^2 \tau_m^2} \right) d \ln \tau_m, \tag{1.66}$$

$$\eta' = \int_{-\infty}^{\infty} \left(\frac{H \tau_m}{1 + \omega^2 \tau_m^2} \right) d \ln \tau_m. \tag{1.67}$$

The loss compliances can be derived by integration from the retardation spectra, in accordance with

$$J' = J_0 + \int_{-\infty}^{\infty} \left(\frac{L}{1+\omega^2\tau_m{}^2} \right) d \ln \tau_m, \tag{1.68}$$

$$J'' = \int_{-\infty}^{\infty} \left(\frac{L\omega\tau_m}{1+\omega^2\tau_m{}^2} \right) d \ln \tau_m + \frac{1}{\omega\eta_N} \tag{1.69}$$

7. Boltzmann Superposition Principle

A linear viscoelastic deformation can be defined as one which obeys the superposition principle, which emphasises that the strain at any time t depends upon the shear history of the sample. It can be explained in the following way. If a stress p_1 is applied at time t_μ, then the shear strain $\varepsilon(t)$ at any later time t will be

$$\varepsilon(t) = p_1 J(t - t_\mu) \tag{1.70}$$

where $J(t)$ is the creep compliance. Now, if instead, various incremental stresses p_i are applied at times t_μ, where $t_\mu < t$, the resultant strain at time t is the sum total of the strains which would be associated with each stress at this time if the stresses were applied independently of each other, so that

$$\varepsilon(t) = \sum_{-\infty}^{t} p_i J(t - t_\mu). \tag{1.71}$$

In integral form, with the initial stress being applied at time $t = 0$, this becomes

$$\varepsilon(t) = J_0 p(t) + \int_0^{\infty} p(t - t_x) \frac{dJ(t_x)}{dt_x} dt_x \tag{1.72}$$

where $t_x = t - t_\mu$

Similar relationships can be developed to explain the influence of the strain history on the stress at any time t

$$p(t) = \sum_{-\infty}^{t} \varepsilon_i H(t - t_\mu) \tag{1.73}$$

where $H(t)$ is the relaxation modulus. The integral form of this equation is

$$p(t) = G_0 \varepsilon(t) + \int_0^{\infty} \varepsilon(t - t_x) \frac{dH(t_x)}{dt_x} dt_x. \tag{1.74}$$

If this concept is applied to creep compliance tests, which will be discussed in experimental detail elsewhere, then, in accordance with Section C.1, at any time t_1 in which bond rupture has led to Newtonian flow

$$\varepsilon(t_1) = p_0 \left(J_0 + J_R + \frac{t_1}{\eta_N} \right). \tag{1.75}$$

The stress is now removed (stress $= -p_0$), and the strain at a later time t_2 is the sum of the original stress p_0 at time t_1 plus the strain due to stress $-p_0$ at time t_2.

In accordance with Eq. (1.71)

$$\varepsilon(t_2) = p_0 \left[J_0 + J_R + \frac{t_1 + t_2}{\eta_N} - J(t) \right] \tag{1.76}$$

so that

$$\varepsilon(t_1) - \varepsilon(t_2) = p_0 \left[J(t) - \frac{t_2}{\eta_N} \right] \tag{1.77}$$

where $[J(t) - (t_2/\eta_N)]$ is the elastic part of the creep compliance curve (Fig. 1.9).

8. Weissenberg Effect

Some viscoelastic fluids show a tendency to flow in a direction normal to that in which the shear stress is applied. This is due to the stress component $(p_{xx} - p_{zz})$ along the line of flow.

In section A.1 it was shown that p_{ij} can be defined by six independent stress components, but the absolute values of the normal components are not important now because of the incompressible nature of the fluid. The state of stress depends, in fact, only on three independent rheological quantities, viz. two differences between pairs of normal stress components $p_{yy} - p_{zz}$ and $p_{xx} - p_{yy}$, and one tangential component of stress (p_{yx}). The other difference between normal stress components $p_{xx} - p_{zz}$ is the sum of $p_{yy} - p_{zz}$ and $p_{xx} - p_{yy}$, and as shown previously, $p_{xy} = p_{yx}$.

Roberts (1953) measured the stress components while subjecting a polymer solution to uniform shear in a Weissenberg rheogoniometer, which is essentially a cone-plate viscometer (see Chapter 2). The stress components p_{yy} and p_{zz} were determined by balancing them against the height of liquid in a series of capillaries and against the pressures exerted by Newtonian fluids respectively. By finding the magnitudes of p_{zz} at various distances (r)

from the axis of rotation the function $\partial p_{zz}/\partial \ln (r/R)$, where R is the value of r at the boundary, was derived. Then, by the principle of force equilibrium

$$p_{xx} = \frac{\partial p_{zz}}{\partial \ln (r/R)} + 2p_{zz} - p_{yy}. \qquad (1.78)$$

It was found for the materials examined that

$$p_{yy} = p_{xx} \qquad (1.79)$$

and, therefore, $\qquad p_{xx} - p_{yy} = p_{xx} - p_{zz} = \dfrac{\partial p_{zz}}{\partial \ln (r/R)} \qquad (1.80)$

For a coaxial cylinder geometry, with the outer cylinder rotating, it can be shown (Lodge, 1964) that

$$r \frac{dp_{zz}}{dr} = p_{xx} - p_{yy} + v_c D \frac{d}{dD} (p_{yy} - p_{zz}) \qquad (1.81)$$

where $\qquad\qquad\qquad v_c \equiv 2 \dfrac{d \log D}{d \log p_{yx}}.$

In steady shear flow $p_{xx} - p_{yy} > 0$, and $p_{yy} - p_{zz} = 0$, so that from Eq. (1.81) it follows that

$$\frac{d}{dr} (-p_{zz}) < 0 \qquad (1.82)$$

i.e. the pressure is greater near the inner cylinder. However, this will not occur in practice as the pressure should be uniform over the horizontal surface of the liquid. In order to produce a pressure distribution along lower horizontal planes, the liquid climbs up the inner cylinder. This, of course, makes viscosity measurements extremely difficult, because the height of liquid between the two cylinders gradually decreases.

D. Large Deformations

The theories of elasticity which have been discussed so far relate to very small strains, so that the distance between the basic structural units, or particles, in the material under test does not change significantly. When the strain is large the stress and strain components can no longer be defined by Eqs (1.1) and (1.4) respectively.

Murnaghan (1937, 1951) made one of the first attempts to develop relationships for finite deformations, and his work gave rise to the second order theory of elasticity. He regarded the classical theory of elasticity as part of a generalised theory which deals with small, though finite, deformations. The location of a particle within the material, according to Murnaghan, can be defined in two alternative ways. In the first, the coordinates are defined in terms of its original unstrained position, whereas by the second way the coordinates are based on the strained positions. According to the latter the strain components ε_{xx}, ε_{yy}, and ε_{yz} now become

$$\varepsilon_{xx} = \frac{\partial \xi}{\partial x} - \frac{1}{2}\left[\left(\frac{\partial \xi}{\partial x}\right)^2 + \left(\frac{\partial \delta}{\partial x}\right)^2 + \left(\frac{\partial \zeta}{\partial x}\right)^2\right] \tag{1.83}$$

$$\varepsilon_{yy} = \frac{\partial \delta}{\partial y} - \frac{1}{2}\left[\left(\frac{\partial \delta}{\partial y}\right)^2 + \left(\frac{\partial \zeta}{\partial y}\right)^2 + \left(\frac{\partial \xi}{\partial y}\right)^2\right] \tag{1.84}$$

$$\varepsilon_{yz} = \frac{1}{2}\left(\frac{\partial \delta}{\partial z} + \frac{\partial \zeta}{\partial y}\right) - \frac{1}{2}\left(\frac{\partial \xi}{\partial y}\frac{\partial \xi}{\partial z} + \frac{\partial \delta}{\partial y}\frac{\partial \delta}{\partial z} + \frac{\partial \zeta}{\partial y}\frac{\partial \zeta}{\partial z}\right) \tag{1.85}$$

These equations are far more complex than Eq. (1.4), but they reduce to the latter when the strain is infinitesimal if one allows for the particular notation employed by Murnaghan (1937).

The stress tensor components are given by

$$p_{ij} = \rho\left(\frac{\partial \theta}{\partial \varepsilon_{ji}} - 2\Sigma \varepsilon_{ik}\frac{\partial \theta}{\partial \varepsilon_{kj}}\right) \tag{1.86}$$

where ρ is the density in the strained condition, and θ is the free energy of deformation. The latter is a function of the components of the strain tensor through its dependency on the strain invariants I_1, I_2, and I_3. If V_0 is the original volume of the sample, and V_s is the volume in the strained condition, then provided the strain is homogeneous

$$\frac{dV_0}{dV_s} = \frac{V_0}{V_s} = \sqrt{(1 - 2I_1 + 4I_2 - 8I_3)}. \tag{1.87}$$

If the deformation of a cube shaped sample is defined by

$$\xi = (\alpha_1 - 1)x, \quad \delta = (\alpha_2 - 1)y, \quad \zeta = (\alpha_3 - 1)z \tag{1.88}$$

then

$$I_1 = \alpha_1{}^2 + \alpha_2{}^2 + \alpha_3{}^2 \qquad (1.89)$$

$$I_2 = \alpha_2{}^2 \alpha_3{}^2 + \alpha_3{}^2 \alpha_1{}^2 + \alpha_1{}^2 \alpha_2{}^2 \qquad (1.90)$$

$$I_3 = \alpha_1{}^2 \alpha_2{}^2 \alpha_3{}^2. \qquad (1.91)$$

The dependency of the function θ on I_1, I_2, and I_3, will vary with different materials, and this relationship can be obtained from experimental stress–strain data.

The majority of quality control tests employed in laboratories for texture or consistency measurement give rise to finite deformations or strains, so that the interpretation of the results from such tests is often difficult. In Chapter 2 the discussion will centre around the replacement of such techniques, wherever possible, by more sensitive ones which depend in their operation upon the development of very small strains. Interpretation of the data then follows from classical theory.

2. Nutting's General Law of Deformation

According to Hooke's law the deformation (s) of a purely elastic body is proportional to the applied force. In fluid flow it is the rate of deformation which is proportional to the applied force. Nutting (1921) pointed out that the former law does not allow for plastic or viscous yield or any other form of energy dissipation, while the latter law assumes complete dissipation of the work done in deformation with no elastic storage of energy. In both cases any second order effects are ignored.

Nutting (1921) found that the materials he examined obeyed the relationship

$$s' = a_N t^B F^M \qquad (1.92)$$

where N and M are functions of temperature, but do not depend on s', t, or F. They are also independent of the dimensions of the sample and the test method, provided it involves only pure shear. The constant a_N is also independent of s', t, and F but it depends on the units and method employed.

When $B = 0$, and $M = 1$, Eq. (1.92) reduces to Hooke's law, while the law of flow is a special case with $B = 1$ and $M = 1$. Experimental values of B found by Nutting (1921) ranged between 0·20–0·91, and M between 0·74–3·5.

When $M > 1$ internal rupture of the structure results in a plastic yield which is more than proportional to F. Conversely, when $M < 1$ the consistency resembles that of quicksand. For low viscosity liquids $B \approx 1$, and when $B \to 0$ it signifies the beginning of the solid state.

Equation (1.92) can be written in two other ways

$$\frac{ds'}{s'} = B \frac{dt}{t} + M \frac{dF}{F} \tag{1.93}$$

or
$$s' \left(\frac{l}{s'} \frac{ds'}{dt} \right)^B = (a_N B)^B F^M \tag{1.94}$$

and Scott Blair (1949) has pointed out that these two equations are "... the simplest hypothesis which we can make to relate expressions for deformations, times, and forces which express all these entities in dimensionless form".

Equation (1.93) pinpoints the errors which may arise by ignoring large second order time effects. Nutting (1921) suggests that the concepts of elasticity and fluid flow can both be incorporated into a generalised law which states that the "percentage deformation is proportional to the percentage change in the force, or log deformation is a linear function of the log of the force producing it".

BIBLIOGRAPHY

Alfrey, Turner Jr. (1948). "Mechanical Behavior of High Polymers" pp. 181-182, Interscience, New York.
Bingham, E. (1922). "Fluidity and Plasticity", McGraw-Hill, New York.
de Waele, A. (1935). *Kolloidzeitschrift* **36**, 332.
Ferry, J. D. (1961). "Viscoelastic Properties of Polymers", Wiley, New York
Frederickson, A. G. (1964). "Principles and Applications of Rheology", Prentice-Hall, New Jersey.
Freundlich, H. (1935). "Thixotropy", Hermann et Cie, Paris.
Gross, B. (1953). "Mathematical Structure of the Theories of Viscoelasticity", Hermann et Cie, Paris.
Houwink, R. (1958). "Elasticity, Plasticity, and Structure of Matter", Dover Publications, New York.
Inokuchi, K. (1955). *Bull. Chem. Soc. (Japan)* **28**, 453.
Leadermann, H. (1958) in "Rheology. Theory and Applications" Vol. 2 (ed. F. Eirich), Academic Press, New York.
Lodge, A. S. (1964). "Elastic Liquids", Academic Press, New York.
Murnaghan, F. D. (1937). *Am. J. Math.* **59**, 235.
Nutting, P. G. (1921). *J. Franklin Inst.* **191**, 679.
Reiner, M. (1960). "Deformation, Strain, and Flow", 2nd edn., Lewis, London.
Reiner, M. (1963). 2nd Annual Report on U.S. Dept. Agric. Project No. UR-A10-(10)-22, p. 20 (Israel Inst. Technol.).
Roberts, J. E. (1953). *Proc. 2nd Intern. Congr. Rheology (Oxford)*, p. 91.
Scott Blair, G. W. (1949). "A Survey of General and Applied Rheology", Pitman, London.
Shama, F. and Sherman, P. (1966). *J. Food Sci.* **31**, 699.
Umstätter, H. (1935). *Kolloidzeitschrift* **70**, 174.
Van Wazer, J. R., Lyons, J. W., Kim, K. Y., and Colwell, R. E. (1963). "Viscosity and Flow Measurement", Interscience, New York.

CHAPTER 2

Experimental Methods

1. Introductory Comments on Rheological Test Methods 33
2. Quantitative Rheological Test Methods 35
   A. General Comments on Viscometers 35
   B. Limitations of the Four Principle Categories of Commercially Available
     Viscometers. 38
   C. Comparison of Flow Data obtained with the Principle Types of
     Viscometer 54
   D. Vibrating Reed Viscometer 55
   E. Viscometric Studies at Very Low Shear Rates, and/or Very Low
     Shearing Stresses 55
   F. Parallel Plate Plastometer 59
   G. Viscometric Measurement of Dynamic Rheological Properties . . 60
   H. Viscometric Study of Stress Relaxation 66
   I. Modulus of Rigidity of Gels 67
   J. Creep Compliance–Time Studies on Solids at Low Shearing Stresses . 69
   K. Dynamic Test Methods for Solids 76
3. Qualitative Rheological Test Methods 81
   A. Viscometers for Fluids 81
   B. Penetrometers 83
   C. Hesion Meter 85
   D. Sectilometer 88
   E. FIRA-NIRD Extruder for Soft Solids 88
   F. Bread and Cereal T.N.O. Panimeter for Solids 90
   G. Rotary Cutter 93
   H. Identation of Hard Solids by a Falling Sphere 94
Bibliography 95

1. INTRODUCTORY COMMENTS ON RHEOLOGICAL TEST METHODS

Rheological measurements are made on materials of industrial interest for one, or more, of the following reasons.

(a) Quality control of the raw materials or ingredients which are used to manufacture a product.

(b) Quality control of a manufacturing process, and/or of the final product.

(c) To study the influence of recipe ingredients, recipe modifications, etc. on rheological properties. Such data may also provide information about relationships between product structure and its rheological characteristics.

(d) To supplement the information provided by panel assessment of product behaviour during practical usage. It will be shown elsewhere, in the last chapter, that suitable rheological measurements could eventually replace the more expert assessments obtained using specially trained panels.

The conditions under which the measurements are made depend on the information which is required. For example, if this relates to the consistency of a semi-solid food product, or a paint, immediately after manufacture then the test should be made under conditions which ensure minimal structure alteration. This is achieved by employing very low shear rates or shearing stresses. Conversely, if the point of interest is the ease with which the food, or paint, spreads on a solid surface then the test should be carried out at much higher rates of shear, or shearing stresses, such that the product suffers appreciable structural alteration. The main problem with the latter type of test is that the actual shear conditions which operate during practical usage of the product are not known, nor can they be calculated at present with any degree of accuracy, so that the shearing stress–rate of shear interrelationship has to be studied over a wide range of medium to large values of the two variables.

Studies on a salad cream or a paint at very low rates of shear, or at low shear stresses, are little more time consuming than those which are made at much higher rates of shear. When dealing with a hard solid product, however, the time difference between tests at low shear stresses and at high shear stresses can be as much as one hour or more. Consequently, the general tendency has been to use high shear stresses irrespective of the type of information which is required. The situation has been aggravated further by the frequent use of empirical test procedures which simulate the mechanical action to which the material is subjected in practice, but which give the result as an arbitrary figure which has no value other than that of indicating whether samples differ in their response to the test conditions.

The published literature contains references to a vast number of techniques. It would appear that there has been an unnecessary proliferation of test methods as new materials or products appeared, with the emphasis generally on the acquisition of qualitative rather than quantitative data. Due to space limitations it is not possible to describe many of these test procedures, nor is it desirable. The approach which will be adopted is one which will attempt a rationalization of test methods, in order to show that the whole range of rheological properties can be covered with a minimum of experimental techniques. For this purpose materials will be regarded as falling within one of three categories—fluid, semi-solid, or solid—although it is readily acknowledged that there is no sharp demarcation between them. Certain materials may not fall within the scope of this classification, but they should be few in number compared with those which do. Table 2.1 enumerates the test methods which are more generally used to investigate the rheological properties of the three categories of the proposed classification. A distinction is drawn between methods which provide merely qualitative information about the material under test and those methods which provide quantitative rheological data, as outlined in Chapter 1. It is readily apparent from Table 2.1 that a single technique can often be used to study more than one category of materials.

2. QUANTITATIVE RHEOLOGICAL TEST METHODS

A. General Comments on Viscometers

A wide range of both commercially available and non-commercial instruments for viscosity measurement have been described in the literature. It is not the purpose of this treatise to describe their construction, or operation, in any detail since this information is already available in excellent texts by, for example, Ferry (1961) and Van Wazer *et al.* (1963). Table 2.2 lists some of the commercially available viscometers, the general principles of their operation, and their limitations. Detailed discussion will be restricted to viscometers which can be built in the laboratory, and which are used to study the more complex rheological properties of materials.

Most viscometers, with the exception of the falling or rolling sphere and the vibrating reed viscometers, will measure both Newtonian and non-Newtonian viscosities. However, the cone-plate viscometer is the only one, due to a small cone angle, which provides uniform shearing conditions throughout the whole of the test sample. In a capillary viscometer the rate of shear varies from zero at the axis of the capillary to a maximum at the capillary wall. When measuring viscosity with a coaxial cylinder instrument one of the two cylinders is rotated. If the outer cylinder, i.e. the cup containing the sample, is rotated the torque transmitted to the inner solid cylinder is measured. In this case the rate of shear varies from a maximum at the rotating cylinder surface to a minimum at the inner cylinder surface, but this shear gradient can be minimized by ensuring that the width of the gap between the two cylinders is as small as possible. If this instrument is to be used to study semi-solid materials then the gap width should not be too small otherwise difficulties will be encountered when trying to introduce a sample into the gap. When the outer cylinder is stationary and the inner cylinder is rotated, the viscosity is derived from the viscous drag which is exerted on the inner cylinder by the sample.

One of the main disadvantages of capillary viscometers is that they cannot

TABLE 2.1

Test methods

Consistency	Quantitative	Qualitative
Fluid		
(a) Newtonian	Viscometers—capillary, coaxial cylinder, cone-plate, variable pressure capillary, vibrating reed, rolling or falling sphere.	Viscometers—rotating spindle, orifice.
(b) non-Newtonian	Viscometers—coaxial cylinder, cone-plate, variable pressure capillary.	
(c) Viscoelastic	Viscometers—coaxial cylinder in normal way, or with appropriate modifications to permit study of creep compliance-time response, stress relaxation, dynamic rheological properties; cone-plate in simple or oscillatory shear.	
Semi-solid	Viscometers—coaxial cylinder in normal way, or with appropriate modifications to permit study of creep compliance-time response, stress relaxation, dynamic rheological properties; cone-plate in simple or oscillatory shear; extrusion capillary.	Penetrometers—cone, rod, sphere.

TABLE 2.1 (continued)

Consistency	Quantitative	Qualitative
Semi-solid	Tensile strength by extrusion.	Hesion meter.
	Modulus of rigidity (weak gels).	
Solid	Creep compliance-time response by parallel plate viscoelastometer, or compression between parallel plates; stress relaxation with same equipment.	Penetrometers—cone, rode, sphere
	Creep compliance-time response by torsion of hollow cylinder.	Sectilometer.
	Stress-strain relationship in elongation.	B.F.M.I.R.A.-N.I.R.D. extruder.
	Modulus of rigidity (gels).	Bread and Cereal T.N.O. panimeter.
	Dynamic rheological properties by torsional vibration of solid cylinder; or vibrating plate viscometer.	Rotary Cutter.
	Micro-tester for tensile strength of crystals.	Indentation by falling sphere.

be used to study the effect on viscosity of the time for which any rate of shear is applied. The sample passing through the capillary changes continuously so that it is not sheared for a significant time. It will be shown elsewhere (Chapter 3) that information of this kind is valuable for elucidating the structure of concentrated dispersions or emulsions. Alternatively, this shortcoming can be put to advantage when dealing with dispersed systems showing time-dependent structure breakdown and recovery if interest centres on the viscosity prior to structure alteration.

Table 2.3 shows the equations which are used to calculate the various parameters associated with viscosity for the four principle classes of viscometer; also the main corrections which have to be made.

B. Limitations of the Four Principle Categories of Commercially Available Viscometers

1. Capillary Viscometers

Corrections have to be made for various phenomena which occur during flow through capillaries when analysing capillary viscometer data.

(a) As the sample passes from the wide reservoir, in which it is held initially, into the very much narrower diameter capillary, the sample is deformed around the shoulders of the reservoir. This effect can be equated with an increase in the effective length of the capillary from L to $L + \Delta L$, where $\Delta L = n_c R_c$, and R_c is the capillary radius, so that the effective shear stress at the capillary wall is given by

$$p = \frac{PR_c}{2(L + \Delta L)} = \frac{PR_c}{2(L + n_c R_c)}. \tag{2.1}$$

Many values of n_c have been proposed within the range 0–1·2. When the ratio of capillary length to radius is very high viz. 200/1 or greater, then the correction for this end effect, or more correctly the entrance effect, is minimized. This effect is then unimportant compared with the pressure drop due to flow through the capillary.

(b) When a dispersed system enters the narrow capillary the dispersed phase may show a tendency to migrate towards the capillary axis. If this occurs there will be a concentration fluctuation across the capillary with the major reduction in concentration in the layers adjacent to the capillary wall. The continuous phase fluid in the latter region then behaves at low rates of shear as a lubricating layer, and the dispersed phase slips along it. Vand (1948) suggested that the net effect was an increase in the volume concentration of dispersed phase by $1/(1 - r/R_c)^2$, where r is the radius of dispersed phase

TABLE 2.2

Commercial viscometers; principles of operation

Type	Principles of operation	Applications and limitations
1. Rotational viscometers **(a) Coaxial cylinders**		
Stormer	Stationary outer cup, inner rotor. Other geometric designs of rotor available. Rotor driven by weights and pulley. Stress varied by applying different weights.	End and edge effects. Also distorted streamlines of flow. Instrument calibrated with fluid of known viscosity. Derived instrument constant includes factors for these effects.
Haake Rotovisko	Fixed outer cup and inner rotor. Several combinations of cups and rotors of different dimensions available. Rotor driven through torque dynamometer. Ten basic speeds from 3·6–582 r.p.m.; using reduction gears can be reduced to 1/10th or 1/100th of these values. No provision for end effects other than overflow from annulus into cup-shaped top of rotor. Some rotors ribbed to prevent slippage, others made of plastic for high temperature studies.	Within restricted viscosity range rates of shear down to 10^{-2} secs^{-1} available. Newtonian and non-Newtonian viscosities can be measured. Also, structure recovery at low shear rate after subjecting to high shear rate. Temperature control not too satisfactory when operating at high shear.
Epprecht (Drage) Rheomat	Fixed outer cup and inner rotating bob of cylindrical shape, but with conical top and bottom. Several combinations of cups and rotors of different dimensions. Viscous drag of inner rotating bob measured via rotation of motor assembly suspended by wire free to rotate against torque spring. Angular deflection of spring recorded. Fifteen driving speeds available. Conical ends reduce turbulent end effects.	End effects present. Width of gap between bob and cup rather large for many combinations so large velocity gradients. Suspension system has a large moment of inertia, so difficult to measure time dependent effects or yield value.

TABLE 2.2 (continued)

Type	Principles of operation	Applications and limitations
Portable Ferranti	Inverted outer rotating cylinder driven by synchronous motor. Viscous drag on inner cylinder measured through calibrated spring with pointer. Nine cylinder combinations of different dimensions available. Five gear speed box gives speeds from 1–300 r.p.m. End effect almost eliminated. Drag on inner cylinder minimised since both cylinders end in same plane.	End effect almost eliminated. Drag on base of inner cylinder negligible. Disagreement between viscosity data obtained with different cylinder combinations when testing low viscosity dispersions at same mean rate of shear. Not a very robust instrument.
(b) Rotating spindle Brookfield Synchro-Lectric	Measure viscous traction on spindle rotating in sample. Spindle (up to 7 models available) driven through a beryllium-copper spring by synchronous motor. Geometry of spindle results in wide range of shear rates in sample.	Geometry of spindle makes it extremely difficult to calculate shear rate. Not suitable for absolute measurements of viscosity. End and edge effects; also distorted flow streamlines.
(c) Cone Plate Ferranti-Shirley	Rotating small-angled cone and stationary lower flat plate. Three cones available—large, medium and small. Apex of cone just touches plate, and sample sheared in gap between. Viscous drag on cone exerts torque on electro-mechanical torque dynamometer. Less than 0·5 ml. sample required.	Uniform shear rate throughout sample provided cone angle small. Suitable for measuring Newtonian and non-Newtonian flow at shear rates exceeding few secs^{-1}. Particularly suited to high shear rate measurements. Practically no end effect: sample held in gap by surface tension.
Weissenberg Rheogoniometer	Cone rigidly fixed while flat lower plate rotates. Selection of cone angles and platen diameters. Flat plate rotates and torsion (i.e. tangential stress) imparted to cone measured; also normal force acting on platen. Two synchronous motors drive 60 speed gearbox. Very small volume of sample necessary. In oscillatory tests platen oscillates about its axis, and oscillatory motion transmitted to cone via sample.	Uniform shear rate throughout sample. Suitable for measuring Newtonian and non-Newtonian flow. Wide range of shear rates from about 10^{-4}–18×10^{-3} secs.$^{-1}$ Oscillatory shear can be used to determine elastic and viscous components.

TABLE 2.2 (continued)

Type	Principles of operation	Applications and limitations
Haake Rotovisko	Rotary cone and stationary plate. Three cones in stainless steel available. Torque developed when cone rotates measured with dynamometer.	Suitable for measuring Newtonian and non-Newtonian flow, particularly at very high rates of shear. Under these conditions danger of frictional heating effects.
2. Capillary viscometers (a) Glass capillary Ostwald U-tube	Reservoir bulb from which fixed volume of sample flows down through capillary to receiver bulb at lower level in other arm of U-tube.	Small driving force since operates by force of gravity. Low shear stress. Suitable for measuring Newtonian viscosity. Wide range of viscosities covered by capillaries of different dimensions.
Cannon-Fenske	Reservoir and receiving bulbs lie in same vertical axis. Sample of fixed volume flows from reservoir bulb through capillary to reservoir bulb. Range of capillary lengths and diameters available.	Design reduces errors introduced when viscometer not aligned correctly in vertical plane. Only suited in this form to Newtonian viscosity measurement. In conjunction with external source of pressure non-Newtonian viscosities can be measured.
Ubbelohde	U-tube viscometer with third arm. Principles of flow as above. Range of capillary lengths and diameters available.	When liquid emerges from capillary flows only over walls of lower bulb forming a "suspended level". As result not necessary to use fixed volume of sample for test. Suited to Newtonian viscosity measurements.

TABLE 2.2 (continued)

Type	Principles of operation	Applications and limitations
(b) Variable pressure		
Instron Rheometer	Sample forced out of container chamber through capillary by plunger fastened to moving crosshead which can move at various speeds. Series of capillaries with different diameters and lengths. Force required to move plunger at each speed detected by load cell situated on crosshead.	Wide range (5 decades) of shear-rates and shear stress available. Suitable for measuring Newtonian and non-Newtonian viscosities. Suitable for high shear rate studies.
Techne Printing Viscometer	Sample forced by air pressure through horizontal capillary tube. Air displaced by this motion moves drop of water along another horizontal tube, and time taken to pass between two graduations noted. Air pressure, or suction, produced by vibrated dead weight gauge. In automated model photo-electric cells control electric timer and printer.	Suitable for Newtonian viscosities within range 0·003 — to 2000 poises. Only small volumes of sample necessary. Using alternately air pressure and suction repeated tests can be made on the same sample.
Bingham	Sample extruded through capillary by air pressure, or other pressure source, into receiver. Rate of flow measured by air displacement from receiver which is recorded by flowmeter. Flow rates noted over wide range of pressure.	Suitable particularly for high shear studies, but not for low shear work.

TABLE 2.2 (continued)

Type	Principles of operation	Applications and limitations
3. Orifice viscometers Redwood Saybolt Engler	Consists essentially of reservoir, orifice, and receiver. Time measured for fixed volume of sample to flow through orifice.	Efflux time taken as arbitrary measure of viscosity, although they are not related in a simple way. Cannot be used for absolute measurements.
4. Rolling sphere Höppler	Cylindrical glass tube contains sample. Time taken for steel ball bearing to roll through middle, graduated, section of tube. This ensures that ball has steady velocity before reaches test section. Set of 6 ball bearings of different diameters available. If liquid sample is opaque time can be determined electrically.	Difficult to define shear stress and rate of shear. Suited only to Newtonian viscosity measurements.
5. Vibrating reed Ultraviscoson	Thin alloy steel blade at end of probe vibrates at ultrasonic frequency. Ultrasonic shear waves develop in sample surrounding blade. Computer translates energy required to produce this motion into viscosity. Automatic temperature compensator.	Only one rate of shear, so suited only to Newtonian viscosities or for detecting deviation from fixed viscosity. Multi probe computer available for taking several viscosity measurements at turn of switch.
6. Penetrometers Cone (Hutchinson) or needle	Rod with needle, or cone, attached drops vertically into sample, after release from stationary position above sample. Graduated scale records penetration depth. With cone penetrometer smooth angled cones recommended rather than A.S.T.M. type.	Geometry of penetrating body complex. No theoretical treatment of even Newtonian viscosity determinations attempted.

TABLE 2.3

Equations for calculating the relevant parameters for the four principle classes of viscometer

Viscometer	Capillary	Coaxial cylinder	Cone-plate	Rolling, or falling, sphere
Newtonian viscosity η	$\eta = \pi P R_c^4 / 8LQ$	$\eta = G_T \dfrac{\left(\dfrac{1}{R_1^2} - \dfrac{1}{R_2^2}\right)}{4\pi h_c \Omega_a}$	For small values of ψ_a $$\eta = \frac{3G_T}{2\pi R_p^3}\Big/ \frac{\Omega}{\psi_a}$$	$\eta_s = \dfrac{2}{9}\left(\dfrac{\rho_1 - \rho_2}{v_x}\right) gR_s^2$
Non-Newtonian viscosity η_∞	$\eta_\infty = \dfrac{\pi R_c^4 g\sigma}{8Lk} \; \dfrac{1}{\dfrac{dQ}{dP} + \dfrac{P}{4}\cdot\dfrac{d^2Q}{dP^2}}$	$\eta_\infty = \dfrac{M}{4h_c \Omega_a}\left(\dfrac{1}{R_1^2} - \dfrac{1}{R_2^2}\right)$ $- (D/\Omega_a)\ln(R_2/R_1)$		No mathematical theory developed

TABLE 2.3 (continued)

	Capillary	Coaxial cylinder	Cone-plate	Rolling, or falling, sphere
Rate of shear D	$D_{capillary\ wall} = \dfrac{4Q}{\pi R_c^{3}}$ $D_{axis} = 0$	$D_{max} = \dfrac{2\Omega_a}{R_1^{2}\left(\dfrac{1}{R_1^{2}} - \dfrac{1}{R_2^{2}}\right)}$ $D_{min} = \dfrac{2\Omega_a}{R_2^{2}\left(\dfrac{1}{R_1^{2}} - \dfrac{1}{R_2^{2}}\right)}$	$D = \dfrac{\Omega}{\psi_a}$	$D_{max} = \dfrac{3v_x}{2R_s}(\sigma_a = 90^{0})$ $D_{min} = 0\ (\sigma_a = 0°)$
Shear stress p	$\dfrac{PR_c}{2L}$	$\dfrac{\left(\dfrac{1}{R_1^{2}} - \dfrac{1}{R_2^{2}}\right)}{4\pi h_c}$	$\dfrac{3G_T}{2\pi R_p^{3}}$	
Yield value p_0	$\dfrac{P_0\, g\sigma R_c}{2L}$	$\dfrac{pG_2}{\ln (R_2/R_1)}$	pG_2	
Principal corrections required	Entrance (end) effect Kinetic energy imparted to sample Wall effect Turbulent flow Plug flow	End effect Turbulent flow in end region		End effect Wall effect

G_2 = Extrapolated value of torque for $\Omega_a = 0$; σ = Density of mercury; k = Constant for converting flowmeter readings into ccs/sec efflux; $P_0 = P_{D\to 0}$.

particles, but this correction may define the concentration change only in the central part of the capillary with radius $R_c - r$, and *not* the concentration change over the whole of the capillary (Higginbotham *et al.* 1958).

The true relative viscosity (η_{rel}) is given by

$$\eta_{rel} = \left[1 + H\left(\frac{1}{\eta_{rel\,(a)}} - 1\right)\right]^{-1} \tag{2.2}$$

where H is the correction factor which has to be applied for the layer of fluid near the capillary wall, and $\eta_{rel\,(a)}$ is the apparent relative viscosity. The factor H is given by

$$H = \left(1 - \frac{\Delta}{R_c}\right)^{-4} \tag{2.3}$$

where Δ is the thickness of the hypothetical layer of fluid, and R_c is the capillary radius. The value of Δ is derived by making viscosity measurements in two different capillary viscometers with capillary radii R_c^x and R_c^y respectively, since

$$\frac{H_x}{H_y} = \frac{1 - (\eta_{rel\,(a)}^y)^{-1}}{1 - (\eta_{rel\,(a)}^x)^{-1}} \tag{2.4}$$

and,

$$\left(\frac{H_x}{H_y}\right)^{-4} = \frac{1 - (\Delta/R_c^y)}{1 - (\Delta/R_c^x)} \tag{2.5}$$

and H_x, H_y, R_c^x, and R_c^y are known.

(c) Not all the applied pressure is used in shearing the sample, since some of it imparts kinetic energy to the sample as it enters the capillary. This effect exerts a greater influence than the entrance effect. The actual pressure (P_a') used to shear the sample is given by

$$P_a' = P - \frac{\rho Q^2}{\alpha_k \pi^2 R_c^4} \tag{2.6}$$

where ρ is the sample density, α_k is the kinetic energy correction factor, and Q is the volume of sample which flows through the capillary per second. When the sample exhibits Newtonian flow α_k is 1, provided that the velocity distribution is parabolic. Published values of α_k range between 0·7 and 2·0, and this may be due to irregularities in capillary bore, and end effects, producing deviations in the form of the velocity distribution (Van Wazer *et al.* 1963).

If flow through the capillary is of the non-Newtonian type which can be defined by a power law type equation (Chapter 3, Section 2A), then

$$\alpha_k = \frac{(4c_a+2)\,(5c_a+3)}{[3(3c_a+1)^2]} \tag{2.7}$$

where c_a is a constant.

Alternatively, if the flow approximates to a Bingham plastic

$$\alpha_k = \frac{2}{(2-p_0/p)} \tag{2.8}$$

where p_0 and p are equal to $PR_c^0/2L$ and $PR_c/2L$ respectively, and R_c^0 is the distance from the capillary axis over which plug flow occurs. The stress p_0 is that required to induce flow at a distance R_c^0 from the capillary axis.

(d) Close to the capillary axis R_c^0 will, of course, be very small, so that P will have to be extremely large if p_0 is to be exceeded. This is not possible with normal experimental conditions, and consequently there is always a thin layer of sample around the capillary axis which moves through the capillary as a solid plug (plug flow). Both Buckingham (1921) and Reiner (1926) proposed correction factors for this phenomenon in their analyses of non-Newtonian flow through capillaries,

$$\frac{1}{\mu_v} = \eta_p = \frac{\pi R_c^4}{8LQ}\left[p - \frac{4(P')}{3} - \frac{(P')^4}{3(P')^3}\right] \tag{2.9}$$

where (P') is the pressure which is required to attain p_0. When $P \gg P'$, the term $(P')^4/3(P')^3$ can be neglected, so that

$$\frac{1}{\mu_v} = \eta_p = \frac{\pi R_c^3}{4Q}\left(p - \frac{4}{3}p_0\right) \tag{2.10}$$

(e) The derivation of Poiseuille's (1840) equation for Newtonian fluids (Table 2.3) is based upon the assumption that flow is laminar throughout the capillary. Reynolds (1883) showed that the deviations which appear at high rates of flow are due to a change from laminar, or streamline, flow to turbulent flow. The change occurs when the Reynolds number (Re) exceeds 2000, where

$$(Re) = \frac{2v\rho R_c}{\eta} \tag{2.11}$$

and v is the mean velocity of the sample $(= Q/(\pi R_c^3))$. Flow is laminar when $(Re) < 1000$, and also in the intermediate region of $(Re) = 1000–2000$

provided the sample undergoes no disturbance when entering the capillary. In non-Newtonian flow the transition to turbulent flow often occurs at about the same (Re) as for Newtonian flow.

(f) Tests on a dispersed system using viscometers with capillaries of widely different dimensions may yield quite different stress–rate of shear curves. It appears to be the capillary radius which is of primary importance in some cases, and then variations in capillary length have little effect (Scott Blair, 1958). As the capillary radius decreases, so does the experimentally determined viscosity. One explanation offered for this effect suggests that the Poiseuille, Buckingham–Reiner, etc. equations are based on integrations which assume that the shearing layers are infinitesimally thin. This assumption is not valid when the particles in a dispersed system are relatively large compared with the radius of the capillary (Din and Scott Blair, 1940).

(g) The rate of shear at the capillary wall is given as $4Q/(\pi R_c^3)$ in Table 2.3, but this is true only for Newtonian flow. For all other conditions the true rate of shear is derived (Rabinowitsch, 1929) by multiplying the Newtonian value by a correction factor $(3+b)/4$, where b is the slope of a plot of $\log [4Q/(\pi R_c^3)]$ against $\log [PR_c/(2L)]$ (capillary wall). If b does not change over a wide range of shear stress a single calculation provides the information required; otherwise, the plot must cover many values of shear stress.

(h) Wellman et al. (1966) point out that for those capillary viscometers in which all, or part, of the driving force is a varying hydrostatic head allowance must be made for the interaction of the hydrostatic head and kinetic energy effects, i.e. for the variation in the fraction of total pressure which is lost as kinetic energy of the sample with the changing hydrostatic head. This correction is especially important when the viscosity of the sample differs greatly from the viscosity of the material used to calibrate the viscometer. They derived a rather complex equation for the effective pressure, but unfortunately its use is limited to capillary viscometers with cylindrical reservoirs. The derivation is based on a differential equation for the change in volume with time, with an additional term to allow for that part of the pressure which is not used to overcome viscous resistance.

The average pressure (P_a') which has to be introduced into the Poisseuille equation (Table 2.3) is given by

$$P_a' = \frac{C}{Z_0 - Z + \ln[(Z_0 - 1)/(Z - 1)]} \qquad (2.12)$$

where

$$Z = \left[1 + \frac{\alpha_k \rho R_c^4}{16L^2 \eta^2}(P_s - C)\right]^{\frac{1}{2}}$$

and
$$Z_0 = \left[1 + \frac{\alpha_k \rho R_c^4}{16 \, L^2 \eta^2} P_s \right]^{\frac{1}{2}}$$

where C is the total change in hydrostatic head, and P_s is the algebraic sum of the applied pressure and the initial hydrostatic pressure.

2. Coaxial Cylinder Viscometers

(a) An examination of the forces exerted by the sample on the curved surfaces of the rotating cylinder leads to the equation for flow between coaxial cylinders (Table 2.3). However, if the inner cylinder is wholly immersed, forces are exerted also on both ends. If the upper end of the inner cylinder is not covered with sample, the effect due to the forces acting on the lower end can be determined by varying the length of the inner cylinder which is immersed in the sample and finding the torque/angular velocity ratio. When this ratio is plotted against the depth of immersion (h_c) a linear relationship is obtained which gives a negative intercept on the h_c axis. This represents the correction Δh_c which has to be added to h_c in the equation for Newtonian flow (Table 2.3). The end effect may assume greater importance in non-Newtonian flow since the rate of shear at the ends of the inner cylinder is lower than in the gap between the two cylinders. Thus, the viscosity is higher at the ends of the inner cylinder. In this case, an end correction should be determined for each sample at every value of Ω which is used.

(b) The end effect increases when flow at the ends of the inner cylinder is turbulent. For Newtonian viscosities greater than 1 poise the correction is almost constant, but when the viscosity is less than 1 poise the effect increases (Lindsley and Fischer, 1947).

(c) Dispersed systems may show slippage at the cylinder surfaces due to phase separation (Van Wazer *et al.* 1963). If this occurs then the viscosity of the fluid phase is measured and not the viscosity of the homogeneous dispersion. To prevent slippage, the surfaces of the cylinders should be either roughened or ribbed depending on the size of the particles in the dispersion. When studying the flow properties of dispersions with large sized particles the surfaces of the cylinders should be ribbed; when the particles are small, roughened surfaces are quite adequate.

Van Wazer *et al.* (1963) suggested that smooth cylinder surfaces contain many very small crevices at the atomic level. If small particles are homogeneously dispersed in a fluid medium many of them are entrapped in the cylinder surfaces when the dispersion is introduced into the viscometer, so that slippage cannot occur. However, if the particles are relatively large they cannot be entrapped within the crevices. Slippage then ensues because there is a separation of fluid at the cylinder surfaces which prevents particles from being carried around with the rotating cylinder. The width of the gap in a coaxial cylinder viscometer should be 10–100 times the largest diameter particles in the dispersion. Thus, when dealing with slurries of paper pulp, cellulose fibres etc., the gap width should be several centimetres. Morrison and Harper (1965) studied the flow behaviour of suspensions of fibrous particles in aqueous glycerol and lubricating oil, and concluded from their data that the fluid film adjacent to a smooth cylinder wall was approximately 10–40 μ thick, the thickness depending to some extent on the shearing stress. Another interesting fact emerged from the observed change in shearing stress with time. "The usual concept of thixotropy relates time effects to breaking down a fluid structure by shear. The time effects observed were clearly related to establishment or breakup of a wall film rather than to any gross change in structure. The rough surface actually produced an increase in stress with time, completely opposite to the effect of thixotropy".

Kuno and Senna (1967) studied the flow properties in a MacMichael viscometer of dry spherical particles of polymethylmethacrylate in the size ranges 125–210 μ, 210–350 μ, and 70–300 μ after compaction. The surface of the sample was marked with carbon black or a black dye, and the cup was then rotated. The powder divided into two parts, an inner one containing particles which flowed continuously or at least moved relative to each other, and an outer region, separated by a slipping plane from the inner region, which moved as a whole with the cup.

3. Rolling Sphere Viscometer

(a) Newtonian viscosity is calculated using Stokes' law from data relating to the descent of a sphere through the sample at a constant speed v_x cm/sec (Table 2.3). If the sphere has a radius (r_s) which is relatively large in comparison with the radius (R_s) of the tube through which it falls a more complex equation has to be used because of the drag imposed by the tube wall. As a first approximation (Ladenburg, 1907) a correction may be applied to v_x so that

$$v_c = v_x \left(1 + 2 \cdot 4 \frac{r_s}{R_s}\right) \qquad (2.13)$$

where v_c is the corrected value of v_x. Faxén (1922) suggested that the constant in the bracket should be 2·1 rather than 2·4. In addition, he amended Eq. (2.13) to a form which for Newtonian fluids gave viscosities (η_F) in close agreement with viscosities obtained by capillary viscometers,

$$\eta_F = \eta_T \left[1 - 2 \cdot 104 \left(\frac{r_s}{R_s}\right) + 2 \cdot 09 \left(\frac{r_s}{R_s}\right)^3 - 0 \cdot 95 \left(\frac{r_s}{R_s}\right)^5\right] \qquad (2.14)$$

where η_T is the viscosity according to Stokes' law. Eq. (2.14) is valid for r_s/R_s ratios up to 0·32 (Bacon, 1936).

Recently Scott Blair and Oosthuizen (1960) proposed that structured and thixotropic systems could be studied conveniently by using spheres which had much smaller diameters than the cylindrical tube. Several tests could then be made on a single sample by rotating the tube carefully between individual tests so that unsheared material became available for further tests. Viscosities so determined agreed well with viscosities derived by U-tube and other viscometers. Nylon spheres were normally used, but when the difference between the density of the sphere and the density of the sample was very small, thereby introducing the possibility of serious errors, metal spheres were used. When flow behaviour deviates only slightly from Newtonian then spheres of different radii and densities can be used (Williams and Fulmer, 1938).

(b) One of the basic assumptions employed in the development of Stokes' law is that the tube into which the sample is introduced has an infinite length. In practice, of course, this is not possible so that the effect due to the two ends of the tube must be taken into account. This effect is proportional to r_s/d_s, where d_s is the distance from the sphere to the base of the tube (Ladenburg, 1907; Altrichter and Lustig, 1937).

Maude (1961) approached this problem by considering the forces which act on two spheres when they move with different velocities. If one of the spheres is stationary, and its radius is very large, then the situation corresponds to that of a sphere approaching an infinite rigid plane, such as the base of the rolling sphere viscometer tube. When the moving sphere has a velocity v_x, the force (F) on the stationary sphere is given by

$$F = 6\pi\eta v_x r_s \lambda' \qquad (2.15)$$

where r_s is the radius of the moving sphere, η is the viscosity of the fluid through which it moves, and

$$\lambda' \approx \left(1 - \frac{9}{8} \cdot \frac{r_s}{d_s}\right)^{-1}$$

On expansion, this gives

$$\lambda' = 1 + \frac{9}{8} \cdot \frac{r_s}{d_s} + \left(\frac{9}{8}\right)^2 \left(\frac{r_s}{d_s}\right)^2 + \dots$$
$$= 1 + 1·125\, r_s/d_s + 1·266\, r_s^2/d_s^2 + \dots . \qquad (2.16)$$

This theoretical derivation agrees satisfactorily with the experimental finding of Hopper and Grant (1948) that

$$\lambda' = 1 + 1·08\, r_s/d_s + 1·4\, r_s^2/d_s^2. \qquad (2.17)$$

Additional useful information about the end effect has been derived by studying the pressure changes induced by a moving sphere (Morgan, 1961) Glass containers of different diameters were used so as to give a wide range of sphere diameter/container diameter ratios. An oil–water interface was developed in each tube, and after introducing a steel sphere its descent through the oil was observed. Particular note was taken of interface movement as this provides a sensitive way for following any variation in the hydrostatic pressure which is associated with the depth of liquid. As the sphere approached the oil–water interface a depression formed in the latter, and it reached a maximum as the sphere passed through the interface. The rate of sphere descent was not affected by the interface. Disturbance of the pressure field ahead of the moving sphere was confined to a region within about five sphere diameters of the interface, but pressure variations extended over a distance of about eight sphere diameters.

Morgan (1961) also determined liquid viscosities by releasing steel spheres at different intervals of time. When the spheres were released at 2–3 min. intervals into glycerine (viscosity 18 poise at 12°C) there was an apparent reduction in viscosity of about 4 per cent. With castor oil (viscosity 10 poise at 12°C) there was no change in viscosity until the interval between successive spheres was 40 sec or less. The apparent change in viscosity has been attributed (Lemin, 1931) to the first sphere setting up currents in the fluid which then affect the descent times of all spheres which follow, but Morgan (1961) questions whether this can account for viscosity variations over very short time periods.

(c) A curvilinear relationship exists between the speed (v_x) of descent of the sphere and the angle (θ_a) at which the tube is tilted. The force of resistance (F_r) is related to v_x by

$$F_r = A_F \eta v_x + \beta_F \rho_2 v_x^2 \qquad (2.18)$$

where A_F and B_F are constants with dimensions $[L]$ and $[L^2]$ respectively; B_F approximately equals $\frac{1}{2}\pi r_s^2$. The viscous resistance $F_\eta (= AF\eta v_x)$ is given by F_r minus the inertia term $B_F \rho_2 v_x^2$

$$F_\eta = \frac{4}{3}\pi r_s^3 (\rho_1 - \rho_2) g \sin \theta_a - B_F \rho_2 v_x^2 \qquad (2.19)$$

where ρ_1 and ρ_2 are the densities of the sphere and fluid respectively.

4. Cone-plate viscometer

(a) When the angle (ψ_a) between the cone and plate does not exceed 6° the shear rate in the gap does not vary by more than 0·35 per cent (McKennell, 1956). At values of ψ_a greater than about 4°, however, edge effects may appear, and at high shear rates the temperature of the sample may rise. It is important, therefore, to use a small cone angle, e.g. $0.3°$, with an average gap width of ~ 0.05 mm.

The increase in temperature (ΔT) in time t is given by

$$\Delta T(°C) = \frac{R\psi_a H}{\kappa_c}\left(\frac{d_f t}{\pi}\right)^{\frac{1}{2}} \qquad (2.20)$$

for an annular element of width dR and radius R (Carslaw and Jaeger, 1947), where H is the amount of heat which is generated in one second for unit volume of liquid. For the Ferranti cone-plate viscometer (McKennell, 1953)

$$H = \frac{3.58 \times 10^{-8} G_T \Omega_a}{\pi R^3 \psi_a} \qquad (2.21)$$

(b) Inertial terms can be neglected with respect to the viscous forces only when ψ_a is very small (Slattery, 1961).

(c) Dilatant materials may be thrown out from the gap between the cone and plate at high rates of shear. If the required information cannot be obtained by testing at low rates of shear then a coaxial cylinder viscometer should be used.

Fluid and viscoelastic materials are not usually thrown out of the gap by the centrifugal force developed on rotation of the cone because it is exceeded by the centripetal force due to surface tension in fluids, and because of the normal component of stress (Weissenberg effect) generated in visco-elastic materials.

(d) Since the gap width between cone and plate is very small this method may not be suitable for studying the flow properties of highly structured materials. Much of the structure may be destroyed when aligning the cone and plate before the test is begun.

When dealing with samples containing particles which are relatively large compared with the gap width, the cone apex should be converted into a small flat zone (McKennell, 1953) so as to enlarge the gap. A "spike" is retained at the apex to assist correct positioning of the cone, and thus ensure constant stress and shear rate conditions.

(e) There is evidence to suggest that certain materials e.g. flocculated china clay suspensions, show pseudoplastic behavior in a cone-plate viscometer whereas they show Bingham flow in a coaxial cylinder viscometer. This has been attributed to the presence of wall layers adjacent to the cone and plate which are deficient in dispersed material (Boardman and Whitmore, 1963).

C. Comparison of Flow Data obtained with the Principle Types of Viscometer

Brodnyan *et al.* (1961) studied the non-Newtonian flow properties of lubricating oil, solutions of ethyl cellulose in cyclohexanone, viscose solutions, and solutions of nitrocellulose in n-butyl acetate, using a capillary viscometer, a cone-plate viscometer, and a viscometer with interchangeable biconical cylinders. The rate of shear–shear stress curves derived for these materials were identical for all three viscometers, so that "For a wide variety of materials, agreement of the viscosities at different shearing stresses measured by instruments of diverse design is shown to be good when all necessary corrections are applied to the experimental data. Among these corrections is one for the energy stored elastically in a fluid. Investigations of the response of fluids to harmonic vibrations have shown that fluids have elastic storage mechanisms: but only recently have there been investigations of the more general problem of the measurement of elastic stresses in laminar flow. For some materials a large fraction of the energy input is stored elastically; therefore the storage correction may be quite important".

Brodnyan *et al.* (1961) further point out that "As a result of the interchangeability of the results of different instruments, each instrument can be used in its own best range, thus simplifying design and experimentation and decreasing costs. Rotational viscometers should be used for low and medium rates of shear, and capillary viscometers for medium and high rates of shear".

Krieger and Woods (1966) found very good agreement between non-Newtonian viscosity–rate of shear data for a 0·7 per cent hydroxyethyl cellulose solution obtained with capillary viscometers, a coaxial cylinder viscometer, and a cone-plate viscometer. Viscosity was measured over four decades of rate of shear. Also included in this comparison were data obtained with a Ferranti–Shirley viscometer in which the cone and plate had been replaced with parallel disk platens. The upper platen, radius 3·75 cm, acted as rotor, and the lower platen was the stator, The apparent viscosity (η_{app}) of a non-Newtonian material at any rate of shear was shown to be

$$\eta_{app} = \frac{4}{D_1{}^4} \int_0^{D_1} D^2 p \, dD \qquad (2.22)$$

where D_1 is the maximum rate of shear ($\Omega_a R/L_a$) at the edge of the disks of radius R, and L_a is the thickness of sample layer between the platens. Eq. (2.22) shows that for any sample "the apparent viscosity depends only upon the maximum shear rate, and is otherwise independent of the dimensional and mechanical variables of the apparatus".

The parallel plane rotational viscometer gave viscosity–rate of shear data in close agreement with the data obtained with the other viscometers. Advantages claimed for the parallel plane instrument included "greater convenience in construction and alignment, and the ability to change the shear rate by simply varying the gap width. Parallel-plane instruments are preferable to cone-and-plate viscometers for the study of coarse mixtures and of suspensions which tend to destabilize when sheared in narrow gaps. They should also be advantageous for measurements at high shear rates, because of the superior heat transfer and lesser tendencies toward centrifugation and secondary flows. Their principal disadvantage lies in the need to correct for nonuniformity of the shear rate. However, the correction is rigorously valid and easily calculable, and is usually small".

D. Vibrating Reed Viscometer

The ultrasonic viscometer operates on quite different principles from the four classes of viscometer which have been described already. For Newtonian materials their viscosity is related to the damping coefficient (β) by the relationship

$$\eta = \frac{4(\beta \rho_m d_l)^2}{\pi \delta_a \rho} \tag{2.23}$$

where ρ and ρ_m are the respective densities of the sample and metal resonator, d_l is the width ($\frac{1}{8}$ in.) of the lateral faces of the thin metal blade which vibrates, and δ_a is the resonance frequency of 28 kH$_z$ (Rich and Roth, 1953). The blade is kept intentionally small so that its lateral faces subject the sample to tangential shear, and any effect due to its edges is negligible. Also, dilation and contraction of the blade are minimized as waves are propagated along its length.

The effective shear rate is a function of the resonance frequency, the amplitude 1μ, and the depth of the sheared layer of sample adjacent to the probe. In practice, a very high shear rate of approximately 10^4–10^5 sec^{-1} is achieved.

If the sample has a high degree of shear elasticity (G) the damping response given by Eq. (2.23) is modified in accordance with

$$\beta = \frac{(\rho)^{\frac{1}{2}}}{2\rho_m d_l} \left[\frac{(G^2 + \omega_a \eta^2)^{\frac{1}{2}}}{2} + G \right]^{\frac{1}{2}} \tag{2.24}$$

where ω_a is the angular velocity. When $\eta = 10$ poise, and $\omega = 1\cdot8 \times 10^5$, then G must exceed $\sim 10^6$ dyne/cm^2 before damping is affected.

E. Viscometric Studies at Very Low Shear Rates, and/or Very Low Shearing Stresses

Rheological studies are not restricted to measuring viscosity. Many materials of practical interest exhibit viscoelastic behavior at very low shear rates, or when subjected to low shearing stresses, i.e. they behave like solids in the initial stages of shear and subsequently they exhibit fluid behavior. Rheological tests under these conditions, which cause minimal structural alteration, provide valuable information about the structure of the sample.

A coaxial cylinder viscometer is easily adapted for this kind of study. In one form of the instrument as used in the author's laboratory (Fig. 2.1) the surfaces of both cylinders are finely ribbed to prevent the sample from slipping along the walls. The outer cylinder remains stationary while the inner

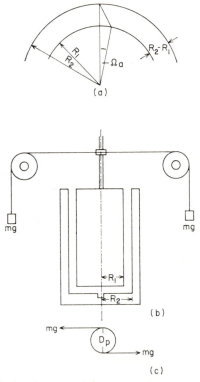

Fig. 2.1. Coaxial cylinder viscometer for creep studies on emulsions (Sherman, 1966). (a) angular displacement of inner cylinder, (b) diagrammatic representation of coaxial cylinder viscometer, (c) weight acting on pulley

cylinder, which is suspended from the torsion head by a thin torsion wire, is rotated slowly by a pulley–weight combination, or alternatively by means of an electromagnet. The angular rotation (Ω_a) of the inner cylinder is then followed optically by reflection of a light beam, which is projected from a lamp via a mirror affixed to the torsion wire on to a graduated scale. Scale readings are taken frequently at short time intervals. A more sensitive method for following the angular rotation of the inner cylinder is to follow the motion of a small metal rod, which is attached horizontally to the torsion wire, by means of a displacement transducer. The latter is linked up with a transducer meter and T–Y recorder, so that a continuous record is obtained of Ω.

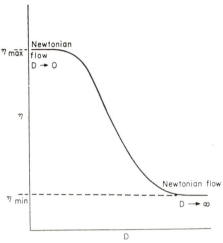

FIG. 2.2. Theoretical $\eta - D$ curve showing the two Newtonian regions

The strain (ε) is given by

$$\varepsilon = \frac{R_1 \, \Omega_a}{R_2 - R_1} \tag{2.25}$$

where R_1 and R_2 are the radii of the inner and outer cylinders respectively. In addition,

$$\text{shear stress } (p) = \frac{mgD_p}{\pi(R_1 + R_2)\, h_c} \tag{2.26}$$

where m is the weight acting over each pulley of diameter D_p, and h_c is the height to which the sample extends in the viscometer.

Since creep compliance is the ratio strain/shear stress, its value at any time t is given by

$$J(t) = \frac{R_1 \, \Omega_a \, \pi(R_1 + R_2)\, h_c}{(R_2 - R_1)\, mg \, D_p} = \text{constant} \times \frac{\Omega_a}{m} \tag{2.27}$$

because Ω_a and m are the only two variables. The creep compliance–time response curve is analysed in accordance with the method discussed in Section 1.C.1. of Chapter 1.

All structured materials should theoretically show two regions in their viscosity–rate of shear curves where viscosity is independent of rate of shear (Fig. 2.2). In practice only one of these regions is often observed viz. the region of minimal viscosity (η_{min}) at high rates of shear, because the structural network of the sample is so weak that a sufficiently low rate of shear is not available with any type of commercially produced viscometer to detect the region of maximum viscosity. With many polymer solutions, and also with bentonite suspensions (Rehbinder, 1954; Rehbinder and Michajlow, 1961), however, the upper Newtonian regions can be identified at rates of shear from 10–100 sec^{-1}. Similar curves are obtained when viscosity is plotted against shear stress. Their form can be semi-quantitatively defined by

$$\eta(p) = \eta_{min} + \frac{\eta_{max} - \eta_{min}}{1 + (p/p_0)^2} \qquad (2.28)$$

where $\eta(p)$ is the viscosity at any shear stress p.

When $\eta_{max} \gg \eta_{min}$, which usually applies,

$$\eta(p) \approx \eta_{min} + \frac{\eta_{max}}{1 + (p/p_0)^2} \qquad (2.29)$$

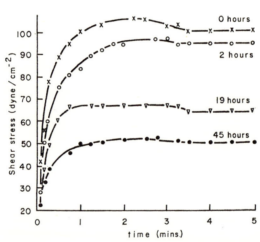

FIG. 2.3. Stress development in an emulsion at 0·12 sec^{-1} after it had aged for different times (Sherman, 1967)

Very low shear rates ($\ll 1$ sec^{-1}) are achieved in a coaxial cylinder viscometer by incorporating a synchronous motor and reduction gears to slow down the speed of rotation. Under these conditions a finite time interval may elapse before a steady-state stress value is achieved, the time interval increasing as the rate of shear decreases. Furthermore, the stress may initially rise to a maximum (Fig. 2.3), and then fall away to an equilibrium value where linkages between structural units are breaking and reforming at the same rate.

Shear rates lower than 10^{-2} sec^{-1} have been achieved with a conventional coaxial cylinder viscometer (Portable Ferranti, model VL) by measuring torque decay (Benis, 1967). In the normal mode of operation one of the cylinders is rotated at a constant angular vecloity, and this exerts a torque on the other cylinder so that it is displaced. If the latter is displaced manually to give a finite torque reading, and it is then allowed to return to its original position, i.e. zero torque, the decay curve provides information about shear stress (p) at the cylinder wall v. dp/dt. Since dp/dt is proportional to the instantaneous angular velocity, the data can be transformed into shear stress v. angular velocity. It is claimed by Benis (1967) that provided the rate of torque decay is reasonably slow the shear rate at the wall, calculated from the angular velocity, is the same as the shear rate which would be achieved if the cylinder had a constant angular vecloity. For a Newtonian fluid

$$D = \frac{2(R_2/R_1)^2}{(R_2/R_1)^2 - 1} k_1 \frac{dM_F}{dt} \qquad (2.30)$$

where M_F is the scale reading, and k_1 is a constant. The shear stress at the cylinder wall is

$$p = k_2 M_F \qquad (2.31)$$

where k_2 is a constant. When the sample exhibits non-Newtonian flow behavior the expression for the shear rate assumes much greater complexity

$$D = \frac{\Omega}{\ln (R_2/R_1)} \left[1 + \ln \left(\frac{R_2}{R_1} \right) \frac{d \ln \Omega_a}{d \ln p} + \frac{[\ln (R_2/R_1)]^2}{3\Omega_a} \cdot \frac{d^2\Omega_a}{d(\ln p)^2} \right.$$
$$\left. - \frac{[\ln (R_2/R_1)]^4}{45\Omega_a} \cdot \frac{d^4\Omega_a}{d(\ln p)^4} + ... \right] \qquad (2.32)$$

For both sugar syrup and Carbopol 934 k_1 had a value of 0·0429, but k_2 was 1·470 for the former and 3·32 for the latter.

One of the principal advantages of this technique is that the time delay in obtaining steady-state readings is eliminated. Errors that may arise when dealing with a viscoelastic fluid were not considered, but it was assumed that they could be ignored provided the rate of decay is very slow.

F. Parallel Plate Plastometer

The parallel plate plastometer consists of two parallel discs of equal radius, one of which remains stationary while the other moves toward it along a vertical axis which is normal to the planes of the discs, and passing through their centres (Landel et al.; 1965; Steggles, 1966). A sample of fluid

is held in the space between the two discs. Provided that the fluid is Newtonian its viscosity (η) can be calculated from the rate of approach of the upper
plate to the lower plate, and from the applied force (mg/A). The theory was
developed by Dienes and Klemm (1946) using three basic assumptions
viz. that the distance between the plates is much smaller than their radius,
that the area of compression is constant, and that the motion of the upper
plate is slow. On this basis,

$$\frac{mg}{A} \cdot \frac{t}{\eta} = \frac{3V_0^2}{8\pi}\left(\frac{1}{h_b^4} - \frac{1}{h_a^4}\right) \qquad (2.33)$$

where V_0 is the volume of the sample, and h_a and h_b are the initial sample
height and the height at any time t.

If the first assumption quoted above is eliminated, then the following
relationship ensues (Gent, 1960).

$$\frac{mg}{A} \cdot \frac{t}{\eta} = 3V_0\left[\frac{V_0}{8\pi}\left(\frac{1}{h_b^4} - \frac{1}{h_a^4}\right) + \left(\frac{1}{h_b} - \frac{1}{h_a}\right)\right] \qquad (2.34)$$

For a non-Newtonian fluid which exhibits a power law relationship
(Eq. 3.1) between stress and rate of shear, a differential equation (Scott,
1931) relates the rate of change in sample height ($-dh_b/dt$) with the plastic
viscosity (η_p)

$$-\frac{dh_b}{dt} = \pi^{(n_s+1)/2}(n_s+2)^{-1}\left[\frac{3n_s+1}{2n_s}\right]^{n_s}h_b^{5(n_s+1)/2}\left[\frac{(mg/A)^{n_s}}{\eta_p V_0^{(3n_s+1)/2}}\right] \qquad (2.35)$$

When the fluid is Newtonian $n_s = 1$, so that a plot of $\log(-dh_b/dt)$
against $\log h_b$ should have a gradient equal to 5; for non-Newtonian systems
obeying the power law relationship the gradient is $5(n_s+1)/2$.

Steggles (1966) compared viscosity data derived by a parallel plate plastometer modified to give higher shear rates than are normally possible with
data obtained on a Ferranti–Shirley cone-plate viscometer. He found good
agreement between the two instruments when the fluids tested were Newtonian. With non-Newtonian fluids the agreement was still good provided the
lower plate was made to rotate at 26 rev/min, and a load of 330 g. acted
on the upper disc. When the lower disc of the plastometer was not rotated
this instrument always gave very much higher viscosities than the Ferranti–
Shirley viscometer (250 and 320 sec^{-1}).

G. Viscometric Measurement of Dynamic Rheological Properties

It is curious that a large number of commercial instruments are available
for measuring the behavior of fluids and semi-solids in simple shear, and

yet the only instrument readily available for testing in dynamic shear is the Weissenberg rheogoniometer. Dynamic techniques are particularly useful for studying viscoelastic and other structured systems. When small amplitudes are applied, the structure alters less than when rotational shear is used, so that the information derived corresponds more closely to that of the zero shear state. As the amplitude increases the structure breaks down to a greater degree as with increasing rotational shear. Many structured systems show a linear viscoelastic response, i.e. their behavior is characterized by a linear differential equation with constant coefficients between the shear strains and shear stresses and their time derivatives (Weissenberg, 1964), only at very low stress, and for these systems dynamic measurements offer distinct advantages.

When the coaxial cylinder viscometer used for creep compliance–time studies, as described in the previous section, is appropriately modified the inner cylinder can be made to undergo oscillatory motion instead of, or in addition to, rotational motion. One way of doing this is to make the outer cylinder oscillate sinusoidally by means of a metal shaft which is connected to its base, and also to a variable speed motor, by an eccentric unit (Van Wazer et al., 1963). The inner cylinder oscillates with the same frequency as the outer cylinder but there is a phase difference since the former is subjected to a torque arising from the viscous drag exerted by the sample between the cylinders.

Motion of the inner cylinder is defined in terms of the torsion pendulum principle (Morrison et al., 1955; Shoulberg et al., 1959).

$$(W_1 + X_1 + I_m) \frac{d^2\Omega_a}{dt^2} + \left(\frac{\eta'}{H_s} + Y_1\right) \frac{d\Omega_a}{dt} + \left(\frac{E'}{H_s} + M_s\right) \Omega_a = Z_t i \quad (2.36)$$

where I_m is the moment of inertia of the pendulum, W_1 and X_1 are instrument constants, H_s is a geometrical shape factor, Y_1 is the instrument damping coefficient, M_s is an instrument torsional stiffness coefficient proportional to the square of the stiffness current, η' is the dynamic viscosity, E' is the dynamic modulus, Z_t is the torque on the inner cylinder due to unit current passing through the torque motor, and i is the current through the torque motor. Eq. (2.36) is valid when the wavelength of transverse waves in the sample is large compared with sample dimensions. When I_m is very small as compared to $W_1 + X_1$, and i changes sinusoidally, then

$$\eta' = H_s \left\{ \frac{Z'(f_i/f_\Omega) \sin j}{\omega_0} - Y' \right\} \quad (2.37)$$

$$E' = H_s [Z'(f_i/f_\Omega) \cos j - M_s + (W_1 + X_1) \omega_0^2] \quad (2.38)$$

where f_i and f_Ω are the amplitudes of the current i and of the angular deflection Ω_a respectively, Z' is an instrument constant proportional to the stiffness current, j is the phase lag between the pendulum deflection and the current and ω_0 is the angular velocity of the outer cylinder. The angular deflection of the inner cylinder is varied by changing the amplitude of the activating current.

The shape factor H_s is given by

$$H_s = \left(\frac{1}{4\pi h}\right) \left[\left(\frac{1}{R_1}\right)^2 - \left(\frac{1}{R_2}\right)^2 \right] \tag{2.39}$$

and the phase lag j by

$$j = \cos^{-1} \left[\frac{f_i^2 + f_\Omega^2 - f_r^2}{2 f_i f_\Omega} \right] \tag{2.40}$$

where f_r is the amplitude of the phase difference between current and the pendulum deflection.

Markovitz et al. (1952) employed a coaxial cylinder viscometer in which coils, situated in a magnetic field, were rigidly attached to both cylinders. The outer cylinder was made to oscillate with a small fixed amplitude and a frequency that could be varied. The dynamic rheological properties of the sample were derived from the motion of the inner cylinder, which depended on its inertia, the properties of the sample, and the restoring torque of the suspension wire. The coils on the outer cylinder consisted of about 1500 turns of General Electric Formex wire wound on a stainless steel form which was rigidly attached to the cylinder. In order to ensure that the coils were in uniform magnetic fields, and parallel to the direction of the field, two large horseshoe-type magnets (1800 gauss) with parallel pole faces were located in the appropriate positions. Motion of the cylinders were reproduced as voltage signals generated in the coils, and these were picked up by very fine gold leads. The voltages were proportional to the angular velocities of the two cylinders. It was found that the inner cylinder could not be made from metal because of considerable damping due to induced eddy currents. Stypol Resin No. 507E cast on to $\frac{5}{16}$ in quartz rod proved most suitable.

The instrument of Markovitz et al. (1952) was used to study both fluids and solids. When studying the latter the outer cylinder was replaced by a stainless steel plug to which the outer coil was rigidly attached. The sample, in the form of a cylinder, was cemented at its base into a small hole in the centre of a Stypol Resin adaptor in the plug, while the upper surface of the sample was similarly cemented to another Stypol Resin adaptor which was attached to a coil suspended from a copper wire.

The principle of operation for liquids involved measuring the angular amplitude ratio (a_r), i.e. the ratio of the displacements of the inner and outer cylinders, and the phase angle (α_p) lag between the inner and the outer cylinders.

$$(A_2+B_2\rho)\,\rho\left(\frac{\omega_0}{\eta^*}\right)^2-i(A_1+B_1\rho)\left(\frac{\omega_0}{\eta^*}\right)-1+\cos\frac{\alpha_p}{\alpha_r}+i\sin\frac{\alpha_p}{\alpha_r}=0 \tag{2.41}$$

where

$$A_1=\frac{I_m(R_2{}^2-R_1{}^2)}{4\pi h_c\,R_1{}^2R_2{}^2}$$

$$A_2=\left(\frac{I_m}{32\pi h_c}\right)\left[4\ln\left(\frac{R_1}{R_2}\right)+\left(\frac{R_2}{R_1}\right)^2-\left(\frac{R_1}{R_2}\right)^2\right]$$

$$B_1=\frac{(R_2{}^2-R_1{}^2)^2}{8R_2{}^2}$$

$$B_2=\frac{(R_2{}^2-R_1{}^2)\,(R_2{}^4-5R_2{}^2R_1{}^2-2R_1{}^4)+12R_2{}^2R_1{}^4\,\ln(R_2/R_1)}{192\,(R_2{}^2)}.$$

When the equipment was used to measure the properties of Vistanex polyisobutylene Eq. (2.41) reduced to the much simpler form

$$\frac{\omega_0}{\eta^*}=3{\cdot}88+11{\cdot}9i, \tag{2.42}$$

When testing solid materials α_r is the ratio of the displacements of the upper suspended end and the lower driven end, and α_p is the phase angle lag between the upper and lower ends. For solids,

$$E'=\frac{2(I_m\omega_0{}^2-k_t)l}{\pi R_2{}^4}\cdot\frac{\alpha_r(\alpha_r-\cos\alpha_p)}{1+\alpha_r{}^2-2\alpha_r\cos\alpha_p}, \tag{2.43}$$

and

$$\eta'=\frac{2(I_m\omega^2-k_t)l}{\pi R_2{}^4\omega_0}\cdot\frac{\alpha_r\sin\alpha_p}{1+\alpha_r{}^2-2\alpha_r\cos\alpha_p}, \tag{2.44}$$

where k_t is the torsional constant of the suspension wire, and l is the length of the sample.

Dynamic rheological properties can be studied also with a suitably modified cone-plate viscometer, or with the Weissenberg Rheogoniometer. Kambe and Takano (1963; 1965) used both a cone-plate viscometer and a coaxial cylinder viscometer. In the former instrument oscillatory torque was supplied to the shaft of the cone by a torque motor. Deflection of the cone was followed with a specially designed electrical detector, and was registered on an electrical recorder. Theoretical treatment of the data followed the treatment of Morrison *et al.* (1955) and Shoulberg *et al.* (1959) apart from a few minor modifications. These resulted in slightly amended versions of Eqs (2.37) and (2.38) for η' and E' respectively.

$$\eta' = H_s \left[\frac{Z'(i_d/f_i) \, (f_\Omega/\Omega) \, (f_i/f_\Omega) \, \sin j}{\omega_0} - Y' \right] \qquad (2.45)$$

$$E' = H_s \{ [Z'(i_d/f_i) \, (f_\Omega/\Omega) \, (f_i/f_\Omega) \cos j] - M_s + (W_1 + X_1 + Z_1) \, \omega_0^2 \} \qquad (2.46)$$

where i_d is the amplitude of the driving current, and Z_1 is the instrument constant.

Using this equipment Takano (1964) compared rheological data for dispersed systems obtained in both simple shear and oscillatory shear. The apparent viscosities of pseudoplastic systems obtained at low rates of shear closely resembled their dynamic viscosities as measured at low frequencies. With dispersed systems exhibiting plasticity there was no agreement between the two sets of data, the apparent viscosity at low rates of shear sometimes exceeding and sometimes falling below the dynamic viscosity at low frequencies. The deviations for plastic systems were attributed to the different ways in which the flocculated particle networks broke down and reformed under the influence of simple shear and oscillatory shear: " . . . the dependence of the apparent viscosity on the rate of shear is connected with the structural changes of the network caused by shearing forces, whereas the frequency dependence of the dynamic viscosity originates mainly from the relaxation of the network structures formed by particles in the medium".

Billington (1965) pointed out that in the technique used by Markovitz *et al.* (1952) for fluids the amplitude of oscillation of the inner cylinder, for a given frequency of oscillation of the outer cylinder, is independent of the time of shearing. Consequently it is not possible to differentiate between fluids which show a time dependent change in their resistance to flow and those fluids which do not. Billington (1965) devised an ingenious method for the dynamic testing of fluids by using two Portable Ferranti viscometers in conjunction. The outer cylinders of the viscometers were replaced by cylinders made from Duralumin, which had Tufnol cups partly filled with mercury attached to their outsides. By making contact with the lower edge of the inner cylinders, the mercury sealed the gap between the cylinders. The two viscometers were placed about 1 metre apart. "The gear wheel attached to the external shaft driving the outer cylinder of the first viscometer is replaced by a circular disc to which is attached at its circumference a rod one metre in length; rotation of the disc causes the extremity of the rod to execute near simple harmonic motion. The end of the metre rod, remote from the external drive, is connected to an arm attached to the internal driving shaft of the second viscometer; in this way the rotary motion of the external drive is made to oscillate the outer cylinder of the viscometer, the displacement of the outer cylinder approximating very closely to simple harmonic motion. The rotation of the inner cylinder of the viscometer is indicated in normal use by the movement of the needle over a scale graduated in 100 equal divisions; the neutral position of the inner cylinder is adjusted to coincide with the 50 division reading on the scale . . . ". Movement of the needle was recorded either photographically; or alternatively using a parallel plate condenser with one plate attached to the inner cylinder.

A comparison of viscosities for Newtonian fluids obtained with this combination of two Portable Ferranti viscometers and also with a capillary viscometer showed very good agreement with a maximum discrepancy of 1·9 per cent, the average discrepancy being 1·4 per cent. Motion of the inner cylinder was defined in a similar way to Eq. (2.36). If it is assumed that damping of this motion by all effects other than the viscous damping of the liquid is proportional to the angular velocity ($\omega_i t$) of the inner cylinder, then

$$\Omega_i^2 = \frac{v'}{X^2 + v_0^2} \Omega_0^2 \tag{2.47}$$

where,

$$v_0 = v' + v_e = \frac{\beta' + \beta_e}{M_0} \; ; \quad X = \frac{f_N^2 - f_0^2}{2\pi f_N^2 f_0} \; ; \quad f_N = \frac{1}{2\pi} \left(\frac{M_0}{I_m} \right)^{\frac{1}{2}}$$

and

$$\beta' = \frac{4\pi R_2^2 R_1^2 H_i}{(R_2^2 - R_1^2)} \eta$$

Ω_i and Ω_0 are the angular displacements of the inner and outer cylinders respectively, β_e is an instrumental damping factor, H_i is the height of the inner cylinder, f_0 and f_N are the applied frequency of oscillation for the outer cylinder and the natural undamped frequency of oscillation for the inner cylinder respectively, and M_0 is the restoring couple per unit angular displacement.

When $f_0 = f_N$, then $X = 0$, and the amplitude of the inner cylinder is a maximum $\Omega_m < \Omega_0$, so that Eq. (2.47) reduces to

$$\frac{\Omega_m}{\Omega_0} v_0 = v' \tag{2.48}$$

and, in conjunction with Eq. (2.44) this leads to

$$\Omega_i^2 = \frac{(v')^2}{X^2 + v_0^2} \Omega_m^2 \tag{2.49}$$

D

Lamb and his co-workers (Barlow *et al.*, 1961) have reviewed the very sophisticated techniques which they employ to study the viscoelastic properties of liquids. These include study of the damping of resonant torsional vibration of quartz crystals immersed in the fluid, change in amplitude and phase of a torsional wave pulse which is propagated along a long thin rod immersed in the fluid by a torsional crystal fixed to the end of the rod, and shear wave propagation in fused quartz and the measurement of the reflection coefficient at the interface between the fused quartz and the fluid. Such techniques are of great academic interest but they are not suited to industrial laboratory studies.

H. Viscometric Study of Stress Relaxation

Stress relaxation measurements following steady-state flow are particularly useful for determining the shape of the distribution function in the region of long relaxation times.

Stress relaxation has been studied both with the coaxial cylinder geometry and the cone-plate geometry.

Schramp *et al.* (1951), used a coaxial cylinder viscometer in which the cylinders were made from stainless steel. After steady state flow had been achieved movement of the outer cylinder was stopped, and the inner cylinder slowly turned back from its final position to its original position. The first few minutes of relaxation were recorded with an automatic camera which exposed a moving film to a deflected light beam at one second intervals. Subsequently, when the relaxation continued more slowly the displacement was followed visually by taking readings of the light beam on a ground glass scale. The total rotation of the inner cylinder back to its rest position approximated to 0·01 radian, so that strictly speaking the relaxation did not proceed at constant strain. However, provided displacement readings were taken at time (t) intervals which were much longer than a critical time t_a ($= b_a \eta / M_s$, where b_a is an apparatus constant) relaxation proceeded at approximately constant strain. Relaxation readings were commenced when $t = 10 t_a$. M_s was very large so that t_a was but a few seconds, and only the very beginning of the relaxation process had to be discarded.

In dynamic studies relaxation can be followed from the rate of decay of the oscillations at a particular frequency after the exciting current has been switched off (Zimm, 1957). Shearing the sample against the container walls is mainly responsible for the damping effect. Hence, the effect depends on the viscosity of the sample.

Weissenberg and Freeman (1948) studied stress relaxation phenomena in viscoelastic liquids e.g. saponified oils, starch pastes, and cellulose acetate, with a cone-plate viscometer. They concentrated on "the relaxation of the various components of stress during periods of stationary rates of laminar shearing displacements". Stress relaxation was initiated by halting the motion of the cone and maintaining the cone and disc at rest. From a continuous photographic record it was deduced that the "Stress components are delayed and show anisotropic time effects, the percentage rates of build-up and relaxation being different as between the components tangential and normal to the plane of the disc".

I. Modulus of Rigidity of Gels

The consistency of solid gels can be studied in creep compliance–time tests at low shearing stress, as will be described later in Section J of this chapter. Their modulus of rigidity can be derived, however, using much simpler equipment. This consists of a glass capillary U-tube (Saunders and Ward, 1954), one arm of which opens up into a wider bore tube, while the other capillary arm is calibrated in millimetres (Fig. 2.4). The tube is filled

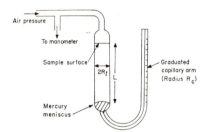

FIG. 2.4. Saunders and Ward's (1954) method for determining rigidity of stiff gels

with mercury to a height corresponding with the base of the wide-bore section, a standard volume of warm sample is then introduced into the latter, and the tube is inserted in a constant temperature bath (10°C for gelatin gels). Originally this tube was designed for testing gelatin gels, and two versions were proposed depending on the concentration of gelatin which was to be used. When 1·4 g of gelatin was used at a concentration of 5·5 g/100 ml soln., the mercury displacement (h_g) was about 6·0 cm, which corresponded to a wide arm radius $R_t = 0.75$ cm, gel column length $L = 13.0$ cm, and a capillary radius $R_c = 0.05$ cm. At a gelatin level of 0·1 g, the solution concentration remaining the same, h_g was restricted to about 0·5 cm to achieve the same strain as previously, which corresponded to $R_t = 0.305$ cm, $L = 6.0$ cm, and $R_c = 0.06$ cm. The warm sample solution is allowed to cool and set to a gel overnight. On the following morning the glass tube is carefully connected up to a source of air pressure and a mercury manometer. The gel is subjected to a shearing stress by applying a known air pressure; this causes a small volume displacement of the gel which results in a substantial movement of the mercury up the adjacent graduated capillary arm. Mercury displacements are determined for several values of air pressure. In order to ensure free movement of the mercury in the capillary arm a small volume of methanol is introduced on to its surface.

Provided $L \gg R_t$, the volume (Q) displacement of gel can be derived from an equation resembling Poiseuille's relationship, and

$$G_R = \frac{PR_t^4}{8LR_c^2 h_g} \qquad (2.50)$$

where G is the modulus of rigidity, and P is the net applied pressure i.e. after correction for the back pressure exerted by the mercury.

The maximum strain at the wide arm wall is $4Q/(\pi R_t^3)$, and the maximum stress is $PR_t/(2L)$.

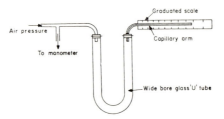

FIG. 2.5. Modified technique for determining the rigidity of soft gels (Scott Blair and Burnett, 1957)

In the preliminary work on gelatin gels (Saunders and Ward, 1954) it was found that for a particular gel the position of the mercury meniscus both under stress, and after its removal, depended on the time for stress application and for its removal. A loading time of 30 sec and a relaxation time of 3·5 min caused no significant residual deformation provided the gelatin concentration did not exceed 20 per cent.

The apparatus described above is suitable for measuring high values of G_R, and for use at relatively high shearing stresses (10^3–5×10^5 dyne/cm^2). It cannot be used on soft gels having much lower values of G_R than gelatin gels, particularly when the deformation under stress may not be all elastic. In this case a glass U-tube (Fig. 2.5) of different construction is used (Scott Blair and Burnett, 1957). The whole tube now consists of wide bore ($\sim 1·5$ cm diameter) glass, and the arm to which pressure is not applied is attached to a nearly horizontal capillary containing a drop of coloured alcohol. Movement of alcohol along the capillary is proportional to the displacement of the fixed volume of gel in the U-tube when subjected to different air pressures for 1 min intervals. It is important to ensure that movement of the alcohol results only from *deformation* of the gel, and not from its slipping along the walls of the tube. This experimental set-up produces a movement of the coloured alcohol which is 100 times greater than the gel displacement. The shear modulus is calculated in a similar way to Eq. (2.50), the curvature of the test sample in the Scott Blair and Burnett (1957) modification having no appreciable effect on the derivation of G_R, so that

$$G_R = \frac{P R_t^{\,4} \rho g}{8 L R_c^{\,2} h_g} \qquad (2.51)$$

where R_t is the radius of the wide U-tube, R_c is the radius of the capillary arm containing the drop of coloured alcohol, L is the length of the gel column, h_g is the movement of the drop of coloured alcohol, and ρ is the density of the liquid in the manometer. Since very much smaller air pressures are applied to weak gels than to gelatin gels, kerosene is used in the mano-meter tube.

The tensile strength of gels, and materials of a similar consistency, can be estimated by extruding them from a variable pressure viscometer (cylinder diameter of 6 cm) fitted with a nozzle of 5–12 mm diameter, and determining the quantity which falls off under its own weight (Ben Arie, 1955). The tensile strength is given by the ratio of sample weight extruded/nozzle area, since the drop breaks away from the nozzle when its weight/unit area equals its tensile strength. Several drops are collected in a weighed bottle, and from the increase in weight of the latter the mean weight per drop can be calculated. When testing very fluid samples much narrower nozzle diameters (3–4 mm) are used. Tests on viscoelastic gels of 2–6 per cent Napalm in gasolene (boiling range 60–90), and associated computations of applied stress and volume of sample extruded, indicated that tensile strength depends on both Young's modulus and the viscosity. Thus, if it was possible to determine the interrelationship of these two parameters with the slope of the graph relating the applied stress to the extrusion velocity from the nozzle, tensile strength, Young's modulus, and the viscosity could be calculated when only two of the three parameters were known.

J. Creep Compliance–Time Studies on Solids at Low Shearing Stresses

1. Parallel Plate Viscoelastometer

This technique can be applied to any material which will retain a rectangular shape prior to and during test. The viscoelastometer was designed originally for studies on solid polymers (Van Holde and Williams, 1953), and it was subsequently modified for low temperature studies on frozen ice cream (Shaw, 1963). The version to be described here contains other modifications and improvements so as to improve the sensitivity, and also to permit automatic operation following insertion of the sample and initial application of the shearing stress (Shama and Sherman, 1968).

In the older version the general principles of operation were as follows. A rectangular block (1 in thick, 2 in × $2\frac{1}{16}$ in horizontal cross-sectional area) was cut from the centre of each of two similar samples using a special sampling device (Fig. 2.6b). The sampler was fitted with a plunger for easy removal of samples. The two samples S_1 and S_2 were placed on either side of a

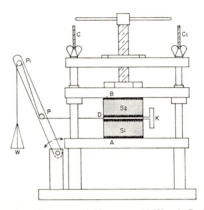

FIG. 2.6a. Creep apparatus (Shama and Sherman, 1968); A, B, D, Ribbed plates; C, C_1 Constant compression device; S_1, S_2 Samples; P, P_1, Pulleys (low friction); W, Weight pan; K, Knife edge:

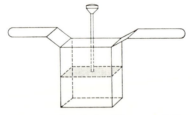

FIG. 2.6b. Sample cutting device (Shama and Sherman, 1968)

thin ribbed metal plate D (Fig. 2.6a), and the combination was then inserted between the two ribbed plates A and B of the viscoelastometer frame. Plate A is stationary, while plate B moves downward into contact with the upper surface of sample S_2 by carefully tightening the screw device attached to its upper surface. One end of plate D was connected to a balance pan W by a cord which passed over two low-friction pulleys P and P_1. When a weight was placed on pan W both samples S_1 and S_2 were subjected to shear, due to movement of plate D in the direction in which the weight was applied. The rate at which plate D moved, which was originally followed with a travelling microscope focussed on the knife edge, K, depended on the weight used and also on the rheological properties of the samples S_1 and S_2. The surfaces of S_1 and S_2 in contact with plates A and B underwent minimum shear, while the sample surfaces in contact with either side of plate D experienced maximum shear. To ensure minimal structural alteration to the samples the maximum movement of plate D did not exceed a few mm. When temperature control of the equipment was required the apparatus was placed in a large box made from polystyrene foam, $1\frac{1}{2}$–2 in thick, fitted with a heating or cooling unit.

In the improved version of the parallel plate viscoelastometer a system of ball bearing pulleys was introduced. Also, a device to prevent samples S_1 and S_2 from being compressed more than $\frac{1}{64}$ in when plate B made contact with the upper surface of sample S_1. The travelling microscope used originally to measure the displacement of plate D on the application of stress could

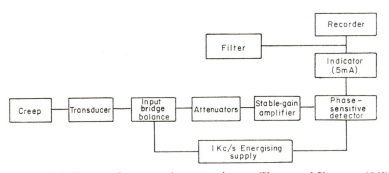

Fig. 2.7. Block diagram of automated creep equipment (Shama and Sherman, 1968)

not follow the instantaneous initial displacement accurately. This limitation was overcome by using a differential inductance displacement transducer with a volume of 0.11 in^3, and an armature of 0.75 g, which covered a range of ± 0.1 in. The transducer was insulated with P.T.F.E. and mounted on one of the stationary plates of the viscoelastometer frame, the free armature being screwed into the plate D and aligned so that it fitted the transducer bore without any friction. The transducer was connected to a transducer meter which was able to register displacements as low as 2.5×10^{-6} in with an accuracy of ± 1 per cent.

Figure 2.7 shows a block diagram of the equipment. All ribbed plates in contact with samples S_1 and S_2 were made from duraluminum. The transducer forming the active arms of the bridge circuit was energised by a 1 kHz supply from a lamp bridge oscillator. The output from the bridge was fed with appropriate attenuators and a stable gain amplifier into a phase-sensitive detector, the reference voltage being derived from the same 1 kHz

source as the energising circuit. The D.C. output from the detector was fed
to a moving coil meter, and also to a jack-socket which was used to drive
a direct inking recorder using a "plug in" filter unit. A T-Y recorder was
used with interchangeable time axis motors, so that the recording speed
could be readjusted after the instantaneous elastic compliance had been
recorded in detail.

The strain at any time t is given by

$$\varepsilon(t) = \frac{\beta_d f_d}{h_d} \qquad (2.52)$$

where β_d is the displacement of plate D, h_d is the thickness of samples S_1
and S_2, and f_d is the reciprocal magnification factor which is introduced only
when the displacement is followed by a travelling microscope.

The stress is

$$p = \frac{mg}{2A} \qquad (2.53)$$

where m is the weight placed on the balance pan W, and A is the cross-
section area of samples S_1 and S_2.

Therefore, the creep compliance at any time t is

$$J(t) = \frac{\varepsilon}{p} = \frac{2\beta_d f_d A}{mg h_d} \qquad (2.54)$$

Subsequent treatment of the data is in accordance with Section C.1 of
Chapter 1 (Eqs 1.26–1.35).

2. Compression Under Constant Stress

The creep compliance–time response of solids has also been studied by
techniques other than the parallel plate viscoelastometer. One of the simplest
of these alternative methods is to introduce a sample of cylindrical form
between a metal plate and a metal core (Fig. 2.8), and to apply a constant
force perpendicular to the upper plate. The two plates should be of the same
diameter, and this should be slightly larger than the diameter of the sample.
The upper plate is connected by a metal rod to a platform on which weights
are placed when the force is to be applied. Slotted weights are employed,
and these are gently lowered around a holder plate by means of a pulley
device. In this way minimal damage is incurred by the sample at the moment
that the force is applied. A travelling microscope, which is focussed on a
horizontal indicator arm attached to the vertical rod support of the platform,
follows the rate of compression on loading and the subsequent recovery

when the load is removed. The sensitivity of the technique can be improved by following the motion of the indicator arm with a displacement transducer, transducer meter, and T–Y recorder. In this case much smaller loads can be applied, and this will result in much smaller strains.

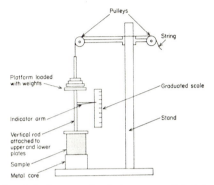

Pulleys

String

Platform loaded with weights

Graduated scale

Indicator arm

Stand

Vertical rod attached to upper and lower plates

Sample

Metal core

FIG. 2.8. Compression between parallel plates

The applied load (mg) is converted into stress/unit area (S_a) by the relationship

$$S_a = \frac{mg}{A} \cdot \frac{h_a}{h_b} \qquad (2.55)$$

where h_a and h_b are the height of the sample before and after deformation respectively, and A is the cross-sectional area. Since the non-recoverable part of the deformation (d_n) is small compared with the height of the sample $h_a \approx h_b$, so that

$$S_a = \frac{mg}{A}. \qquad (2.56)$$

Assuming a Poisson's ratio of 0·5, the shearing stress is

$$p = \frac{S_a}{3} = \frac{mg}{3A}. \qquad (2.57)$$

The total compression is defined by

$$\int_{h_a}^{h_b} \frac{dh}{h} = \ln\left(\frac{h_b}{h_a}\right) \qquad (2.58)$$

and, since $h_b = h_a - d_n$, with $d_n = \Delta h$, this becomes

$$\ln\left(\frac{h_a - d_n}{h_a}\right) = \ln\left(1 - \frac{d_n}{h_a}\right).$$

This expression expands into a series of the form

$$-\frac{d_n}{h_a} - \left(\frac{d_n}{h_a}\right)^2 - \left(\frac{d_n}{h_a}\right)^3 - \left(\frac{d_n}{h_a}\right)^4 \ldots$$

and, because $d \ll h_a$, the shearing strain (ε) is d_n/h_a.

The Newtonian viscosity (η_N) in the linear region of the creep compliance time curve is calculated from the gradient in the usual way. Strictly speaking this accounts for flow in one plane only, and is valid only when the compression is very small. Under all other conditions significant viscous flow occurs in all planes (Dienes and Klemm, 1946), and η_N has to be multiplied by a correction factor $2\pi h_a^2/A$, but even this correction is really valid only when the radius of the sample is much greater ($>10\times$) than the height of the sample.

Following the application of large deformations to plastic solids Sone (1961) determined their viscosity using Oka and Ogawa's (1960) theory for Bingham bodies with a yield value. The change in the height of the sample with time is

$$-\frac{dh}{dt} = \left(\frac{mg}{A} - \frac{p_0 V_0^{3/2}}{3\pi^{1/2} h_b^{5/2}}\right) \frac{2\pi h_b^5}{3\eta V_0^2} \qquad (2.59)$$

where V_0 is the volume of the sample, and p_0 is the yield value.

When $dh/dt = 0$, the limiting height of the sample (h_l) is given by

$$h_l = \left(\frac{V_0^{3/2} p_0}{3\pi^{1/2} mg/A}\right)^{2/5}$$

so that the yield is obtained from the height of the specimen as the rate of decrease in height approaches zero.

Equation (2.59) can be rearranged to

$$\log\left(-\frac{dh/dt}{h_b^5}\right) = \log\left(\frac{mg}{A} - \frac{p_0 V_0^{3/2}}{3\pi^{1/2} h_b^{5/2}}\right) + \log\left(\frac{2\pi}{3\eta V_0^2}\right)$$

and a plot of the left hand side of the rearranged equation (abscissa) against the first term on the right hand side (ordinate) is linear, giving an intercept on the ordinate axis (the second term on the right hand side) from which the viscosity can be calculated.

3. Extension Under Constant Stress

A sample, in the form of a long strip or fibre, is clamped at either end. One of the jaws, usually the one holding the lower end of the sample, remains stationary while the other is made to move slowly upward at a constant speed so as to stretch the sample. The Instron tensile testing machine (Instron Engineering Co.) is eminently suitable for this purpose. Analysis of the data resembles the procedure described in the previous section with rate of extension replacing rate of compression. The total longitudinal deformation is defined as in Eq. (2.58) if l_a and l_b, the initial and the extended lengths respectively, replace h_a and h_b. Similarly, the shear strain (ε) is d_n/l_a, with d_n now equal to Δl.

A micro-tensile testing machine, described by Marsh (1961), which measures the tensile strength of crystal whiskers could probably be modified to carry out similar tests on small samples. This instrument applies loads of 1 mgf–100 gf to samples with cross-sectional areas of 10^{-7}–10^{-2} mm^2. Extensions ranging from a few Angstroms to 15 mm are measured by a mirror and telescope system. The sample is glued to two silica rods which have flat ends, and the load is applied by means of a torsion balance. Extension of the sample upsets the null condition of the extension detector, and this condition is restored after applying each load increment by means of a sensitive micrometer which operates through a reduction lever mechanism. Consequently the readings of the loading mechanism, following calibration against known weights, and of the micrometer give the load–extension data directly.

4. *Torsion of Hollow Cylinder*

A hollow cylinder of 3–4 cm height, 1 cm wall thickness, 2 cm inner radius and 3 cm outer radius, is cut from a block of sample, and placed between roughened, or ribbed, metal discs (Van den Tempel, 1958). The lower disc is held stationary while a couple is applied to the upper disc. Rotation of the upper disc is followed by determining the motion of a horizontal arm either optically or electronically.

The shear at a distance r from the axis of the cylinder is

$$\varepsilon_r = \frac{\alpha r}{h_a} \qquad (2.60)$$

where h_a is the height of the cylinder, and α is the angle through which the upper disc rotates.

The shearing stress at this point is given by

$$p_r = \frac{Mr}{I_p} \qquad (2.61)$$

where M is the moment to which the upper surface of the sample is subjected, and I_p is the polar moment of inertia of the sample. The latter is equal to

$\pi/2\,(r_u^4-r_i^4)$, where r_u and r_i are the outer and inner radii respectively of the hollow cylinder.

Using gels of triglycerides in paraffin oil shear variations of 3×10^{-5} could be detected. A drawback of the technique is that during preparation of the hollow cylinder a great deal of surface damage is incurred with respect to its bulk.

K. Dynamic Test Methods for Solids

1. Free Torsional Vibrations

A sample in the form of a square block ($1\times1\times5$ cm) is fixed between two clamps and subjected to periodic torsion (Nederveen, 1963). The lower clamp is rigidly attached to a heavy metal yoke (Fig. 2.9), while the upper clamp

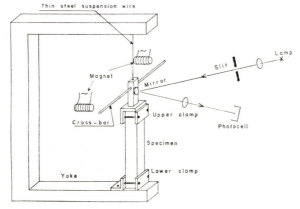

FIG. 2.9. Apparatus for free torsional vibrations (Nederveen, 1963)

is suspended from a thin steel wire which has a mirror and horizontal cross-bar attached to it. Two earphone magnets positioned near the ends of the cross-bar are coupled in opposite phase to an audio frequency oscillator.

This apparatus measures the real, or storage, elastic modulus in the frequency range 5–60 Hz. Vibrations are excited electromagnetically around the resonance frequency v_r, and its amplitude is determined as a function of the driving frequency by projecting the reflected light beam on to a photocell. The illuminated surface of the photo-tube, and therefore its electrical output signal, is proportional to the amplitude of oscillation. Simultaneously, the reflected light beam is projected on to a screen, so that the absolute value of deformation can be measured from the angle of rotation (2α).

G' is calculated from the resonance frequency by

$$G' = \frac{4\pi^2\,v_r^2\,I_r}{0{\cdot}14\,a_c^4/L_r} \tag{2.62}$$

where I_r is the rotational moment of inertia a_c and L_r are the width and length of the sample respectively.

The maximum deformation in the sample, which occurs along vertical lines drawn halfway between, and parallel to, the longitudinal ribs is

$$\left(1 - \frac{8}{\pi^2} \operatorname{sech} \frac{\pi}{2}\right) \frac{\alpha a_c}{L_r} = \frac{0 \cdot 675 \, \alpha a_c}{L_r}.$$

Waterman (1964) used a simpler version of this equipment for fibres in the low frequency range 0·2–20 Hz, and a rather similar principle for the frequency range 50–1000 Hz. In the latter range the vibrating system consisted of the sample with a body composed of two half cylinders of ferroxcube glued to its middle. This affixed body constituted the core of a differential transformer. In contrast to the system used by Nederveen (1963) only the lower end of the sample was clamped, while the upper end was connected to a transducer of the loudspeaker type. The system was put into resonance by adjusting the frequency of the signal generator driving the transducer. The current through the transducer was then chopped by a multivibrator which also triggered an oscilloscope. Tan δ was then calculated from the decay of the signal, and Young's modulus from the sample dimensions and the weight of the body attached to the sample.

$$E' = \frac{L_r^2 \, \rho(\omega')^2}{4Z_d} \left(1 - \frac{\lambda^2}{4\pi^2}\right) \tag{2.63}$$

and

$$\tan \delta = \frac{\lambda}{\pi} \left(1 - \frac{\lambda^2}{4\pi^2}\right)^{-1} \tag{2.64}$$

where

$$Z_d \tan Z_d = \frac{m_t}{m_c}$$

and L_r is now the length of the thread, ρ is its density, ω' is the real component of the angular frequency, and m_t and m_c are the respective weights of the thread and the body attached to it.

2. Dynamic Measurements on Soft Solids using a Horizontal Pendulum, or Tuning Fork

The rheological parameters of soft solids can be derived from their mechanical stiffness and internal dissipation as determined at several frequencies. One method employed for studying these properties at low frequencies

($\sim$ 1–2 Hz) is a horizontal pendulum (Fig. 2.10). This consists of a steel rod of $\frac{1}{2}$ in diameter and 27 in length which is attached at one end to a fixed platform while the other end has a weight attached to it. The weight is attached by a wire to the fixed platform. By varying the weight the pendulum

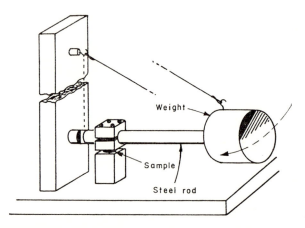

FIG. 2.10. Horizontal pendulum for low frequency shear measurements. (Rorden and Grieco, 1951)

is made to vibrate at different frequencies about its fulcrum. The period and rate of decay are determined microscopically; then the sample, in the form of a circle ($\sim\frac{1}{2}$ in diameter, and 0·05–0·07 in thickness) is placed between the pendulum and the fixed platform near the fulcrum, and the period and rate of decay are determined once again (Rorden and Grieco, 1951).

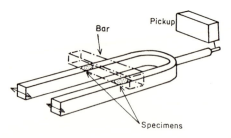

FIG. 2.11. Tuning fork for higher frequency measurements (Rorden and Grieco, 1951)

The horizontal pendulum cannot be made to vibrate at high frequencies so the same workers made use of a vibrating tuning fork (Fig. 2.11) in the frequency range 100–10,000 Hz. The fork was set in motion by a periodic driving force applied to the two ends by two electromagnetic telephone receivers, and the vibrations were picked up by a crystal of Rochelle salt. The amplitude was recorded as a function of the driving frequency. Identical samples were then placed on both prongs of the fork and a bar was placed over them; the bar was of such weight that it remained stationary under its own weight. The changes in vibrational frequency and in decrement due to the sample were then determined. From these data the storage modulus and the loss modulus respectively, can be calculated.

In both techniques the dimensions of the sample are small compared with the wavelength of the elastic wave. Consequently any inertia due to the sample can be ignored (Ferry, 1961).

3. Bending Vibrations

Young's modulus can be determined from the bending vibrations of a soft solid (Nederveen 1963a), or of a long thin rod of a hard solid (Nederveen, 1963b).

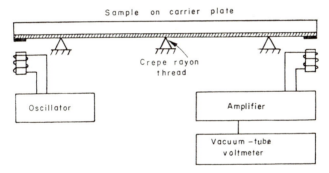

Fig. 2.12. Apparatus for bending vibrations (Nederveen, 1963)

In the former test, if the sample ($6 \times 8 \times 190$ mm^3) is too soft to support its own weight, it can be supported on a carrier plate of phosphor bronze (0·2 mm thickness). The carrier plate rests on three thin crêpe rayon threads (Fig. 2.12), which are located as close as possible to the node lines of the vibration. The thickness of the carrier plate is such that its vibrational behavior will not dominate that of the sample. Bending vibrations are excited by means of pieces of razor blade which are located at each end of the carrier plate. They are detected electromagnetically. The resonance curve is measured with a voltmeter, and the amplitude of vibration with a microscope. Young's modulus is determined from the different responses in the absence and the presence of the sample.

By simplifying the double layer beam theory (Oberst, 1952) the ratio

(B_{sc}/B_c) of the bending stiffness of the sample plus carrier to the bending stiffness of the carrier is

$$\frac{B_{sc}}{B_c} = \frac{1+2(E_s/E_c)\,(2\xi_r+3\xi_r^2+2\xi_r^3)+(E_s/E_c)^2\,\xi_r^4}{1+(E_s/E_c)\,\xi_r} \tag{2.65}$$

and,

$$\frac{\tan\delta_{sc}}{\tan\delta_s} = \frac{(E_s/E_c)\,\xi_r}{1+(E_s/E_c)\,\xi_r} \cdot \frac{3+6\xi_r+4\xi_r^2+2(E_s/E_c)\cdot\xi_r^3+(E_s/E_c)^2\,\xi_r^4}{1+2(E_s/E_c)\,(2\xi_r+3\xi_r^2+2\xi_r^3)+(E_s/E_c)^2\,\xi_r^4} \tag{2.66}$$

where E_s and E_c are the moduli of the sample and carrier respectively, ξ_r is the ratio of the thicknesses of the sample and carrier, δ_{sc} and δ_s are the loss angles of the sample plus carrier and of the sample respectively.

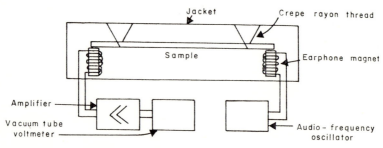

FIG. 2.13. Apparatus for resonance vibrations in flexure (Dekking, 1961)

In Nederveen's (1963a) experiments $1<B_{sc}/B_c<20, \xi = 30$, and $10^{-5}<a<10^{-3}$, so that several terms in Eqs (2.65) and (2.66) could be neglected. Consequently,

$$E_s = E_c \cdot \frac{[(B_{sc}/B_s)-1]}{4\xi_r^2[1+\frac{3}{2}\xi_r-(B_{sc}/B_c)/4\xi_r^2]} \tag{2.67}$$

where

$$\frac{B_{sc}}{B_c} = \frac{v_{r(sc)}^2\,m_{sc}}{v_{r(c)}^2\,m_c}$$

and $v_{r(sc)}$ and $v_{r(c)}$ are the resonance frequencies of the sample plus carrier and of the carrier respectively, and m_{sc} and m_c are the respective masses of the sample plus carrier and of the carrier. The ξ_r values selected were such that $B_{sc}/B_c \gg 1$.

Dekking (1961) used a similar technique for hard solids (Fig. 2.13). A sample of approximately 18 cm length, 7 mm width, and $3\frac{1}{2}$ mm thickness, which had a small piece of steel attached to each end, was supported by two crêpe rayon threads. The latter were located as close as possible to the nodes. Below each piece of steel was an earphone magnet. One of these was coupled to an audio-frequency oscillator (1–15,000 Hz), while the other was coupled to an amplifier and a vacuum tube voltmeter. The frequency of the oscillator was varied over a suitable range, to permit a study of the resonance curves associated with a number of modes of flexural vibration.

In general, the two components of E are derived (Ferry, 1961) from the relationships

$$E' = \left[\frac{|F|}{C_b\,(wt_b{}^3/l^3)\,|X|} \right] \cos \delta \qquad (2.68)$$

and

$$E'' = \left[\frac{|F|}{C_b\,(wt_b{}^3/l^3)\,|X|} \right] \sin \delta \qquad (2.69)$$

where X is the displacement, w, t_b, and l are the width, thickness, and length of the sample respectively, and C_b is a constant whose value depends on the procedure used to produce the bending vibrations. For a bar which is supported on knife edges at both ends, and which is bent by a knife edge in the centre, $C = 2$. It has a value of 16 for a bar which is clamped at both ends and is bent in the middle, and $C = \frac{1}{4}$ for flexure of a bar which is clamped at one end only.

3. QUALITATIVE RHEOLOGICAL TEST METHODS

The methods which will be discussed in this section can be used to obtain qualitative information about the consistency of a wide range of materials, although the publications in which they were originally described may have dealt with their application to a specific material. Some of the more interesting techniques which have been designed to simulate the mechanical action to which a material is subjected during practical usage will be described elsewhere in the appropriate applications chapter.

A. Viscometers for Fluids

1. Rotating Spindle

The Brookfield Synchro-Lectric viscometer is used widely throughout many industries. A synchronous motor drives a cylinder, spindle, or disc element which is immersed in the fluid at a constant speed of rotation. The force required to overcome the resistance offered by the sample to this motion is derived from a beryllium copper spring which drives the immersed element. A dial at the top of the viscometer records the extent to which the spring is wound in surmounting the viscous resistance. Consequently, the dial reading is proportional to the viscosity of the sample.

With the cylindrical element attached to the viscometer the data could be treated as indicated under coaxial cylinder viscometer in Table 2.1 provided the gap between the cylinder and the beaker containing the sample fluid was small. Unfortunately, the gap width is usually quite large, so that the relationships quoted in Table 2.1 are no longer valid. If the spindle or disc elements are attached to the viscometer the geometry of the system becomes more complex and complicated end effects are introduced. No satisfactory theoretical treatment is available for these conditions.

Bowles *et al.* (1955) proposed a method for translating Brookfield viscometer data for plastic viscosity into absolute units of yield value, and viscosity, in the linear region of the shearing stress-rate of shear curve. The apparent viscosity (η_1, η_2) is determined at two widely different rates of shear (D_1, D_2). Then, if η_1 and η_2 are the slopes of the lines drawn through the origin of the curve and points with coordinates (p_1, D_1) and (p_2, D_2) respectively.

$$p_1 = c_d D_1 \times \eta_1 \tag{2.70}$$

and

$$p_2 = c_d D_2 \times \eta_2 \tag{2.71}$$

where c_d is a proportionality factor relating the speed of rotation of the attached element (rev/min) and rate of shear if it can be assumed that " . . . Brookfield viscosities are slopes (of the shearing stress–rate of shear curve) connecting the true flow curve with the origin."

The plastic viscosity (η_∞) is given by

$$\eta_\infty = \frac{p_2 - p_1}{c_d (D_2 - D_1)} = \frac{D_2 \eta_2 - D_1 \eta_1}{D_2 - D_1} \tag{2.72}$$

Provided $D_2/D_1 = P_f$, where P_f is a proportionality factor, then

$$\eta_\infty = \frac{P_f \eta_2 - \eta_1}{P_f - 1} \tag{2.73}$$

Furthermore, $P_f = 2$ for most speed increments, so that

$$\eta_\infty = 2\eta_2 - \eta_1 \tag{2.74}$$

and, the yield value (p_0) is

$$p_0 = p_1 - D_1 \eta_\infty = 2D_1 (\eta_1 - \eta_2) \tag{2.75}$$

Unfortunately, Eqs (2.70)–(2.75) disregard the complexity of the flow geometry due to the rotating elements employed. Consequently, it is unlikely that the readings η_1 and η_2 represent the true apparent viscosities at the rates of shear corresponding to D_1 and D_2, even if it can be assumed that uniform shearing conditions are operative, which is very doubtful. The variation in shear conditions becomes particularly significant when attempting to compare apparent viscosity data obtained with different attachment geometries. Similar criticisms can be levelled against the approach proposed by Fryklöf (1961) and Neuwald (1966) for converting data obtained with a Brookfield viscometer into absolute values of shear rate, shear stress, and viscosity.

2. Orifice Viscometers

Typical examples are the Redwood, Engler, and Saybolt viscometers. Their operation is based upon the Hagen–Poisseuille law that the time required for a given volume of fluid to flow through a capillary is proportional to the viscosity of the fluid. The design of orifice viscometers does not, in fact, meet the requirements of the Hagen–Poisseuille law, so that there is no simple relationship between efflux time and viscosity. The ratio of orifice length to orifice diameter is very much less than for a standard capillary viscometer, and there is a much more pronounced entrance effect than in the latter. In addition, the variation in the hydrostatic head becomes a really serious problem.

While orifice viscometers may be useful for determining the viscosity of Newtonian fluids, due to their simplicity of operation, they cannot be used for non-Newtonian fluids.

B. Penetrometers

A cone, or needle, or sphere, affixed to a short rod which is located within a hollow, vertical, tube is held initially in a stationary position just in contact with the surface of a semi-solid or solid sample. When a release mechanism is brought into operation the cone, needle, or sphere, penetrates the sample with a given force for a predetermined time and the depth of penetration, or the rate of penetration, is measured. The main disadvantages of this technique are that the area of contact between the penetrating body and the sample does not remain constant throughout the test, and that displaced sample moves in a direction opposite to that in which the penetrometer moves.

The yield value (p_0) for a cone penetrometer (Fig. 2.14) is calculated from the depth of penetration (d_p) in accordance with

$$p_{0(g/cm^2)} = \frac{mg\,K_b}{(d_p)^{n_j}} \tag{2.76}$$

where m is the weight (g) of the cone plus the weight of the other moving parts of the penetrometer, n_j is a constant (≈ 2) whose precise value depends on the properties of the sample (Rehbinder and Semenenko, 1949; Agranat and Volarovich, 1957; Mottram, 1961) and

$$K_b = \frac{1}{\pi} \cos^2 a_p \cot a_p$$

where a_p is half the cone angle.

The original Institute of Petroleum grease testing cone (ASTM 5–25) consisted of a small angle (30°) cone superimposed on a wide angle (90°) cone. With this particular design the geometry of the system becomes complicated when the penetration depth exceeds the height of the 30° cone section, and then Eq. (2.76) no longer applies. A smooth, single angle (40°), cone overcomes this difficulty (Haighton, 1959). For long service it should be made of aluminium with a tip of hard steel.

Investigations with cones of different angles (Sambuc and Naudet, 1959) showed that

$$d_p \tan (a_p/2) = K_a \qquad (2.77)$$

where K_a is a constant. Reproducibility was most satisfactory when d_p was restricted to 7.5–20 mm.

Factors which influence the determination of yield value (Eq. 2.76) are the smoothness of the cone, the sharpness of the cone tip, the time of penetration, and the kinetic energy of the cone (Haighton, 1959). The cone should be carefully cleaned after each test so that sample material does not accumulate on its surface. When testing hard solids the cone may show a maximum penetration after 2–3 seconds, but for soft solids or when using small cone angles the penetration may continue for several hours. For most materials, however, a penetration time of a few seconds provides reproducible results. The minimum time required to achieve reproducible readings should be determined for each different material that is tested. When dealing with extremely soft or extremely hard consistencies the most suitable angle and weight of cone for use should be determined for each material.

Haighton (1959) classified the consistency of a wide range of materials in accordance with the arbitrary yield value range into which their yield value falls.

The rod penetrometer may consist of a metal rod surmounted by a small platform. The base of the rod is brought into contact with the surface of the sample, and the depth, or rate, of penetration is determined for different loadings of the platform. If the load is converted into shearing stress, and the penetration into velocity gradient, then in the case of non-Newtonian flow a curve will be obtained resembling that for pseudoplastic or plastic flow (Fig. 1.5), and the behavior will be defined by Eq. (1.23) or Eq. (1.24).

Fig. 2.14. Cone penetrometer with scale for reading yield value directly (Haighton, 1959)

TABLE 2.4

Textural Classification of Materials according to Yield Value (Haighton, 1958)

Yield Value (g/cm²)	Assessment
< 50	Very soft, just pourable
50– 100	Very soft, not spreadable
100– 200	Soft, but already spreadable
200– 800	Plastic, and spreadable
800–1000	Hard, but satisfactory spreadable
1000–1500	Too hard, limits of spreadibility
> 1500	Too hard

Similar curves can be obtained with the cone penetrometer if cones of different weights are used.

Tests on soft solids with a sphere penetrometer, which operated essentially in the same way as the cone and rod penetrometers already described, showed lack of reproducibility (Foley, 1959). In order to achieve greater

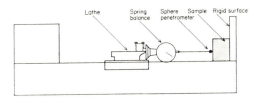

FIG. 2.15. Sphere penetrometer with lathe attachment (Foley, 1959)

control of the speed of penetration, and to eliminate frictional errors, a lathe was linked up to the penetrometer. A variable gear box attached to the lathe provided a wide range of penetration speeds, and the resistance offered by the sample was measured as a thrust using a spring balance which was mounted between the saddle of the lathe and the rod of the penetrometer (Fig. 2.15). The penetrometer was arranged so that the rod moved in a horizontal plane instead of vertically when entering the sample. The resistance to penetration, or the thrust required for penetration, was found to increase rapidly until the sphere was well embedded in the sample. While the sphere moved through the sample the thrust remained almost constant, and it did not decrease until the sphere approached the opposite face.

C. Hesion Meter

Interest in the stickiness, or tackiness, of semi-solids and soft solids led to the design of an instrument to measure the force required to pull a movable plate away from the surface of a thin layer of material which was sandwiched between this plate and another, stationary, plate (Green, 1944, 1949).

Two principal forces are involved in these measurements viz. the cohesion forces between the basic structural units of the sample, and also the forces of adhesion between the sample and the plate. When the cohesion forces are greater than the forces of adhesion, rupture occurs at the surface of the sample in contact with the moveable plate as the latter moves slowly upwards and away from the stationary plate, thus providing a measure of stickiness. Alternatively, if adhesion exceeds cohesion, then the rupture takes place within the sample itself. The term 'hesion' has been proposed (Claassens, 1958) for the combined effects due to adhesion and cohesion.

The consistency of materials in which the forces of adhesion are greater than the cohesion forces can be determined qualitatively using the simple instrument (Autard, 1968) shown in Fig. 2.16. The sample is deposited in a Petri dish, and the surface is carefully levelled and smoothed out with a large spatula. A circular aluminium plate which has a somewhat smaller diameter than the dish, and which is connected by a vertical metal rod to two cords via metal rod junctions and also to an Oldak gauge, is brought carefully into contact with the upper surface of the sample. The cord attached to the upper end of the Oldak gauge is connected at its other end to an axle which rotates through a constant speed motor and applies an upward force to the aluminium plate. This force causes a thin metal plate, which passes through the lower metal rod junction, to bend in the centre since it is clamped towards its extremities over two knife edges. The deformation is recorded by the gauge. The sample draws out into a "neck" Fig. (2.17) as the axle rotates and pulls up the cord connected to the metal plate. This neck progressively thins in the middle and eventually ruptures, leaving a layer of sample on the underside of the metal plate. The maximum gauge reading, i.e. when the upward acting force just overcomes the cohesion forces and the sample breaks following extension, is related to sample consistency.

In this form the apparatus provides only "single point" determinations. To study non-Newtonian flow behavior the constant speed motor is replaced by a balance pan on which different weights are placed. The time (t) taken for sample rupture over a wide range of pan loadings (mg) is determined, and a plot of mg against $1/t$ gives a typical non-Newtonian flow curve. Under these conditions the viscosity equations for the linear portions of pseudoplastic and plastic flow curves are

$$\eta_\infty \text{ (pseudoplastic)} = \frac{(t_s + c_s)^2 \, mg \, 4\cdot6t}{\pi r^4} \qquad (2.78)$$

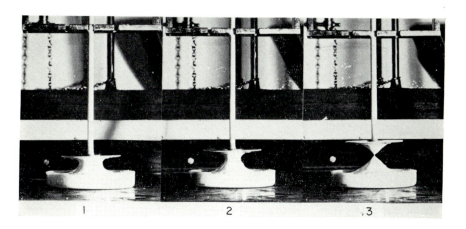

FIG. 2.17. Stages in sample extension in Hesion test.

Facing p. 86

and,

$$\eta_\infty \text{ (plastic)} = \frac{(t_s + c_s)^2 \, (mg - i_s) \, 4 \cdot 6t}{\pi r^4} \qquad (2.79)$$

where t_s is the thickness of the sample, c_s is a constant with a value which has to be determined by trial and error, r is the radius of the aluminium plate, and i_s is the intercept on the mg axis in the mg–$1/t$ plot.

The consistency of semi-solids may be examined in another way by shearing a thin layer between two parallel vertical plates (Kolvanovskaya and Mikhailov, 1961), one of which is stationary while movement of the other is activated by a loading device.

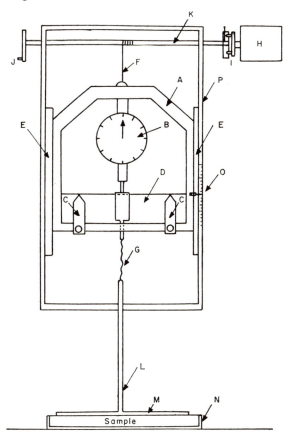

FIG. 2.16. Hesion meter (Autard, 1968).

A Movable framework, B Gauge for measuring deflection of spring, C Adjustable knife edges, D Spring, E Grooves, F Upper cord connecting axle to framework, G Lower cord connecting framework to test body, H Motor, I Gear, J Handle, K Axle, L Rod, M Test body, N Container for sample, O Scale, P Fixed frame.

D. Sectilometer

The principle of this technique is that a fine wire cuts through a sample and the thrust on the wire is measured (Mohr, 1948; Prentice, 1956; Sambuc and Naudet, 1959).

In a typical sectilometer (Fig. 2.18) the wire is held taut in a metal frame. A sample of standard dimensions is placed on a platform, and its surface is brought in contact with the wire. The platform is made to move slowly upwards (e.g. 0·07 cm/sec for a soft solid) at a steady rate through a motor drive so that the wire cuts through the sample. The thrust on the wire is transmitted by the metal frame holding the wire to a lever balance, where it is registered.

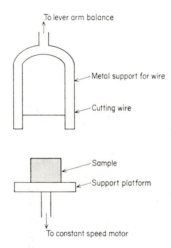

FIG. 2.18. Sectilometer for assessing the consistency of solid materials (Prentice, 1956).

This technique is not suitable for very soft materials because the wire does not cut through the sample until a minimal thrust has been developed (Prentice, 1956). For a sample width of 3 cm this may be as much as 40 g.

In another version of the sectilometer (Sambuc and Naudet, 1959) the sample is placed on a spring balance, and the wire, which is motivated by a constant speed motor, cuts through it at a rate of 0·044 cm/sec. The force exerted on the sample is recorded by the spring balance.

E. FIRA—NIRD Extruder for Soft Solids

This instrument (Fig. 2.19) was designed originally for studying the rheological properties of margarine, butter, and cooking fats (Prentice, 1952, 1954, 1956), but it can be used for a wide range of soft solids also.

It records automatically the force required to extrude a sample through a small diameter orifice ($\frac{1}{8}$ in) at a constant speed. The main components of the extruder are a cylindrical tube ($\frac{3}{8}$ in internal diameter) into which the sample is introduced after cutting from a block by means of a cylindrical borer, a mobile carriage which draws the tube back against a piston at uniform speed, and a leaf spring to which the pressure on the piston is transmitted via a

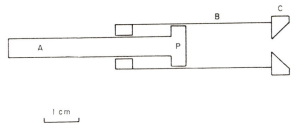

FIG. 2.19. Diagram of the FIRA–NIRD extruder (Prentice, 1954).

thrust rod. Movement of the spring, as a result of this pressure, is magnified by a system of levers which also actuate a pen recorder. In this way a complete trace is derived of the variation in thrust during extrusion of the sample. Five calibrated springs are available with the instrument. A full scale deflection of 3 in on the trace represents a thrust of 750, 1500, 3000, 6000, or 12000 g, depending on which spring is used.

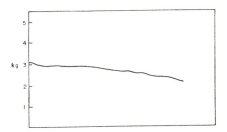

FIG. 2.20. Typical FIRA–NIRD extruder recording (Prentice, 1954).

Figure (2.20) shows a typical trace. After the filled cylinder has been placed in position the instrument is switched on, and the carriage begins to move. This results in a linear increase in thrust on the piston (AB). When the thrust reaches a value sufficient to produce extrusion of the sample (at B) the trace curves over (BC). In the latter region of the trace the recorded thrust is the sum of the force required to extrude the sample through the orifice and the force required to overcome the frictional resistance between the surface of the sample and the walls of the cylinder. The frictional resistance is maximum at the time when extrusion begins. As extrusion proceeds it decreases, since it is proportional to the amount of sample which is left in the cylinder. Consequently, the force which is required to overcome frictional resistance also decreases throughout the extrusion period. The force required for extruding the sample through the orifice remains constant, so that the recorded thrust falls progressively (CD) during the extrusion.

When the thrust rod hits a stop on the mobile carriage (at D) the trace continues in a straight line (DE) parallel to AB until the extruder switches itself off automatically. The minimum thrust value at D is called the extruder thrust. The slope of the trace CD provides a measure of the frictional resistance between the sample and the cylinder walls. For fats this value is related to their stickiness. Since the recording unit travels at four times the speed of the carriage the extruder friction is represented by one quarter of the horizontal distance between the two sloping lines. The general form of the trace indicates the homogeneity of the sample, a very wavy trace indicating poor homogeneity.

F. Bread and Cereal T.N.O. Panimeter for Solids

A technique for quantitative compression tests at constant stress on solid materials was described in Section 2.J.1. of this chapter. The Panimeter, which was designed by Hintzer (1949, 1951) for the Instituut voor Graan, Meel, en Brod T.N.O., Wageningen, Holland, works on a similar principle (Fig. 2.21), but in this case the stress is not constant throughout the test.

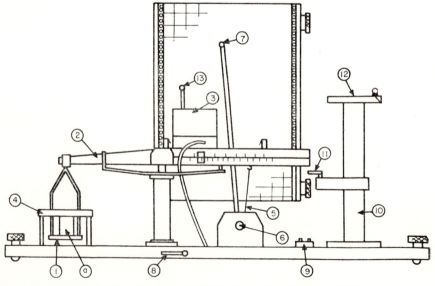

FIG. 2.21. Bread and Cereal T.N.O. Panimeter. (Robson, 1966).

A cylindrical sample (approximately 1·1 in length and 1·2–2 in diameter) is placed on a pan (1) which hangs by the two metal rods connected to one end of a balance beam (2). The rods pass through holes in a rigidly fixed, spherical, plastic disc (4) which is supported by two metal rods affixed to the wooden base of the instrument. The balance beam is raised into its operating position by turning a handle (8), which functions in the same way as the elevating mechanism on a chemical balance, and a weight (3) is then driven along the balance beam from the fulcrum by a built-in motor after switching it on at (9). This raises the end of the beam to which the pan is attached so that the sample is compressed between the pan and the fixed plastic disc. As the weight moves further along the beam the sample is compressed to an increasing degree. A cord (5), which is attached to the right hand side of the balance beam, winds around a pulley (6). To the shaft of the latter is attached an arm which carries a pen (7) at its end so that a continuous record can be obtained on chart paper of the extent of sample compression.

At the beginning of the compression a force of 50 g. is applied to the sample immediately after the motor has been switched on. The motor moves slowly along the balance beam by a rack and pinion mechanism to a predetermined position, where it remains stationary for 2 min. At the end of this time sample compression is at its maximum. The motor is then reversed, and the weight moves slowly back to the fulcrum where it is allowed to rest for 2 min before the motor is switched off.

The % compression and % recovery are calculated from

$$\% \text{ compression} = \frac{\text{maximum chart reading}}{1000} \times \frac{16 \cdot 68}{28} \times 100 \qquad (2.80)$$

$$= 0 \cdot 0594 \times \text{maximum chart reading}$$

$$\% \text{ recovery} = \frac{\text{maximum chart reading} - \text{minimum chart reading}}{\text{maximum chart reading}} \times 100$$

$$(2.81)$$

The chart traces (Fig. 2.22) are similar to creep compliance–time response curves in general form, but the resemblance cannot be carried any further since the Panimeter tests are not made at constant stress.

In its original form the Panimeter suffers from two major disadvantages. The holes in the rigid plastic disc have a much larger diameter than the rods which pass through them and support the pan on which the sample is placed. Consequently, when the weight begins to move along the balance beam the pan tends to move sideways to some extent as well as upwards, so that the sample is subjected to a shearing motion as well as to compression. This shearing action was eliminated (Robson, 1966) by inserting Nylon guides in the holes in the rigid plate. The other point to be made is that interpretation of the chart data is complicated by the variation in the shearing stress during much of the test. Modifications to the operating circuit can be introduced (Robson, 1966) to overcome this and to make the operation fully automatic. A push button now starts the motor and moves the weight along the beam until a switch is tripped so that the weight is maintained at any preselected position on the beam. A cam timer holds the weight there for 2 min, and then by means of a two-way relay circuit the motor is reversed and the weight returns to its original position. The recovery curve is traced until 2 min after the return of the weight when the cam timer switches off the chart drive.

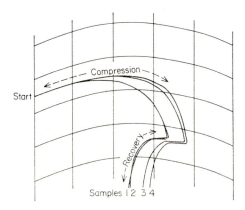

Fig. 2.22. Panimeter chart trace (Robson, 1966)

Other modifications can be introduced so that the Panimeter operates under conditions of constant strain (Robson, 1966) instead of constant stress. Constant strain tests are particularly valuable when dealing with non-linear behavior in food products, for example. This aspect will be discussed in some detail in Chapter 4. A brass pillar (10) is built on the base of the instrument near the right hand end of the beam. A light-action leaf switch (11) connects with a lead screw inside the pillar so that it can be raised or lowered by means of a handwheel (12). The sample is placed on the pan, the beam is raised, and the height of the leaf switch is adjusted so that its contacts are just closed by the end of the beam. The leaf switch is then lowered by an amount equivalent to the compression which the sample is to undergo. Thus, the weight will now move along the beam until the sample has been compressed to the required degree since the leaf switch will then close. Following this, the direction of the motor is reversed by means of a series of relay switches, and the weight now moves down the beam until the leaf switch reopens.

G. *Rotary Cutter*

The mechanical treatment to which many materials are subjected in practical usage can be simulated by cutting a spiral path through a sample with a blunt two-bladed cutting device. The resultant forces which develop depend on the consistency of the sample.

Figure 2.23 shows a useful form of this instrument (Robson, 1966). A rectangular sample (1) is held in a stationary inverted sample holder, and this is brought into contact with the cutting device (2) by raising the platform (4) on which rests the hardened point (3) of the cutting blade shaft. The platform is connected to a rack and pinion device (5) by means of which it, and the cutting blade, are raised at a constant speed. A rod with a

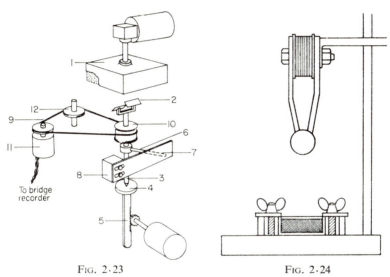

FIG. 2.23   FIG. 2.24

FIG. 2.23. Rotary cutter (Robson, 1966). (1) sample (2) cutter (3) hardened point (4) platform (5) rack and pinion (6) rod (7) leaf spring (8) leaf spring anchorage (9) cord loop (10) drum (11) potentiometer (12) idler pulley

FIG. 2.24. Schematic drawing of falling sphere indenter (Soltøft, 1947)

rounded tip (6), which is attached to the shaft of the cutting blade, makes contact with the surface of a leaf spring (7) through this tip. The leaf spring anchorage (8) permits a certain degree of adjustment to be made to the effective length of the spring according to the consistency of the sample. For hard samples the spring is shortened, and for soft samples it is lengthened. Any movement of the cutting blade, depending on the consistency of the sample and the rigidity of the spring, is transmitted via a cord loop (9) which winds round a drum (10) on the cutter shaft and then round a pulley with frictionless bearings connected to the shaft of a potentiometer (11).

Any slack is taken up by an idler pulley (12). The potentiometer is connected to a Wheatstone bridge circuit, and the out-of-balance signal is fed into a pen recorder.

The "Rotary Cutter coefficient" is derived by cutting out from the paper trace the area between the curve and the base line. This coefficient is expressed as the "area under the graph" (A.U.G.). For many materials the curve has an oscillating outline but as yet it has not been possible to relate this variability to the variations in consistency which might be expected within a single sample.

H. Indentation of Hard Solids by a Falling Sphere

This technique (Lexow, 1940; Soltøft, 1947) can only be applied to relatively hard solids as the sphere is liable to sink through very soft solids. A steel sphere is dropped from a height on to the plane surface of a sample, and the diameter of the indentation, or its depth, depends on the hardness of the sample.

The steel ball (diameter $1\frac{1}{2}$ in, weight approximately 225 g) is initially held by an electromagnet (Fig. 2.24) at a height of not less than 2 in above a receptacle containing the sample. If the sample is of suitable dimensions several tests can be made at different points on its surface. When the current is switched off the sphere is released, and it makes a depression in the surface of the sample the volume of which appears to be proportional to the square root of the distance through which the sphere falls, at least in the case of fats (Soltøft, 1947) and metals. For fats it has been found that a film of fat is deposited on the surface of the steel sphere as it indents the sample, and that the diameter of this film corresponds to the diameter of the indentation in the surface of the sample.

The test conditions should be so arranged that the diameter of the indentation is equivalent to about $\frac{3}{4}$ of the diameter of the steel sphere. If the indentation is too shallow it is very difficult to obtain reproducible results when the surface of the sample is slightly uneven. Lexow (1940) found that satisfactory results were obtained when the sphere was dropped through heights which did not exceed about 3 in.

BIBLIOGRAPHY

Agranat, N. N. and Volarovich, M. P. (1957). *Kolloid Zhur.* **1**, 1.
Altrichter F. and Lustig, A. (1937). *Physik. Z.* **38**, 376.
American Society for Testing Materials ASTM5-25; D217-52T.
Autard, P. S.C.I. Monograph No. 27 "Rheology and Texture of Foodstuffs" p. 236.
Bacon, L. R. (1936). *J. Franklin Inst.* **221**, 251.
Barlow, A. J., Harrison, G., Richter, J., Seguin, H. and Lamb, J. (1961). *Laboratory Practice* **10**, 786.
Ben Arie, M. M. (1955). *J. Polymer Sci.* **17**, 179.

Benis, A. M. (1967). *Biorheol.* **4**, 33.

Billington, E. W. (1965). *J. Sci. Instrum.* **42**, 569.

Boardman, G. and Whitmore, R. L. (1963). *Brit. J. appl. Phys.* **14**, 391.

Bowles, R. L., Davie, R. P. and Todd, W. D. (1955). Modern Plastics 33, 140, 142, 144, 146, 148.

Brodnyan, J. G., Gaskins, F. H. and Philippoff, W. (1961). American Society for Testing Materials, Special Technical Publication No. 299, p. 14.

Buckingham, E. (1921). *Proc. Am. Soc. Testing Materials* **21**, 1154.

Carslaw, W. S. and Jaeger, J. C. (1947). "Conduction of Heat in Solids", Oxford University Press, Oxford, p. 56.

Claassens, J. W. (1958). *S. Afr. J. Agric. Sci.* **4**, 457.

Dekking, P. (1961). Doctoral dissertation, Leiden University.

Dienes, G. J. and Klemm, H. J. (1946). *J. appl. Phys.* **17**, 458.

Din, F. J. and Scott Blair, G. W. (1940). *J. appl. Phys.* **11**, 574.

Faxén, H. (1922). *Ark. Mat. Aftr. Fys.* **17**, 1.

Ferry, J. D. (1961). "Viscoelastic Properties of Polymers". Wiley, New York.

Foley, J. (1959). M.Sc. Thesis, University College of Cork.

Fryklöf, L. E. (1961). *Svensk. Farm. Tidskr.* **65**, 753.

Gent, A. N. (1960). *Brit. J. appl. Phys.* **17**, 458.

Green, H. (1944). *Ind. Eng. Chem. Anal. Edn.* **13**, 632.

Green, H. (1949). "Industrial Rheology and Rheological Structures", Chapman and Hall, London. Chapter 9.

Haighton, A. J. (1959). *J. Am. Oil Chem. Soc.* **36**, 345.

Higginbotham, G. H., Oliver, D. R. and Ward, S. G. (1958). *Brit. J. appl. Phys.* **9**, 372.

Hintzer, H. M. R. (1949). *Bakkers vokblad.* **8**.

Hintzer, H. M. R. (1951). *Bakkers vokblad.* **10**.

Hopper, V. D. and Grant, A. M. (1948). *Aust. J. scient. Res.* **A1**, 28.

Kambe, H. and Takano, M. (1963). Aeronautical Research Inst. Univ. Tokyo, Report No. 378.

Kambe, H. and Takano, M. (1965). Proc. 4th Intern. Cong. Rheology 3, 557.

Kolvanovskaya, A. S. and Mikhailov, V. V. (1961). *Colloid J.* (U.S.S.R.) **23**, 606.

Krieger, I. M. and Woods, M. F. (1966). *J. appl. Phys.* **37**, 4703.

Kuno, H. and Senna, M. (1967). *Rheol. Acta.* **6**, 284.

Ladenburg, R. (1907). *Annln Phys.* **23**, 447.

Landel, R. F., Moser, B. G. and Bauman, A. J. (1965). Proc. 4th Intern. Congr. Rheology 2, 663.

Lemin, C. E. (1931). *Phil. Magazine* **12**, 589.

Lexow, T. (1940). *Fette u. Seifen* **47**, 334.

Lindsley, C. H. and Fischer, E. R. (1947). *J. appl. Phys.* **18**, 988.

Markovitz, H., Yavorsky, P. M., Harper, R. C., Jr., Zapas, L. J. and De Witt, T. W. (1952). *Rev. Scient. Instrum.* **23**, 430.

Marsh, D. M. (1961). J. Scient. Instr. **38**, 229.

McKennell, R. (1953). Proc. 2nd Intern. Congr. Rheology, p. 350, Butterworths, London.

McKennell, R. (1956). *Anal. Chem.* **28**, 1710.

Maude, A. D. (1961). *Br. J. appl. Phys.* **12**, 293.

Mohr, W. (1948). *Milchwissenschaft* 3, 234.

Morgan, P. G. (1961). *Chem. Eng. Sci.* **15**, 144.

Morrison, F. R. and Harper, J. C. (1965). *Ind E. Chem. Fund.* **4**, 176.

Morrison, T. E., Zapas, L. J. and De Witt, T. W. (1955). *Rev. scient. Instrum.* **26**, 357.

Mottram, F. J. (1961). *Lab. Practice* **10**, 767.
Nederveen, C. J. (1963a). *J. Colloid Sci.* **18**, 276.
Nederveen, C. J. (1963b). *Rheol. Acta* **3**, 2.
Neuwald, F. (1966). *J. Soc. Cosmetic Chem.* **17**, 213.
Oberst, H. (1952). *Akust. Beih.* **4**, 181.
Oka, S. and Ogawa, S. (1960). *J. Japan. Soc. Test. Mat.* **9**, 321.
Poiseuille, J. L. M. (1840). *C.r. hebd. Acad. Sci. Séance Paris* **11**, 961.
Prentice, J. H. (1952). B.F.M.I.R.A. Research Report No. 37.
Prentice, J. H. (1954). *Lab. Practice* **3**, 186.
Prentice, J. H. (1956). B.F.M.I.R.A. Research Report No. 69.
Rabinowitsch, R. (1929). *Z. phys. Chem.* **A145**, 1.
Rehbinder, P. (1954). *General Disc. Faraday Soc.* **18**, 151.
Rehbinder, P. A. and Semenenko, N. N. (1949). *Dokl. Akad. Nauk. SSSR* **64**, 835.
Rehbinder, P. and Michajlow, N. W. (1961). *Rheol. Acta* **4/6**, 361.
Reiner, M. (1926). *Kolloidzeitschrift* **39**, 80.
Reynolds, O. (1883). *Phil. Trans. R. Soc.* **174**, 1935.
Rich, S. R. and Roth, W. (1953). *J. appl. Phys.* **24**, 940.
Robson, A. H. (1966). *J.Fd. Technol.* **1**, 291.
Rorden, H. C. and Grieco, A. (1951). *J. appl. Phys.* **22**, 842.
Sambuc, E. and Naudet, M. (1959). *Rev. Franc. Corps Gras* **6**, 10, 18.
Saunders, P. R. and Ward, A. G. (1954). Proc. 2nd Intern. Congr. Rheol. p.284, Butterworths, London.
Schramp, F. W., Ferry, J. D. and Evans, W. W. (1951). *J. appl. Phys.* **22**, 711.
Scott, J. R. (1931). *Trans. Inst. Rubber Ind.* **7**, 169.
Scott Blair, G. W. (1958). *Rheol. Acta* **213**, 123.
Scott Blair, G. W. and Burnett, J. (1957). *Lab. Practices* **6**, 570.
Scott Blair, G. W. and Oosthuizen, J. C. (1960). *Br. J. appl. Phys.* **11**, 332.
Shama, F. and Sherman, P. (1968). S.C.I. Monograph No. 27 "Rheology and Texture of Foodstuffs" p. 77.
Shaw, D. J. (1963). *In* "Rheology of Emulsions", (P. Sherman, Ed.), p. 125. Pergamon Press, London.
Sherman, P. (1967). *J. Coll. Interface Sci.* **24**, 97.
Shoulberg, R. H., Zimmerli, F. H. and Kohler, O. C. (1959). *Trans. Soc. Rheol.* **3**, 27.
Slattery, J. C. (1961). *J. Colloid Sci.* **16**, 431.
Soltøft, P. (1947). Doctoral dissertation, Danish Techniske Hojskole, Copenhagen.
Sone, T. (1961). *J. Phys. Soc.* (Japan). **16**, 961.
Steggles, J. S. (1966). *Chem. Ind.* p. 976.
Takano, M. (1964). *Bull. chem. Soc. Japan* **37**, 78.
Tempel, M. van den (1958). *Rheol. Acta* **1**, 115.
Vand, V. (1948). *J. Phys. Colloid Chem.* **52**, 277.
Van Holde, K. E. and Williams, J. W. (1953). *J. Polymer Sci.* **11**, 243.
Van Wazer, J. R., Lyons, J. W., Kim, K. Y. and Colwell, R. E. (1963). "Viscosity and Flow Measurement", Interscience, New York.
Waterman, H. A. (1964). *Kolloid Z. u. Zeitschr. für Polymere* **196**, 18.
Weissenberg, K. (1964). in "The Testing of Materials by means of the Rheogonio-meter", Farol Research Engineers, Bognor, England.
Weissenberg, K. and Freeman, S. M. (1948). *Nature* **161**, 324.
Wellman, R. E., de Witt, R. and Ellis, R. B. (1966). *J. chem. Phys.* **44**, 3070.
Williams, J. C. and Fulmer, E. J. (1938). *J. appl. Phys.* **9**, 760.
Zimm, B. H. (1957). *J. Polymer Sci.* **26**, 101.

CHAPTER 3

Rheology of Dispersed Systems

1. Introduction 97
2. Rheology of Flow 98
   A. Power Laws Relating Stress and Rate of Shear in non-Newtonian Flow 98
   B. Theories of non-Newtonian Flow Based on Structure Breakdown by
      Shear 101
   C. Rheological Behavior at Low Rates of Shear 125
3. Factors Arising From the Constituents of a Dispersed System Which
   Influence Its Viscosity 127
   A. Internal Phase 130
   B. Continuous Phase 157
   C. Emulsifying Agent 158
   D. Rheological Properties of the Adsorbed Film of Emulsifier . . 162
   E. Electroviscous Effect 164
   F. Hydrocolloids, Pigments, and Crystals 168
4. Rheological Properties of Dispersed Systems at Very Low Shearing Stresses 169
5. Rheological Changes in Dispersed Systems During Aging . . . 172
Bibliography 180

1. INTRODUCTION

Many of the non-solid systems of industrial interest are essentially dispersions of solid particles in fluid media or of liquid droplets in fluid media. The former will be called dispersions in future discussion, and the latter emulsions. In the stationary state the particles or droplets may come together and form a network structure which exhibits viscoelastic behavior at low shear. With increasing shear the interlinked structure gradually breaks down until eventually, at sufficiently high shear, these systems flow like viscous liquids. Both low and high shear studies are of interest, the former probably in relation to production quality control, and the latter to practical usage conditions.

Dispersions have been studied in much greater detail than emulsions, probably because of the greater complexity of emulsions. The latter are not thermodynamically stable systems, and the globules begin to coalesce with one another soon after the emulsion has been prepared, although the rate may be very substantially reduced by suitable formulation. Coalescence therefore leads to a progressive change from the original state of dispersion. Study of emulsion rheology is aggravated further by the possibility of the globules deforming in shear, or when they are packed tightly together. If significant deformation occurs then it is no longer possible to apply the theories which have been developed for solid spherical particles, or for that matter, the theories for any other well defined geometry of shape.

We will discuss the flow properties of dispersed systems before dealing with their viscoelastic properties. This may appear illogical at first sight, but as much more is known about the flow properties it will enable us later to extrapolate back, as it were, to the stationary state and draw some relevant conclusions. Two main approaches have been used to study the flow properties. One of these is a phenomenonological approach which only considers the stress–strain relationship without regard for the structure or composition of the system, while the other approach attempts to explain behavior primarily in terms of particle–particle interactions.

2. RHEOLOGY OF FLOW

A. Power Laws Relating Stress and Rate of Shear in non-Newtonian Flow

For many dispersed systems Eq. (1.23) is not valid for pseudoplastic flow, and the D–p relationship is more satisfactorily defined by (Ostwald, 1925; de Waele, 1925)

$$D = \frac{1}{\eta_p}(p)^{n_s} \tag{3.1}$$

where n_s is a measure of the internal structure, and η_p is the parameter corresponding to viscosity. A Newtonian fluid has $n_s = 1$, and any other value therefore indicates a deviation from Newtonian flow. Metzner and Reed (1955) refer to n_s as the "flow behavior index". From Eq. (3.1) it is apparent that when $n_s > 1$ the apparent viscosity increases with increasing shear stress, and when $n_s < 1$ the apparent viscosity decreases with increasing shear stress. If a pronounced yield value is found in the D–p curve then Eq. (1.24) can be modified to give a power law equation (Herschel and Bulkley, 1926; Scott, 1931) of the form

$$D = \frac{1}{\eta}(p-p_0)^{n_s} \tag{3.2}$$

Much criticism has been levelled against these power laws on dimensional grounds. According to Houwink (1958) η_p "... is distinguished from the coefficient of viscosity, since $n_s \neq 1$, and therefore the dimensions of η_p are different from those of η, and its value may not be expressed in poises. When $n_s = 1$ we shall speak of true flow without taking into consideration the value of p_0. When $n_s \neq 1$ we shall speak of quasi-flow whatever may be the value of p_0". Reiner (1960) questioned the general validity of power laws. He claimed that a power equation "... is not one law which only for different materials, i.e. different values of n_s, gives different numerical results: what we have before us are as many different laws as there are values of n_s, or for every material a different law. They are what would be called 'individual' laws—if there was such a thing".

Van Wazer et al. (1963) suggest that the dimensional opposition can be overcome if the apparent viscosity at any rate of shear is regarded as having arisen by some change from a reference viscosity (η_r) at unit rate of shear

$$\eta_p = \frac{\delta p}{\delta D} = \eta_r \, D^{n_s - 1} \tag{3.3}$$

and, by integrating this equation with respect to stress (dp) and rate of shear (dD), the power law is derived

$$p = \left[\frac{\eta_r}{n_s - 1} \right] D^{n_s} = k_c \, D^{n_s}. \tag{3.4}$$

Many structural dispersed systems show two regions in which the viscosity is independent of shear rate. One of these regions is at very low rates of shear, and the other is at very high rates of shear. For such systems a power law can only hold, at best, for that part of the viscosity-rate of shear curve where the viscosity decreases with increasing rate of shear.

Many p–D curves which are only slightly curvilinear will give a reasonably straight line on a double logarithmic scale, and in such cases the power law cannot be regarded as valid. When the p–D curve shows pronounced curvature, and a double logarithmic plot is linear over several decades of shear rate, then the power law relationship is significant. In this case, by using exponential equations to define D and p, the validity of a power law can be proved (Scott Blair, 1965; Scott Blair and Prentice, 1966, Scott Blair, 1967). An interlinked dispersed system will show a fall in apparent viscosity as either p or D increases. If the incremental increase in stress (Δp) is assumed to be inversely related to the number (Δn) of linkages which are broken, then

$$-\frac{\Delta n}{\Delta p} = \frac{1}{p} \qquad (3.5)$$

or

$$\Delta n = -\frac{\Delta p}{p} \qquad (3.6)$$

For every increase in stress Δp there is a corresponding increase in ΔD. It is assumed (Scott Blair, 1967) that when D is doubled the number of linkages which can reform is halved

$$-n_s \frac{\Delta n}{\Delta D} = \frac{1}{D} \qquad (3.7)$$

where $n_s > 1$ when viscosity decreases with increasing rate of shear

From Eqs (3.6) and (3.7), and replacing Δ by d so as to cover the experimental range of p and D,

$$\frac{dp}{dD} = \frac{1}{n_s} \cdot \frac{p}{D} \qquad (3.8)$$

which is another way of formulating the power law. When the system has a finite yield value Eq. (3.8) becomes

$$\frac{dp}{dD} = \frac{1}{n_s} \left(\frac{p - p_0}{D} \right) \qquad (3.9)$$

which is equivalent to Eq. (3.2).

An alternative proof has also been proposed (Scott Blair and Prentice, 1966). In this case Eq. (3.5) is written as

$$-\frac{dn}{dp} = \frac{C_1}{p} \qquad (3.10)$$

where C_1 is a constant, so that the number of linkages which are now broken depends upon the stress in a rather more complicated way. When integrated this gives

$$n = C_2 - C_1 \ln p. \qquad (3.11)$$

When $\ln p = 0$, i.e. $p = 1$ dyne/cm^2, let $n = C_3$, so that $C_2 = C_3$, and

$$C_1 \ln p = C_3 - n \qquad (3.12)$$

Now, the influence of rate of shear on the breakage of linkages can be given in a form similar to Eq. (3.10)

$$-\frac{dn}{dD} = \frac{C_4}{D} \tag{3.13}$$

which, when integrated, becomes

$$C_4 \ln D = C_5 - n \tag{3.14}$$

where C_5 is the value of n when $D = 1 \text{ sec}^{-1}$.

Combining Eqs (3.11) and (3.14)

$$C_1 \ln p - C_4 \ln D = C_3 - C_5. \tag{3.15}$$

Therefore,

$$D = \frac{p/C_1 C_4}{\exp\left[(C_3 - C_5)/C_4\right]} \tag{3.16}$$

which is yet another way for presenting the power law.

B. Theories of non-Newtonian Flow Based on Structure Breakdown by Shear

No completely satisfactory theory has been proposed yet to explain non-Newtonian flow. Those theories which are now available can be fitted to the limited amount of experimental data for which they were developed, but they cannot successfully predict much of the experimental data produced by other workers. With the exception of the Williamson (1929) theory for pseudoplasticity most theories are based on the kinetics of the flocculation–deflocculation reaction which proceeds under the influence of Brownian motion and rate of shear. Flocculation due to Brownian motion is known as perikinetic flocculation; that due to shear is called orthokinetic flocculation. The probability (J_c) of collisions due to shear between solid particles, thus leading to flocculation, is given (von Smoluchowski, 1917; Tuorila, 1927) by

$$J_c = \tfrac{4}{3} N_i N_j (R_{ij})^3 D \tag{3.17}$$

where N_i and N_j are the number of particles of types i and j per cm^3, R_{ij} is the radius of collision of a central particle i with a particle j.

The probability (I) of collisions due to Brownian motion between type i and type j particles is

$$I = 4\pi \mathscr{D}_{ij} R_{ij} N_i N_j \tag{3.18}$$

where $\mathscr{D}_{ij}$ is the mutual diffusion constant.

Thus, the ratio of the two collision probabilities is

$$\frac{J}{I} = \frac{(R_{ij})^2 D}{3\pi D_{ij}}. \tag{3.19}$$

But

$$R_{ij} = r_i + r_j \tag{3.20}$$

where r_i and r_j are the respective radii of particles i and j, and

$$D_{ij} = \frac{2kT}{3\pi\eta_0 R_{ij}} \tag{3.21}$$

where k is Boltzmann's constant, T is the temperature on the absolute scale, and η_0 is the viscosity of the fluid medium in which the particles are dispersed. Therefore,

$$\frac{J}{I} = \frac{\eta_0 (r_i + r_j) D}{2kT}. \tag{3.22}$$

The ratio of the two collision probabilities is strongly dependent on the size of the particles. At a rate of shear of 1 $\sec^{-1}$ the ratio rises above unity when one of the particles has a radius greater than 0·5 μ. In dispersions with larger particles flocculation is mainly due to shear.

The precise influence of shear on flocculation is not known, and in most theories of non-Newtonian flow a proportionality factor is adopted, based upon the assumption that the particles form long chains when they come in contact. If the particles were to assume a spherical configuration the influence of shear could well be in the form of an exponential in D rather than a product of D and the proportionality factor. Another limitation in most theories is the total disregard for the influence of shear on aggregate size, and also for the possible influence of aggregate size on the rate constants for flocculation and deflocculation.

Only two theories offer alternative concepts to the above. One of these, based on the theory of rate processes, assumes a structure composed of several flow units, whereas the other, based on reaction kinetics, makes no assumption regarding the flocculate structure other than that shear causes some structural change. It is not easy to apply either of these theories to experimental data.

1. Williamson's Theory of Pseudoplasticity

Williamson (1929) suggested that in pseudoplastic flow part of the shearing force is used to break down the flocculated structure, while the remainder is

used to produce flow at the higher rates of shear. His graphical procedure
for calculating the total power (pD) required for linear flow from the p–D
curve is given in some detail because his approach was subsequently extended by both Goodeve (1938) and Gillespie (1960). OGCB in Fig. 3.1
represents a pseudoplastic flow curve, with the area enclosed by RDOT
defining the power required to overcome the plastic resistance when $D = B$,
and the area enclosed by BRDK representing the power associated with
viscous deformation. The procedure is now as follows. From G, which is a
point on the curve below where it becomes linear, a line GF is drawn parallel
to BD. The perpendicular lines GE and FM are drawn so as to enclose a
rectangle GMFE, which represents the power used in overcoming viscous
resistance ($p_l D$). XMFO is the power used to overcome plastic resistance

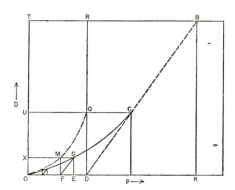

FIG. 3.1. Williamson's (1929) graphical analysis of a pseudoplastic flow curve

($p_r D$). In similar fashion a number of other points are taken on the flow
curve below C, and the procedure is repeated. This produces another curve
OMQ which depicts the shear stress required to overcome plastic resistance
as the rate of shear alters.

The total power (pD) is given by

$$pD = p_r D + p_l D. \tag{3.23}$$

Equation (3.23) can be written as

$$p = \frac{p_\infty D}{C_w + D} + \eta_\infty D \tag{3.24}$$

where C_w is a measure of the curvature of OMQ, and p_∞ is the value of p_r as
D approaches infinity, and η_∞ is the viscosity at infinitely high rate of shear.
When $p_\infty = 0$, p and D will be linearly related as in Poiseuille's equation;
when $C_w = 0$, flow is defined by the Bingham expression (Eq. 1.24).

Williamson's analysis gives satisfactory results only when the linear and curved portions of the p–D curve are well defined.

Equation (3.24) can be rewritten as

$$\frac{1}{p-\eta_\infty D} = \frac{1}{p_\infty} + \left(\frac{C_w}{p_\infty}\right)\left(\frac{1}{D}\right) \tag{3.25}$$

so that a straight line should be obtained when the left hand side of the equation is plotted against $1/D$. Shangraw et al. (1961) found this not to be the case either at low or high rates of shear, the experimental data falling below the theoretical equivalents. An additional equation for plastic flow was then introduced

$$p_2 = p_\infty + \eta_\infty D \tag{3.26}$$

It was assumed that the viscoelastic resistance, which is inversely proportional to the rate of shear, decreases at a rate which is proportional to the residual resistance

$$\frac{dS_1}{dD} = -b_1 p_1 \tag{3.27}$$

where b_1 is a constant of proportionality. After rearrangement, integration gives

$$p_1 = b_v \exp(-b_1 D) \tag{3.28}$$

where b_v is the coefficient of viscoelastic resistance.

Combining Eqs (3.26) and (3.28)

$$p_{corr} = p_2 - p_1 = p_\infty + \eta_\infty D - b_v \exp(-b_1 D). \tag{3.29}$$

Since $\exp(-b_1 D) \to 1$, the yield value is given by

$$\underset{D \to 0}{p_{corr}} = p_\infty - b_v \tag{3.30}$$

but it is difficult to apply Eq. (3.29) to experimental data because b_1 is not known. An arbitrary value of 0·001 was used by Shangraw et al. (1961).

2. The "Impulse" Theory of Pseudoplasticity

Goodeve (1939) extended the concept that two independent effects contribute to pseudoplasticity, viz. a Newtonian effect where the shearing force is proportional to the rate of shear, and a thixotropic effect where the shearing force is constant irrespective of rate of shear. When particles in dispersions, or globules in emulsions, make contact, links are formed between them. Under the influence of shear, these links are stretched, distorted and ruptured. Redevelopment may then follow. Simultaneously momentum is transferred from a moving layer to an adjacent slower moving layer, and

$$\frac{F}{A} = b_G D + w. \tag{3.31}$$

Dividing through by D gives

$$\eta_\infty = b_G + (w/D) \tag{3.32}$$

where b_G is the residual viscosity, and w is the coefficient of thixotropy which is equivalent to the momentum transferred by particle or globule interaction/sec/unit area from one moving layer to the next.

The Newtonian and thixotropic factors quoted in Eqs (3.31) and (3.32) were derived by considering the force (F) resisting link distortion in the thixotropic contribution.

$$F = E_m x \tag{3.33}$$

where E_m is the modulus of elasticity, and the extension x is approximately $l_c Dt$, where l_c is the distance through which F is transmitted. The mean life of each link (t_c) is

$$t_c = \frac{F_c}{E_m l_c D} \tag{3.34}$$

where F_c is the critical value of F at which a link ruptures.

The impulse (I_f) which is transmitted from one moving layer to the slower moving adjacent layer during shear is inversely proportional to D.

$$I_f = \int_0^{F_c} F \, dt \tag{3.35}$$

and substituting for F_c from Eq. (3.34)

$$I_f = \int_0^{t_c} E_m l_c Dt \, dt = \frac{E_m l_c v t_c^2}{2} = \frac{F_c^2}{2E_m l_c D}. \tag{3.36}$$

The total force (F_T) between adjacent moving layers is the product of the average impulse $(F_c/2)$ and the number of impulses/sec $(= N_l l_c/t_c)$

$$F_T = \frac{F_c N_l l_c}{2} \tag{3.37}$$

where N_l is the number of links/cm^3.

The time span of an impulse in the thixotropic contribution to non-Newtonian flow was assumed to be controlled by the rate of shear. However, if the links are so weak that the energy required to break them is less than $1\,kT$, then thermal energy will be mainly responsible for their rupture, and the flow is Newtonian. Under these conditions the probability of link destruction is

$$\frac{1}{t_{N_c}} = f\exp\left(-E/kT\right) \tag{3.38}$$

where f is the frequency with which energy redistribution occurs between a link and its surroundings, and E is the energy required to break a link. Shear now pulls on a link only for a time t_N, so Eq. (3.36) becomes

$$I_f = \frac{E_m\, l_c\, D\, t_N^{\,2}}{2} \tag{3.39}$$

and the total force is

$$F_N = \frac{N_l\, E_m\, l_c^{\,2} D \exp\left(+E/kT\right)}{2f} \tag{3.40}$$

so that F_N is directly proportional to the rate of shear. Equation (3.40) shows the well established exponential relationship between Newtonian viscosity and temperature. Finally,

$$F = F_T + F_N. \tag{3.41}$$

Goodeve (1939) considered that in simple thixotropic systems at high rates of shear, and in systems with low Brownian motion, link formation and rupture proceed mainly under the influence of shear. Gillespie (1960) applied a similar approach to that given by Eqs (3.31)–(3.41), but he now included the effect of Brownian motion on the kinetics, which was disregarded by Goodeve. Thus, link formation and rupture have two rate constants each of which is controlled by Brownian motion and by shear.

The generalised equation for the rate of link formation is

$$\frac{dN_g}{dt} = K_1 N_0 - K_2 N_g \tag{3.42}$$

where N_g is the average number of links/particle, N_0 is the original number of particles before flocculation takes place, and K_1 and K_2 are the rate constants for link formation and rupture respectively. In view of the influence of both Brownian motion and shear on K_1 and K_2, the latter can be defined by

$$K_1 = (K_3 + K_4\, D)\, N_b \tag{3.43}$$

and
$$K_2 = K_5 + K_6 D \qquad (3.44)$$

where K_3 and $K_4 D$ are the rate constants for flocculation by Brownian motion and shear respectively, and K_5 and $K_6 D$ are the rate constants for deflocculation. N_b is the number of links which are formed between two contacting particles.

Now, in the equilibrium state

$$\frac{dN_g}{dt} = 0 \qquad (3.45)$$

so that,

$$N_g = \frac{K_1 N_0}{K_2}. \qquad (3.46)$$

Two particles must share a link, so Eq. (3.37) yields

$$N_l = \frac{K_1 N_0^2}{K_2}. \qquad (3.47)$$

It is assumed that $t_c = 1/K_2$, so Eqs (3.34), (3.37) and (3.47) give

$$F_T = \frac{E_m l_c^2 K_1 N_0^2 D}{4K_2^2}. \qquad (3.48)$$

Provided the particles do not make contact at more than one point, then (von Smoluchowski, 1916, 1917)

$$K_3 = 8\pi \mathscr{D}_{ij} r_{eff} W \qquad (3.49)$$

where r_{eff} is the effective radius of contact, and W is the probability of collision between two particles. Furthermore, it can be shown that

$$K_4 = \frac{32 r_{eff}^3}{3} \qquad (3.50)$$

and

$$K_b = \frac{\bar{l}_c}{l_v} \qquad (3.51)$$

where $\bar{l}_c$ is the mean value of l_c, and l_v is the critical distance separating two particles when the link between them ruptures.

The shear force is derived by combining Eqs (3.31), (3.43), (3.48), and (3.51)

$$\frac{F}{A} = D + \frac{E_m l_v^2 N_b N_0^2 (K_4 D + K_3) D}{4(K_5/K_6 + D)^2} \qquad (3.52)$$

and the total binding energy between two particles is $\frac{1}{2} E_m l_v^2 N_b$.

At high rates of shear link formation and rupture proceed under the influence of shear only, and

$$\frac{F}{A} = \eta_\infty D + \frac{E_m l_v^2 N_b N_0^2 K_4 D}{4(K_5/K_6) + 2D} \tag{3.53}$$

which closely resembles Eq. (3.24).

Most of the parameters in Eqs (3.52) and (3.53) can be determined graphically, with the exception of K_5 and K_6 for which one has to resort to first principles. Equation (3.53) can be divided through by D to give

$$\eta_{eff} = \eta_\infty + \frac{E_m l_v^2 N_b N_0^2 K_4}{4(K_5/K_6) + 2D} \tag{3.54}$$

so that a plot of η_{eff} against $1/D$ should be linear at high values of D, thus giving an intercept with a value of η_∞. Equation (3.53) can be written still another way

$$\frac{1}{\eta_{eff} - \eta_\infty} = \frac{4(K_5/K_6) + 2D}{E_m l_v^2 N_b N_0^2 K_4} \tag{3.55}$$

so that the denominator on the right hand side is derived from the slope of the straight line obtained at high rates of shear in the plot between $1/(\eta_{eff} - \eta_\infty)$ and D.

For pseudoplastic polystyrene latex dispersions with volume fractions of latex ranging from 0·13 to 0·28 the rate constants ratio K_5/K_6 ranged between 51 and 107, showing no relationship between volume fraction and the K_5/K_6 ratio. The rate constants ratio K_3/K_4 ranged between 19 and 55.

3. The Casson Equation

Casson (1959) assumed that flocculated particles form long chains due to attraction forces which bind the particles together. At very low rates of shear the chains move as single entities, but at higher rates of shear they break down into smaller units. Eventually, at very high rates of shear the flocculated structure is completely broken down and the viscosity is due to the flow and interaction of individual particles. To simplify the mathematical treatment he adopted a model of rigid cylindrical rods with a high length to width ratio for the chains of particles. Casson (1959) justified his assumption of a chain like structure for the flocculated particles on the grounds that a chain has a greater number of possible configurations than other more compact models.

By considering the flow of the fluid with respect to individual rods, and the total rate of energy dissipation due to all the rods, the following viscosity equation was obtained for very dilute dispersions

$$\frac{\eta}{\eta_0} = 1 + (a_x a_3 - 1)\,\phi + \eta_0^{\frac{1}{2}} a_x\,a_4\,\phi/D^{\frac{1}{2}} \tag{3.56}$$

where η is the viscosity of the dispersion, η_0 is the viscosity of the fluid in which the particles are dispersed, a_3 and a_4 are constants such that the axial ratio

$$J = a_3 + a_4\,y_c,$$

where

$$y_c = \frac{1}{(\eta_0\,D)^{\frac{1}{2}}}, \qquad a_x = \frac{(\tan\theta_c\,\cos\theta_p)^2}{(\tan\theta_c\,\cos\theta_p)^2 + 1},$$

where θ_c and θ_p define the orientation of the rods in shear with respect to the x, y, z axes. In developing Eq. (3.56), any effects due to Brownian motion were ignored as they were considered insignificant for rods of finite thickness.

Equation (3.56) was then extended to higher concentrations by considering the effect of adding an additional volume $\delta\phi$ of particles. The original dispersion contained a volume of particles equivalent to $(1-\delta\phi)$. The more concentrated dispersion can be regarded as a very dilute system with a volume concentration $\delta\phi$ dispersed in a continuous medium which is formed by the original dispersion. The new volume (ϕ^*) of particles is now

$$\phi^* = \phi(1-\delta\phi) + \delta\phi \tag{3.57}$$

so that the increase in volume concentration is

$$d\phi = \phi^* - \phi = \delta\phi(1-\phi) \tag{3.58}$$

and

$$\delta\phi = \frac{d\phi}{1-\phi} \tag{3.59}$$

If the viscosity of the original dispersion was η, this becomes the viscosity of the continuous phase in the new dispersion. It follows from Eq. (3.56) that the new viscosity (η^*) is given by

$$\frac{\eta^*}{\eta} = 1 + (a_x a_3 - 1)\,\delta\phi + \eta^{\frac{1}{2}} a_x\,a_4\,\delta\phi/D^{\frac{1}{2}}. \tag{3.60}$$

The small difference $(d\eta)$ in viscosity between the new and old dispersions is $\eta^* - \eta$, so that Eqs (3.56), (3.59) and (3.60) give

$$d\eta = \left[\eta(a_x a_3 - 1) + \frac{\eta^{\frac{1}{2}} a_x a_4}{D^{\frac{1}{2}}} \right] \frac{d\phi}{1 - \phi}. \tag{3.61}$$

Let

$$(a_x a_3 - 1) = A_c$$

and

$$(a_x a_4 / D^{\frac{1}{2}}) = B_c$$

then Eq. (3.61) becomes by rearrangement

$$\frac{d\eta}{(A_c \eta + B_c \eta^{\frac{1}{2}})} = \frac{d\phi}{1 - \phi}. \tag{3.62}$$

When integrated with the boundary condition $\eta = \eta_0$ when $\phi = 0$,

$$\eta^{\frac{1}{2}} = \left[\frac{\eta_0}{(1 - \phi)^{A_c}} \right]^{\frac{1}{2}} + \frac{B_c}{A_c} \left[\left(\frac{1}{1 - \phi} \right)^{A_c/2} - 1 \right] \tag{3.63}$$

To obtain the relationship between shear stress and rate of shear both sides are multiplied by $D^{\frac{1}{2}}$

$$\eta^{\frac{1}{2}} D^{\frac{1}{2}} = p^{\frac{1}{2}} = \left[\frac{\eta_0}{(1 - \phi)^{A_c}} \right]^{\frac{1}{2}} D^{\frac{1}{2}} + \frac{B_c D^{\frac{1}{2}}}{A_c} \left[\left(\frac{1}{1 - \phi} \right)^{A_c/2} - 1 \right] \tag{3.64}$$

or

$$p^{\frac{1}{2}} = K_c D^{\frac{1}{2}} + K_0 \tag{3.65}$$

where

$$K_c = \left[\frac{\eta_0}{(1 - \phi)^{A_c}} \right]^{\frac{1}{2}} \tag{3.66}$$

and

$$K_0 = \frac{B_c D^{\frac{1}{2}}}{A_c} \left[\left(\frac{1}{1 - \phi} \right)^{A_c/2} - 1 \right]. \tag{3.67}$$

A plot of $p^{\frac{1}{2}}$ against $D^{\frac{1}{2}}$ [Eq. (3.65)] should give a straight line with a slope K_c and an intercept K_0. Equations (3.66) and (3.67) then give α_3 and α_4, because of their relationships to A_c and B_c respectively, provided α_x is known. The latter quantity is not known, however, and Casson assumed a value of $(0 \cdot 5)^{\frac{1}{2}}$.

There appear to be two weaknesses in Casson's approach. The first of these is in the linear chain model proposed to define the flocculated structure.

Weymann (1965) points out that, although statistical arguments indicate that a chain has many possible configurations, the most likely one is a coil in which cross-links are formed so as to reduce the potential energy. The second weakness lies in the extension of Eq. (3.56) to more concentrated dispersions. This equation was developed for non-interacting chains, and no allowance is made in the subsequent treatment for chain interaction. Therefore Eq. (3.65) cannot be valid for other than dilute dispersions.

Scott Blair (1966) pointed out that Eq. (3.2) is just as valid as Eq. (3.67) for viscosity data at low rates of shear, and the former is to be preferred in some ways because the concept on which it is based can be more easily justified. The former equation does not fit data obtained on dilatant systems at high rates of shear. If such systems show a finite rate of shear for negligible stresses, which is quite feasible, Eqs (3.2) and (3.67) can be rewritten as

$$p = A^1 (D - D_0)^{n^1} \qquad (3.68)$$

and

$$p^{\frac{1}{2}} = B^1 (D^{\frac{1}{2}} - D_0{}^{\frac{1}{2}}) \qquad (3.69)$$

and then both equations fit the whole range of viscosity data equally well.

4. The Cross Equation

Cross (1965) assumed that flocculated particles form random linked chains, the size of the latter depending upon the rate of shear. His approach resembles that of Gillespie (1960) in that he treated link rupture as a process which is influenced by both Brownian motion and shear, but link formation, on the other hand, was considered to be influenced by Brownian motion only.

In pseudoplastic flow the equilibrium state, at any rate of shear, of a chain having an average number of links L_v was postulated as

$$\frac{dL_v}{dt} = 0 = K_3 N_t - (K_5 + K_6 D^e) L_v \qquad (3.70)$$

with D^e taken as an even function. It follows that under these conditions

$$L_v = \frac{K_3 N_t}{K_5 + K_6 D^e}. \qquad (3.71)$$

If $L_v = 0$, when $D = 0$

$$L_0 = \frac{K_3 N_t}{K_5} \qquad (3.72)$$

and, therefore

$$\frac{L_v}{L_0} = \frac{K_5}{K_5 + K_6 D^e} = \frac{1}{1 + (K_6/K_5) D^e}. \tag{3.73}$$

An analogy was drawn between the dependence of the viscosity of a linear polymer on its molecular weight, i.e. the number of segments in a chain, and the viscosity of a random chain of spherical particles. For a linear polymer

$$\eta_v = \eta_\infty + B_1 L_v \tag{3.74}$$

where B_1 is a constant. If $\eta_v = \eta$ when $L_v = L_0$,

$$\frac{\eta_v - \eta_\infty}{\eta - \eta_\infty} = \frac{L_v}{L_0}. \tag{3.75}$$

Combining Eqs (3.73) and (3.75)

$$\eta - \eta_\infty = (\eta_v - \eta_\infty) \left(1 + \frac{K_6}{K_5} D^e \right). \tag{3.76}$$

This is the basic relationship between equilibrium viscosity and rate of shear. Unfortunately rather tedious graphical procedures are required to determine the values of K_6/K_5 and D, and furthermore, the constant c is not known for all conditions of K_6/K_5. Equation (3.76) can be rewritten in the form

$$\frac{1}{\eta_v - \eta_\infty} = \frac{1 + (K_6/K_5) D^e}{\eta - \eta_\infty} \tag{3.77}$$

In accordance with Eq. (3.77) Cross (1965) plotted $1/(\eta_v - \eta_\infty)$ against D^e for arbitrary values of c. A linear plot was obtained when $c = 2/3$, and this was subsequently found to hold for many other dispersions.

5. Flow and the Kinetics of Globule Aggregation

Van den Tempel (1963) and de Vries (1963) also believed that the viscosity of a dispersion or emulsion at any rate of shear depended upon the flocculation $\rightleftharpoons$ deflocculation reaction. It was assumed that flocculation proceeds according to von Smoluchowski's (1916, 1917) theory of slow flocculation, and that the kinetics of the flocculation $\rightleftharpoons$ deflocculation reaction is the controlling factor rather than the rate of link formation as proposed by Goodeve (1939), Gillespie (1960), Casson (1959), and Cross (1965).

In addition to the probability of collisions due to Brownian motion as defined by Eq. (3.18) the probability of collisions due to a velocity gradient is given by Eq. (3.17). To these two terms, which must be summed over the whole dispersion, must now be added a term for the influence of Brownian motion on deflocculation due to collision between k-type flocculates, which are formed by union of i type and j type flocculates on collision, and flocculates of other sizes, and a term for deflocculation due to the velocity gradient. These terms are collectively

$$-N_k \sum_{i=1}^{\infty} 4\pi \mathcal{D}_{ik} R_{ik} N_i - N_k \sum_{i=1}^{\infty} \tfrac{4}{3}(R_{ik})^3 N_i D$$

Van den Tempel (1963) introduced two additional terms to define the rupture of flocculates by the velocity gradient. These are

$$\sum_{m=k+1}^{\infty} 2N_m(P_f D) - N_k(P_f D)(k-1)$$

where P_f is a proportionality factor which allows for the effect of D. In order to derive these latter terms it was assumed, as did Casson (1959) previously, that the flocculated particles formed long rigid chains, although it was realised that this assumption would probably be valid only when the flocculates did not contain many particles. According to the long straight chain model an m type flocculate would contain $(m-1)$ links.

Under equilibrium conditions the summations of Eqs (3.17) and (3.18) can be combined with the other four terms to give the zero rate of change in the number of k type flocculates. After allowing for Eqs (3.20) and (3.21)

$$\frac{dN_k}{dt} = 0 = \frac{2kT}{3\eta_0} \left[\sum_{i+j=k} N_i N_j - 2N_k \sum_{i=1}^{\infty} N_i \right]$$

$$+ \tfrac{2}{3} D_s^{\ 3} D \left[\sum_{i+j=k} N_i N_j - 2N_k \sum_{i=1}^{\infty} N_i \right]$$

$$+ (P_f D) \left[2 \sum_{m=k+1}^{\infty} N_m - N_k(k-1) \right]. \tag{3.78}$$

The solution of Eq. (3.78) for the number of primary particles, i.e. the particles which remain unassociated with other particles, is obtained by making $k = 1$

$$0 = \frac{2kT}{3\eta_0} (-2N_1 N) + \tfrac{2}{3} DD_s^{\ 3}(-2N_1 N) + 2(P_f D)(N - N_1) \tag{3.79}$$

where

$$N = \sum_{k=1}^{\infty} N_k.$$

For paired particles

$$\left(\frac{2kT}{3\eta_0} + \tfrac{2}{3}DD_s^3\right)(2NN_2 - N_1^2) = (P_f D)(2N - 2N_1 - 3N_2). \quad (3.80)$$

Only the constant P_f remains unknown, but it is claimed that it can be derived in principle from the way in which viscosity decreases with increasing rate of shear.

A dispersion, or emulsion, containing flocculated particles shows a higher viscosity than when the particles are unassociated. This is due to immobilisation of fluid in the interstices between particles within the flocculate structure. When a low shear rate is applied each flocculate moves as a single entity with a volume that is larger than the total volume of particles within the flocculate. The effective volume fraction (ϕ_a) of dispersed phase is given (Mooney, 1946) by

$$\phi_a = \tfrac{4}{3}\pi r^3 [N_1 + f_s(N_0 - N_1)] \quad (3.81)$$

where f_s, the swelling factor, is a measure of the immobilised fluid, and N_0 is the initial number of particles per unit volume. Equation (3.79) in conjunction with Eq. (3.81) gives ϕ_a as a function of N_0 and D, if f_s is known. Discussion in Section 2.C shows how f_s can be determined. For concentrated oil-in-water (O/W) emulsions containing 10^{10} and 10^{11} particles/cm^3, P_f was taken as 20, and f_s as 1·6 over the shear rates 10^{-4}–10^2 sec^{-1}. Values of ϕ_a were then inserted in the viscosity equation proposed by Mooney (1946) so as to calculate the viscosity-rate of shear relationship;

$$\eta_{\text{rel}} = \frac{\surd(1 + 0·5\phi)}{1 - \phi} \exp\left(\frac{1·25\phi}{1 - \phi}\right) \quad (3.82)$$

where η_{rel} is the relative viscosity (ratio of the viscosity of the emulsion to the viscosity of the continuous water phase) when the system is completely deflocculated, and ϕ is replaced by ϕ_a. Figure 3.2 shows double logarithmic plots of the calculated viscosity data against rate of shear for the two values of N_0. There were discrepancies between experimental and calculated viscosity data at the lower end of the shear rate range because the aggregates were probably larger than was assumed in developing the theory. Hence, the linear chain model no longer applies in this region.

Two other comments may help explain these discrepancies. First f_s should not be assumed to be independent of shear rate. The flocculates formed in the stationary emulsion will break down to an increasing extent as the rate of shear rises. Consequently the volume of immobilised fluid, and hence f_s, will decrease. The second, and perhaps more important, comment concerns the validity of Eq. (3.82). A critical discussion on this point is given in

Section 5, so for the moment all that will be said is that its main weakness lies in its having been derived empirically from the analysis of very limited viscosity data.

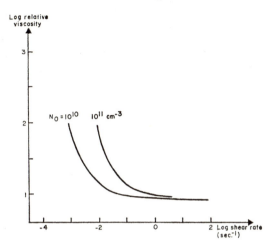

FIG. 3.2. Calculated relation between viscosity and shear rate for O/W emulsion $\phi = 0.50$; $P_f = 20$; $f_s = 1.60$ (Van den Tempel, 1963).

6. *Theory of Rate Processes*

Ree and Eyring (1955) and Kim *et al.* (1960) believed, as did Williamson (1929), Goodeve (1938), and Gillespie (1960), that flow involves two principle types of units, one Newtonian and one or more non-Newtonian units with different characteristics. Following the theory of rate processes (Glasstone *et al.*, 1941) it was assumed that a particle cannot move past its neighbors until it has surmounted a finite potential energy barrier. In their model the surface of each unit is subdivided into localised areas $x_1, x_2, ..., x_n$, with shear stresses per unit area $p_a, p_b, ..., p_n$, acting on these areas. All units in the same shear plane move with identical shear rate D.

Newtonian units are not linked to similar units on the opposite side of a shear surface so that they shear readily, but this is not the case for the non-Newtonian units since they are connected to similar units on the other side of the shear surface.

According to Glasstone *et al.* (1941) the shear rate is given by

$$D = (\lambda^e/\lambda_1) \, 2k' \sinh a_n p_n = \frac{\sinh a_n p_n}{\beta_n} \qquad (3.83)$$

where k' is the rate constant for flow of a unit which belongs to the *n*th group of units, i.e. for a particle to pass over the potential energy barrier, and

$$a_n = \frac{(\lambda^e \lambda_2 \lambda_3)_n}{2kT}$$

where λ^e is the distance that a unit moves between equilibrium positions, λ_1 is the distance between planes of flow units of one type, and $\lambda_2 \lambda_3$ is the cross sectional area of a given flow unit. The denominator β_n is the mean relaxation time, and

$$\beta_n = \frac{1}{(\lambda^e/\lambda_1)_n \, 2k^1}$$

while

$$k^1 = \frac{kT}{h} \, \exp\left(-\Delta F_n/RT\right)$$

with ΔF_n equal to the free energy of activation for flow of the nth unit, h is the Planck constant, k is the Boltzmann constant, and T is the absolute temperature.

The stress is given by

$$p = \sum_{n=1}^{n} x_n p_n \tag{3.84}$$

where $x_n p_n$ is the force acting on the nth group of units. It follows from Eq. (3.83) that

$$p = \sum_{n=1}^{n} \frac{x_n}{a_n} \sinh^{-1}\beta_n D. \tag{3.85}$$

The viscosity at any rate of shear is then

$$\eta(D) = \frac{p}{D} = \sum_{n=1}^{n} \frac{x_n \beta_n}{a_n} \cdot \frac{\sinh^{-1}\beta_n D}{\beta_n D}. \tag{3.86}$$

This is the basic viscosity equation. When $\beta_n D \to 0$

$$\lim \frac{\sinh^{-1}\beta_n D}{\beta_n D} = 1$$

and, when $\beta_n D \to \infty$

$$\lim \frac{\sinh^{-1}\beta_n D}{\beta_n D} = 0$$

Over the whole range of shear rates it is assumed that four conditions are satisfied.

(a) $\beta_1 D \ll 1$

(b) $\beta_2 D \geqslant 1$

(c) $\beta_3 D \gg 1$

(d) $\beta_n D \ggg 1$ (where $n \geqslant 4$)

Units conforming with (a) are Newtonian, while units conforming with conditions (b)–(d) are non-Newtonian. Units falling under (b) differ from those which come under (c) and (d) in that they show Newtonian behavior at low rates of shear, and only become non-Newtonian at high rates of shear.

In the experimental systems studied it was found that n did not usually exceed three. Maron and Pierce (1956) obtained this value for latex emulsions, so that Eq. (3.86) becomes

$$\eta(D) = \frac{x_1 \beta_1}{a_1} + \frac{x_2 \beta_2}{a_2} \cdot \frac{\sinh^{-1}\beta_2 D}{\beta_2 D} + \frac{x_3 \beta_3}{a_3} \cdot \frac{\sinh^{-1}\beta_3 D}{\beta_3 D}$$

with $x_1 \beta_1 / a_1$ representing the Newtonian contribution.

The normal procedure to determine the values of the various parameters is to plot $\eta(D)$ against $\log D$ over a sufficiently wide range of shear rates so that a plateau is obtained at both the upper and lower ends of the range. When Eq. (3.86) is differentiated

$$\frac{d\eta(D)}{d \ln D} = \frac{x_n \beta_n}{a_n} \left\{ \frac{1}{[(\beta_n D)^2 + 1]^{\frac{1}{2}}} - \frac{\sinh^{-1}(\beta_n D)}{\beta_n D} \right\}$$

and further differentiation leads to

$$\frac{d^2\eta(D)}{d (\ln D)^2} = \frac{x_n}{a_n} \cdot \frac{\beta_n}{D^2} \left\{ \sinh^{-1}(\beta_n D) - \frac{\beta_n D}{[(\beta_n D)^2 + 1]^{\frac{1}{2}}} - \frac{(\beta_n D)^2}{[(\beta_n D)^2 + 1]^{3/2}} \right\}$$

$$(3.87)$$

An inflexion point is found in the curve where the left hand side of Eq. (3.87) equals zero, so it follows that $\beta_n D_i = 2.90$, where D_i is the value of D at the inflexion point. From Eq. (3.86)

$$-\left[\frac{d\eta(D)}{d \log D} \right] = 0.290 \times 2.303 \frac{x}{a} \beta_n \qquad (3.88)$$

and

$$\frac{x}{a} = 0.5162D \left[\frac{-\mathrm{d}\eta(D)}{\mathrm{d}\log D} \right]. \tag{3.89}$$

Thus x_1/a_1 and β_n can be determined, and the viscosity η_1 of the principal flow unit is then calculated from Eq. (3.86). Figure 3.3 shows the result of such an analysis. The next stage is to plot $(\eta - \eta_1)$ against $\log D$. This second curve will be flat over a wider range of shear rates than was the original curve, but at high values of D an inflexion point may still be observed. If this proves to be the case the calculations are repeated to obtain η_2. When $[\eta - (\eta_1 + \eta_2)]$ is plotted against $\log D$ in the example given in Fig. 3.3 a straight line is obtained. This line represents the Newtonian viscosity, and no further calculation is required.

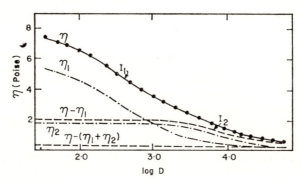

FIG. 3.3. Analysis of non-Newtonian viscosity of a solution of polyacrylonitrile in diethyl formamide (Kim *et al.*, 1960).

In spite of the attractiveness of this theory it is severely limited in its practical application by the difficulties experienced in identifying the inflexion points. If only two types of flow units are involved, the parameter β_n can be derived from viscosity measurements η_α, η_β, η_γ, at three different rates of shear D_α, D_β, and D_γ (Gabrysh *et al.*, 1963), since it can be shown that in this case

$$\beta_n \approx \log^{-1} X_G / 2D_\alpha \tag{3.90}$$

where

$$X_G = \left[\log \frac{D_\beta}{D_\alpha} - k' \left(\frac{D_\beta}{D_\alpha} \log \frac{D_\gamma}{D_\alpha} - \frac{D_\gamma}{D_\alpha} \log \frac{D_\beta}{D_\alpha} \right) \right] \bigg/ k' \left(\frac{D_\beta}{D_\alpha} - \frac{D_\gamma}{D_\alpha} \right) - 1 + \frac{D_\beta}{D_\alpha}.$$

When more than two types of flow units are present a graphical, albeit rather tedious, procedure can be employed to identify them (Maron and

Sisko, 1957). If three flow units are present, for example, Eq. (3.87) can be rewritten as

$$\eta(D) = \frac{x_1 \beta_1}{a_1} + \frac{x_2 \beta_2}{a_2} \Delta_2 + \frac{x_3 \beta_3}{a_3} \Delta_3 \qquad (3.91)$$

where

$$\Delta_2 = \frac{\sinh^{-1}\beta_2 D}{\beta_2 D}, \text{ and } \Delta_3 = \frac{\sinh^{-1}\beta_3 D}{\beta_3 D}.$$

When $\beta_n D \to 0$

$$\eta(D) = \frac{x_1 \beta_1}{a_1} + \frac{x_2 \beta_2}{a_2} + \frac{x_3 \beta_3}{a_3}. \qquad (3.92)$$

Plots of η against Δ_2 are extrapolated to zero rate of shear. Next, values of β_2 are selected and η is plotted against Δ_2. With suitable values of β_2 the graphs are linear at high rates of shear, but they show curvature at low rates of shear. Straight lines drawn through the linear portion of the graphs give $x_2 \beta_2/a_2$ as the slope and $x_1 \beta_1/a_1$ as the intercept. The value of $x_3 \beta_3/a_3$ can then be calculated by Eq. (3.92). Finally Δ_3 is derived for each value of D using Eq. (3.91) in revised form

$$\Delta_3 = \frac{\eta(D) - [(x_1 \beta_1/a_1) + (x_2 \beta_2/a_2) \Delta_2]}{x_3 \beta_3/a_3}. \qquad (3.93)$$

Knowing Δ_3 and D the parameter β_3 can be calculated.

7. Interpretation of Globule Interaction in terms of the Derjaguin-Landau-Verwey-Overbeek Theory

The interaction between globules after flocculation can be explained in terms of the prevailing forces of repulsion and attraction (Derjaguin, 1934, 1939, 1940; Verwey and Overbeek, 1948) as an alternative to the assumption that links are formed (Goodeve, 1939; Gillespie, 1960) between the globules.

In dispersions or emulsions the repulsion forces are usually of electrostatic origin, their magnitude depending on the electric charge, electrolyte concentration, particle size, and the distance between the particles (H_0). When all the particles are spherical with diameter D_s, and they are dispersed in a polar medium, e.g. an oil-in-water emulsion, the repulsion energy (V_R) is derived by considering the interaction between two spheres. Provided the thickness ($1/\chi$) of the diffuse electrical double layer around the particles is much smaller than D_s so that $\chi D_s/2 > 300$, which normally applies,

$$V_R = \frac{\varepsilon_d D_s \psi_p{}^2}{4} \{\ln[1 + \exp(-\chi H_0)]\} \qquad (3.94)$$

where ε_d is the dielectric constant of the continuous phase, and ψ_p is the potential of the outermost part of the diffuse double layer.

If the particles are dispersed in a non-polar medium, e.g. pigment particles in mineral oil or a water-in-liquid paraffin emulsion, the diffuse double layer is several microns thick. In this case it is no longer feasible to consider the interaction between two spheres in isolation, because the repulsion energy now decreases much more slowly with increasing H_0. Allowance must be made for the superimposed interaction effects between many particles. The expression for the repulsion energy becomes much more complicated (Albers and Overbeek, 1959) as a result.

$$V_R = \frac{\varepsilon_d D_s^2 \psi_p^2 \exp(\chi D_s)}{4} \left\{ 8 \cdot 88 \frac{1 \cdot 3 \chi r_h + 1}{\chi^3 r_h^3} \exp(-1 \cdot 3 \chi r_h) \frac{\exp(\chi q_2) - \exp(-\chi q_2)}{q_2} \right.$$

$$+ \frac{\exp[-\chi(r_h - q_2)]}{r_h - q_2}$$

$$\left. - \frac{6}{\chi r_h q_2} \{ \exp[-\chi(r_h + q_2)] - \exp[-\chi \sqrt{(r_h^2 + q_2^2 - \tfrac{5}{3} r_h q_2)}] \} \right\}$$

$$(3.95)$$

where r_h is the radius of a hypothetical sphere on which the twelve particles in closest proximity to a central reference particle are situated, and q_2 is the mean distance from the centre of the sphere at which any of the twelve particles may be found. The limits of this mean distance are 0 and $(r_h - D_s)$, where

$$r_h = \frac{0 \cdot 905 D_s}{\phi^{\frac{1}{3}}}$$

Equation (3.95) gives a value of V_R which is drastically reduced, the precise reduction depending on the value of χ, as compared with the V_R derived by Eq. (3.94). In fact for most non-polar continuous media V_R can be safely assumed to be zero.

The attraction energy (V_A), which operates over greater distances than the repulsion energy, is not affected by the nature of the continuous phase. Irrespective of whether the latter is polar or non-polar

$$V_A = -\frac{A}{12} \left\{ \frac{D_s^2}{(D_s + H_0)^2 - D_s^2} + \frac{D_s^2}{(D_s + H_0)^2} + 2 \ln \left[1 - \frac{D_s^2}{(D_s + H_0^2)} \right] \right\}$$

$$(3.96)$$

for spherical particles, where A is the London–van der Waals constant (Verwey and Overbeek, 1948). When $H_0 \ll D_s$

$$V_A = - \frac{AD_s}{24 H_0}.$$ (3.97)

If H_0 is less than ~ 160 Å the attraction potential is reduced by time (retardation) effects (Schenkel and Kitchener, 1960), and the modified form of Eq. (3.97) is

$$V_A \approx - \frac{AD_s}{24 H_0} \left[\frac{1}{1+1 \cdot 77 \, (2\pi H_0/\lambda_w)} \right]$$ (3.98)

where λ_w ($\approx 10^{-5}$) is the wavelength of the intrinsic electronic oscillations of the atoms. The majority of dispersions and emulsions are rendered stable by the incorporation of a surface active material which is adsorbed at the surface of the particles to form a protective film. If this film is more than a few Angstroms thick V_A is reduced further (Vold, 1961).

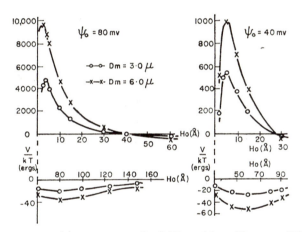

FIG. 3.4. Potential energy curves for O/W emulsions (Sherman, 1963a).

The net potential energy of interaction (V) between the particles is equal to the sum of V_A and V_R. Figures 3.4 and 3.5 illustrate the characteristic shapes of V–H_0 curves for an O/W emulsion stabilised by a non-ionic emulsifying agent containing about 0·5% (wt/wt) soap, and for a W/O emulsion stabilised by a non-ionic emulsifying agent, respectively (Sherman 1963a). All V–H_0 curves show a maximum value (V_{max}) at a certain distance between the particles. If the particles are to move closer together than this critical H_0, then the potential energy barrier V_{max} has to be surmounted. When V_{max} does not exceed a few kT some of the particles are able to cross this barrier, but when it has a value greater than 20–25 kT this is not possible. For the O/W emulsions (Fig. 3.4) the values of V_{max} are very much larger than for the W/O emulsions (Fig. 3.5), which is to be expected from a comparison of Eqs (3.94) and (3.95), and, in addition, V falls away more rapidly from its maximum with increasing H_0 for the former. Because V_{max} is so large for the O/W emulsion it is not possible for particles to pass over this potential energy barrier, and instead they flocculate at a distance of separation corresponding to the position (60–80 Å approx.) of the secondary minimum in the V–H_0 curve.

Although the particles are held together by attraction forces at this point, the attraction is still relatively weak. If the particles were able to pass over the peak in the curve into the primary minimum, where they would be separated only by a few Angstroms, the attraction forces would be very much stronger. Because V_{max} is usually only a few kT for W/O emulsions the particles should be able to pass over the peak. When the particles are not surrounded by an adsorbed layer of surface active material this occurs, and

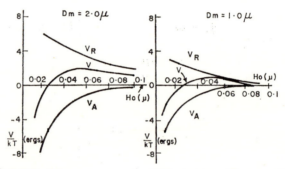

FIG. 3.5. Potential energy curves for W/O emulsions (Sherman, 1963a).

the particles quickly coagulate at the small distance now separating them. More often than not, however, the particles are protected by an adsorbed layer, so that they cannot come closer than twice the thickness of this adsorbed layer. For the data shown in Fig. 3.5 this minimum distance corresponds to 35–40 Å, so that the attraction forces between particles are not large, although they are usually greater than for particles of corresponding size in O/W emulsions which flocculate in the secondary minimum. Dispersions or emulsions containing a non-polar continuous phase do not show a secondary minimum in their V–H_0 curve.

Many practical systems contain particles which are not spherical so that the precise form of the equations to calculate V_R and V_A may not be readily derived. For a suspension of glycerylstearate crystals in groundnut oil or paraffin oil van den Tempel (1961) assumed a value for V_A which was intermediate between that for spherical particles (Eq. (3.97)) and that for cubes with sides equal to the diameter of the former. A previous investigation

(Vold, 1954) had shown that the attraction forces between anisometric particles can generally be calculated in this way.

The attraction potential between well oriented cubes is given by

$$V_A = -\frac{AD_s^2}{12\pi H_0^2} \tag{3.99}$$

in the simplest form. The attraction force for the crystals was taken as intermediate between Eqs (3.97) and (3.99), so that it was proportional to $D_s^{1.5}/H_0^{1.5}$. Because of the non-polar nature of the continuous phase V_R was considered to be insignificant so that V_A also represented the net interaction potential.

Three principal difficulties arise when attempting to apply the Derjaguin–Landau–Verwey–Overbeek theory. First, the zeta potential (ζ), i.e. the potential of the innermost part of the diffuse double layer, is calculated from the electrophoretic mobility of the particles in an electric field, and ψ_p is derived from this potential. At low ion concentrations ζ and ψ_p are very similar, but when the continuous phase contains high ion concentrations they are widely different, with ψ_p often twice as large as ζ (Davies and Rideal, 1963). Sometimes it is not easy to determine the precise relationship between ζ and ψ_p. Another difficulty is that the value of A for insertion in Eqs (3.96)–(3.98) is often not known with any degree of certainty, and the general shape of the $V–H_0$ curve can, in some cases, be drastically altered if the wrong value of A is selected. The procedure for determining A for an emulsion stabilised by a surface active agent is given elsewhere (Sherman, 1967b); for suspensions of solid particles in fluid media A can be calculated from the rate of particle flocculation (Srivastava, 1965) because the particles do not coalesce together as in emulsions. Finally, while it is possible to qualitatively predict rheological behavior from the $V–H_0$ curves, no quantitative relationship has yet been established. According to Houwink (1958) V_{max} is related to the yield value of a dispersion, or in the case of an ideally elastic body to its tensile strength. Elsewhere (Section 4), a method for calculating the modulus of elasticity of a flocculated dispersion from the net attraction forces between particles will be discussed.

8. *The Rate of Shear Required to Overcome the Attraction Forces Between the Particles in a Flocculate*

When a dispersion is subjected to shear the flocculate structure begins to break down. In terms of the theory outlined in the previous section this will initially involve the weakest attraction forces between particles, while the final stage in deflocculation involves the strongest attraction forces. The globules held together by the latter will separate only when these forces are just exceeded by an opposing tension (F_t) which arises from the axial component of the hydrostatic force due to the continuous phase.

An approximate estimate of F_t can be obtained from Stokes' law

$$F_t = 6\pi\eta_0\, r\, (\Delta f/2)\sin\alpha_f = 3\pi\eta_0\, r\, (2r+H_0)\, D\sin\alpha_f\cos\alpha_f \quad (3.100)$$

where Δf is the difference in rate of flow between the centres of the two particles considered (Fig. 3.6), r is the radius of the particles, and α_f is the angle between the line joining the centres of the particles and the direction in which the particles are sheared.

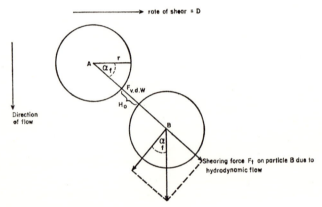

Fig. 3.6. Rotation of a two-particle aggregate under influence of a shearing force. (Albers and Overbeek, 1960).

Since

$$2\sin\alpha_f\cos\alpha_f = \sin 2\alpha_f$$

and the two particles will separate when the attraction force (V_A) is just exceeded by F_t, for a non-polar continuous phase (Albers and Overbeek, 1960)

$$\tfrac{3}{2}\pi\eta_0\, r\, (2r+H_0)\, D\sin 2\alpha_f > V_A \quad (3.101)$$

or, using the uncorrected equation for V_A (Eq. 3.97),

$$\tfrac{3}{2}\pi\eta_0\, r\, (2r+H_0)\, D\sin 2\alpha_f > -\frac{AD}{24H_0^{\,2}}. \quad (3.102)$$

Thus, the value of D at which the particles separate can be derived. Calculations indicated that $\alpha \sim 30°$, but the precise value is of little importance as it will not affect D very much.

C. Rheological Behavior at Low Rates of Shear

When the continuous phase is non-polar particles flocculate very quickly, due to the very diffuse character of the electrical double layer and the absence of a significant potential energy barrier. Under these conditions it can be assumed that the particles flocculate initially according to von Smoluchowski's (1916, 1917) theory of rapid flocculation.

The number of unassociated particles (N_1), flocculates (N_2) containing two particles, ... flocculates (N_k) containing k particles, after a time t is

$$N_1 = \frac{N_0}{[1+(t/t_f)]^2}, \quad N_2 = \frac{N_0(t/t_f)}{[1+(t/t_f)]^3}, \quad ...N_k = \frac{N_0(t/t_f)^{k-1}}{[1+(t/t_f)]^{k+1}}$$

(3.103)

where t_f, the time taken for N_0 unassociated particles to decrease to $N_0/2$, is given by

$$t_f = \frac{1}{4\pi \mathscr{D} D_s N_0}.$$

With non-polar continuous media $t_f \ll t$ usually, so that Eq. (3.103) can be written as

$$N_1 \approx \frac{N_0}{(t/t_f)^2} \approx N_2 ... \approx N_k.$$

(3.104)

In polar continuous media flocculation will occur in the primary minimum provided V_{max} does not exceed $15-25\,kT$, the precise rate depending on V_{max}. The rate of flocculation is then reduced by a factor W, where

$$W = 2 \int_2^\infty \exp\left(\frac{V}{kT}\right) \frac{ds}{s^2} \approx \frac{1}{2\chi r}\exp\left(\frac{V_{max}}{kT}\right)$$

(3.105)

where $S = H_0/r$ (Verwey and Overbeek, 1948).

Following particle flocculation the system behaves at low shear as if it had an apparent volume concentration of dispersed phase as given by Eq. (3.81), where now, for slow flocculation,

$$N_1 = \frac{N_0}{(1+4\pi D_s \mathscr{D} N_0\,W)}.$$

(3.106)

From Eqs (3.81) and (3.106)

$$\phi_a = \phi \left[f_s - \frac{f_s-1}{(1+4\pi D_s \mathscr{D} N_0\,W)} \right].$$

(3.107)

Using the technique described in Section 2.B.5 Mooney (1946) found that $W \sim 10^{-1}$ for a concentrated latex dispersion ($\phi = 0.44$), and that ϕ_a

ranged between 0·790–0·676 for shear stresses of 1·33–10·7 dynes/cm². The systems were agitated very vigorously to break down all structure, and the rate of structure redevelopment was then followed by measuring the viscosity at frequent intervals. ϕ_a was calculated at each time interval from Eq. (3.82) by determining the value of ϕ on the η–ϕ curve which corresponded to the observed viscosity. By inserting these values of ϕ_a in Eq. (3.107) the corresponding values of f_s were derived. They ranged between 1·796 and 1·537, depending on the shear stress, at infinite time after reflocculation.

In most cases $V_{\max}$ is too large in polar continuous media (Section B.7) for the particles to flocculate in the primary minimum of the potential energy curve. Instead, they flocculate in the secondary minimum, where there is no potential barrier to entry. In such cases $W = 0$, so that Eq. (3.107) becomes

$$\phi_a = \phi(f_s), \quad \text{or} \quad f_s = \frac{\phi_a}{\phi}. \tag{3.108}$$

This relationship also applies when the continuous phase is non-polar because in Eq. (3.81) $N_1 \ll N_0$.

During flocculation in a system containing a range of particle sizes collisions initially occur most frequently between the smallest and the largest particles (Verwey and Overbeek, 1948). In effect, the larger particles act as nuclei around which flocculates build up due to deposition of smaller particles. These flocculates then link up at several points, the number of interconnecting points increasing as ϕ increases and/or D_s decreases. Under these conditions the flocculates develop a compact central core with a progressively decreasing density of particles toward the surface (Vold, 1963). As the size of the flocculates increases the proportion of particles found in the core increases.

In the absence of shear random close packing of spheres of equal size gives $\phi = 0·637$ (Yerazanis et al., 1962), so that $f_s = 1·56$, whereas an equiproportional mixture of spheres with sizes in the ratio 1/3·8, gives $\phi = 0·68$, and $f_s = 1·47$. Both values of f_s quoted are higher than the theoretical value for optimum close packing ($\phi = 0·7402$). When increasing shear is applied the flocculates break down to smaller sizes, thus reducing f_s. Before this occurs, however, it is possible that at very low rates of shear the particles roll over one another and rearrange themselves to form more densely packed structures. According to Thomas (1964) the diameter of a flocculate is always within the limits

$$\left(\frac{\phi_a}{\phi}\right)^{5/3} < \frac{D_f}{D_s} \leqslant \left(\frac{\phi_a}{\phi}\right)^2 \tag{3.109}$$

irrespective of shear rate, where, ϕ_a, in this instance, is for $D_s \to \infty$, provided the flocculate does not elongate under the influence of shear.

At very low rates of shear (10^{-5}–10^{-6} sec^{-1}), or at very low shear stresses of the order of a few dyne/cm^2, emulsions and dispersions show viscoelastic behavior which follows the relationships quoted in Section 1.C.1 of Chapter 1. This behavior substantiates the belief that the flocculates interlink after flocculation has continued for some time.

3. FACTORS ARISING FROM THE CONSTITUENTS OF A DISPERSED SYSTEM WHICH INFLUENCE ITS VISCOSITY

The behavior of an emulsion or dispersion when sheared, in flocculation, etc. depends upon its composition and the chemical nature of the ingredients. Table 3.1 lists the principal constituents of a dispersed system, and it also indicates the factors through which they exert an effect. A distinction is drawn between dispersions and emulsions, because in the latter more factors are involved due to the deformable nature of the particles of the dispersed phase. The form of presentation shown in Table 3.1 is convenient but not altogether satisfactory since it suggests that each factor acts independently of the others. In practice, two or more factors operate simultaneously, and the net effect usually differs in magnitude from the sum of the individual contributions.

The possibility that two, or more, factors exert an effect simultaneously has often been ignored in the past when analysing experimental data. A simple example will serve to illustrate this point. Two emulsions are prepared from the same ingredients using exactly the same technique of emulsification and homogenisation in both cases, the only difference being the volume concentration of dispersed phase. Their viscosities are now compared over a wide range of shear rates. If the emulsion with the higher concentration of dispersed phase shows a higher viscosity at each rate of shear the conclusion will be drawn that the higher concentration of dispersed phase is wholly responsible for this. However, it is possible that the more concentrated emulsion will have a larger mean particle size and size distribution than the other emulsion, because the efficiency of dispersion using a standardised technique for emulsification will decrease as the volume concentration of dispersed phase increases. Thus, the simultaneous influence of mean particle size, if not size distribution, on viscosity must also be considered.

Emulsion rheology has received far less attention than the rheology of dispersions because emulsions are much more difficult systems to study. In emulsions a deformable fluid constitutes the internal phase, and the emulsifier establishes a third phase in the form of a layer adsorbed around the particles which modifies the forces of cohesion between the particles,

TABLE 3.1

Factors influencing the rheological properties of dispersed systems

Factors influencing rheological properties	Emulsions	Dispersions of solid particles in fluid media
1. Dispersed Phase		
(a) volume concentration (ϕ)	Influences hydrodynamic interaction between particles; flocculation leading to formation of aggregates.	Influences hydrodynamic interaction between particles; flocculation leading to formation of aggregates.
(b) viscosity (η_i)	Influences particle deformation in shear, and when closely packed in flocculated state; may also influence emulsion viscosity and fluid circulation within particles.	Particles solid, i.e. of infinitely high viscosity, so they do not deform in shear or when closely packed in flocculated state.
(c) particle size, particle size distribution, and particle shape	Influence viscosity over whole range of shear rates, and also the visco-elastic properties of the more concentrated emulsions; since particle size etc. influence the number of particles/unit volume they also influence the rate of flocculation.	Influence viscosity over whole range of shear rates, and also the visco-elastic properties of the more concentrated emulsions; since particle size etc. influence the number of particles/unit volume they also influence the rate of flocculation.
(d) chemical constitution	Influences interaction forces between particles; also partition of surface active agents between the two liquid phases.	Influences interaction forces between particles.
2. Continuous Phase		
(a) viscosity (η_o)	Direct proportionality with emulsion viscosity.	Direct proportionality with dispersion viscosity.

TABLE 3.1 (continued)

Factors influencing rheological properties	Emulsions	Dispersions of solid particles in fluid media
(b) chemical constitution, polarity, pH	Influence viscosity through effect on potential energy of interaction between particles; also partition of surface active agents between the two liquid phases.	Influence viscosity through effect on potential energy of interaction between particles.
(c) electrolyte concentration—if polar medium	Influences surface charge on particles and hence influences viscosity; may also affect solubility of surface active agents in continuous phase.	Influences surface charge on particles, and hence influences viscosity; may also affect solubility of surface active agents in continuous phase.
3. Surface Active Agents		
(a) chemical constitution	Influences solubility in oil and water phases and partition between the two; also adsorption at oil-water interface.	Influences solubility in continuous phase; also adsorption at the solid-liquid interface.
(b) concentration	Influences viscosity, emulsion type, and emulsion inversion; also, solubilization of dispersed phase in micelles.	Influences viscosity through effect on η_0.
(c) adsorbed film at interface	Thickness influences particle dimensions and hence viscosity; it also influences interaction between particles; rheological properties of this film affect particle deformation in shear and when flocculated, fluid circulation within particles, and interfacial slippage.	Thickness influences particle dimensions, and hence viscosity; it also influences interaction between particles.
(d) electroviscous effects	When particles very small the viscosity may be increased by primary and secondary effects.	When particles very small the viscosity may be increased by primary and secondary effects.
4. Additional Stabilising Agents		
pigments, hydrocolloids, hydrous oxides etc.	Modify potential energy of interaction between particles through effect on surfaces of particles, gelation, etc; also modify hydrodynamic interaction.	Modify potential energy of interaction between particles through effect on surfaces of particles, gelation etc; also modify hydrodynamic interaction.

F

and also the forces between the particles and the continuous phase. If the particles in an emulsion are only slightly deformed when sheared the deformation can be calculated from

$$\frac{L_A - B_A}{L_A + B_A} = \frac{DD_s \eta_0}{2\gamma_T} \left(\frac{(19/16)\eta_i + \eta_0}{\eta_i + \eta_0} \right)$$ (3.110)

where L_A and B_A are the dimensions of the major and minor axes respectively, γ_T is the interfacial tension between the oil and water phases, and η_i is the viscosity of the dispersed phase (Taylor, 1934). Particles which are not larger than a few microns generally do not suffer much deformation even at high rates of shear, so that information about their flow behavior can be drawn from data for dispersions of solid particles which have the same size. Interpretation of experimental data relating to larger deformations is more difficult unless the emulsion particles assume a well defined geometrical shape.

A. Internal Phase

1. Volume Concentration

The viscosity (η) of a dilute dispersion or emulsion containing spherical, non-deformable, particles is given by

$$\eta = \eta_0(1 + a\phi)$$

or

$$\frac{\eta}{\eta_0} = \eta_{\text{rel}} = 1 + a\phi$$ (3.111)

provided the volume concentration of dispersed particles is so low that there is no interaction, the distance between the particles is very much greater than their diameter, there is no slippage at the interface between the particles and the fluid of the continuous phase, and η arises only from the dissipation of energy, or viscous drag, which is produced by modification of the fluid motion in the proximity of the particle surfaces (Einstein, 1905, 1906, 1911). If these conditions are fulfilled then the constant $a = 2.5$.

Equation (3.111) can also be written as

$$\frac{\eta}{\eta_0} - 1 = \eta_{\text{sp}} = a\phi$$ (3.112)

where η_{sp} is the specific increase in viscosity. When $\phi \to 0$

$$\frac{\eta_{\text{sp}}}{\phi_{\phi \to 0}} = \eta_{\text{int}} = [\eta] = a$$ (3.113)

where η_{int} is the intrinsic viscosity, and a plot of η_{sp} against ϕ gives a straight line with a gradient equal to a. Since Eq. (3.113) is valid only when $\phi < \sim 0.05$, its application requires very accurate viscosity measurements on very dilute dispersions or emulsions. Sometimes it may be found that viscosity data derived at higher concentrations than 0.05 follow a relationship of the form of Eq. (3.113), but in such cases, of course, $a > 2.5$. Recently it has been suggested (Kynch, 1954) that Eq. (3.111) is not valid when $\phi > 0.01$. It will be shown in Section 3.A.3 that the limiting value of ϕ depends on particle size, for which no allowance is made in Eq. (3.111).

At higher concentrations of dispersed phase the particles begin to interact with one another because the flow patterns in the continuous phase associated with them draw closer together, and eventually in concentrated systems they overlap. When this latter condition is achieved the viscosity is no longer the sum of the effects due to the individual particles. The first quantitative treatment (Guth and Simha, 1936) of this hydrodynamic interaction gave

$$\eta_{rel} = 1 + a\phi + 14.1\phi^2 \qquad (3.114)$$

or

$$\frac{\eta_{sp}}{\phi} = a + 14.1\phi$$

Subsequent analyses suggested that the second constant has a value which is lower than 14.1 if one takes into account the possibility of collisions between globules, slight flocculation (Gillespie, 1963) leading to an apparent value of ϕ which is higher than the theoretical value, etc. All amended forms of Eq. (3.114) can be represented more accurately by a polynomial in ϕ rather than by a quadratic function, since the latter form leads to substantial errors at higher values of ϕ

$$\eta_{sp} = a\phi + b\phi^2 + c\phi^3 + d\phi^4 \ldots \qquad (3.115)$$

where c, d, e, etc. are additional constants. Table 3.2 gives some of the values for a, b, and c quoted in published literature. It is very significant that there is general agreement that $a = 2.5$, in spite of the wide limits of quoted values for b. Few values have been reported for c, or the constants associated with still higher powers of ϕ.

It has been suggested (Thomas, 1965) that Eq. (3.115) can be modified so as to include an exponential term which allows for the probability of a particle being transferred from one plane of shear to another at an infinitely high rate of shear, i.e. when deflocculation is complete

$$\eta_{sp} = a\phi + b\phi^2 + c\phi^3 \ldots + A_T \exp(B_T \phi) \qquad (3.116)$$

where A_T and B_T are constants. No explanation was offered as to the significance of these latter constants, and furthermore, they are not really constant

TABLE 3·2

Published values of the constants in Eq. (3.115)

System	Particle, or globule. diameter (μ)	Constants for Eq. (3.115)			Optimum value for which Eq. (3.115) tested	Reference
		a	b	c		
O/W emulsions	2·5–4·5	1·5–2·3	~2–13 depending on value of a	—	0·17	Nawab and Mason (1958)
W/O emulsions	1·0–3·0	2·3–2·8	0–9·7	—	0·47	Sherman (1950a)
Methyl methacrylate spheres	53–178	2·33–2·46	—	—	0·325	Higginbotham et al. (1958)
Glass spheres	1–10	2·49	12·7	—	0·1838	Manley and Mason (1954)
O/W emulsions	0·7–3·0	2·5	—	—	0·365	Leviton and Leighton (1936)
Latex emulsions	0·1388 0·1216	2·5	—	—	$\phi \to 0$	Maron and Ming Fok (1955)
Theoretical treatment of monodisperse systems of rigid spheres	—	2·5	—	—	≪ 0·05	Roscoe (1952)
Theoretical analysis of Vand's data	100–160	2·5	1·43	—	0·50	Mooney (1951)
Theoretical treatment for rigid spheres	—	2·5	2·5	—	very small spheres undergoing strong Brownian motion	Saito (1950)

TABLE 3.2 (continued)

System	Particle, or globule diameter (μ)	Constants for Eq. (3.115)			Optimum value for which Eq. (3.115) tested	Reference
		a	b	c		
Theoretical treatment of monodisperse systems of rigid spheres	—	2·5	2·64 – 6·4	—	0·09	Simha (1952)
Theoretical treatment	—	2·5	4·7	—	—	de Bruijn (1942, 1948)
Asphalt/water emulsions	mainly ~ 2·7μ	2·5	4·94	8·78	0·707	Eilers (1941)
Latex emulsions	0·0990 – 0·8710	2·5	6·29 – 7·64	26·9 – 36·3	0·24	Saunders (1961)
Rigid spheres (a) Theoretical treatment	152 – 422	2·5	6·75 – 10·0	—	0·25	Kynch (1956)
(b) Glass spheres	—	2·5	7·35	—	—	Vand (1948)
	100 – 160	2·5	7·17	16·2	~ 0·50	Vand (1948)
(a) Glass spheres	125 – 205	—	—	—	0·20	Eirich et al. (1936)
(b) Spores	4 – 5·5	2·5	7·8	—	0·08	Eirich et al. (1936)
Theoretical treatment	2·5	7·8 (uncorrected)	—	—	—	Guth and Simha (1936)
Polystyrene latex	0·2640	2·5	~ 10	~ 50	0·02 – 0·08	Cheng and Schachman (1955)
Review of published data	infinitely large	2·5	10·05	—	< 0·25	Thomas (1965)
O/W emulsions	0·0138 – 0·1025	2·6 – 5·0	—	—	0·33	Van der Waarden (1954)
W/O emulsions	mainly 2 – 3	3·5 – 5·1	—	—	0·33	Albers (1957)

since their values will depend on particle size, and on other factors which are discussed later.

In concentrated dispersions particle size is no longer insignificant with respect to their distance of separation, and interaction may occur between, for example, a particle and its next but one neighbor. This interaction is not of the same magnitude as that between adjacent particles, since in the former instance the interaction is reduced by the shielding effect exerted by the intervening particle. A cage model has been proposed (Simha, 1952) to define this interaction in which a concentric spherical enclosure of diameter D_c is drawn around the reference particle. This enclosure defines the boundary beyond which there is no interaction between the reference particle and other particles. The value of D_c depends on the dispersed phase concentration. In addition, it will also depend on D_s or mean particle size, although this point is not made in the mathematical development and was not pointed out until a later date (Simha and Somcynsky, 1965). Equation (3.111) is modified, in accordance with the aforementioned model, to

$$\eta_{\text{rel}} = 1 + aI_H \phi \qquad (3.117)$$

to allow for interaction between particles which are separated by the intervening particle, where I_H, a complex function of the ratio D_s/D_c, is a factor which defines the interaction between the reference particle and all other particles. The function I_H is given by

$$I_H = \frac{4[1 - (D_s/D_c)^7]}{4[1 + (D_s/D_c)^{10}] - 25\{(D_s/D_c)^3[1 + (D_s/D_c)^4]\} + 42(D_s/D_c)^5} \qquad (3.118)$$

I_H increases slowly until the ratio $(D_s/D_c) \approx 0.5$; between 0.5 and 1.0 it increases much more rapidly.

In dilute systems D_c is proportional to the distance between the particles, but in concentrated systems it is proportional to the distance between particles minus D_s, since the latter becomes more significant as the distance between the particles decreases with increasing dispersed phase volume concentration. When the latter is small

$$\frac{D_s}{D_c} = \frac{\phi^{\frac{1}{3}}}{f_v} \qquad (3.119)$$

where f_v is given by $(8\phi_{\max})^{\frac{1}{3}}$, and in concentrated systems

$$\frac{D_s}{D_c} = \frac{\phi^{\frac{1}{3}}}{f_v(1 - \phi^{\frac{1}{3}}/f_v)} \qquad (3.120)$$

where $\phi_{\max}$ is the maximum concentration of dispersed phase which can be incorporated. For hexagonal, or simple cubical, close packing f_v is 1·81 or 1·61 respectively. When these two values are introduced into Eq. (3.118), Eq. (3.117) can be represented in alternative ways

$$\eta_{rel} = 1 + 2\cdot5\phi \left[1 + \frac{25}{4f_v^3}\phi - \frac{21}{2f_v^5}\phi^{5/3} + \frac{625}{16f_v^6}\phi^2 + \ldots \right] \qquad (3.121)$$

or

$$\eta_{rel} = 1 + 2\cdot5\phi \left[1 + \frac{25}{4f_v^3}\phi + \frac{75}{4f_v^4}\phi^{4/3} + \frac{27}{f_v^5}\phi^{5/3} + \frac{785}{16f_v^6}\phi^2 + \ldots \right]. \qquad (3.122)$$

When $\phi \to \phi_{\max}$

$$\eta_{rel} = \frac{54}{5f_v^3} \left[\frac{\phi^2}{(1-\phi/\phi_{\max})^3} \right]. \qquad (3.123)$$

From $\phi = 0$ to $\phi = \phi_{\max}\,f$ changes only slightly ($\sim$ 1·3 to $<$ 2).

Frankel and Acrivos (1967) also considered the hydrodynamic interaction between particles in concentrated systems, but they did not assume any artificial boundary for the influence of one particle on another. They found that for $\phi \to \phi_{\max}$

$$\eta_{rel} = \frac{9}{8} \left[\frac{(\phi/\phi_{\max})^{1/3}}{1-(\phi/\phi_{\max})^{1/3}} \right]. \qquad (3.124)$$

Specifically for emulsions Hatschek (1911) found that when ϕ exceeded 0·5 the linear portion of the shear stress–rate of shear curve for non-Newtonian flow could be represented by

$$\eta_{rel} = \frac{1}{1-\phi^{\frac{1}{3}}} \qquad (3.125)$$

but Sibree (1930, 1931) obtained higher values than those predicted by this equation. He attributed the discrepancy to an increase in the size of the particles due to hydration of the adsorbed emulsifier layer by continuous phase fluid. This was considered to increase ϕ by a factor h_s ($= 1\cdot3$), so that Eq. (3.125) was amended to

$$\eta_{rel} = \frac{1}{1-(h_s\phi)^{\frac{1}{3}}} \qquad (3.126)$$

Because some of the continuous phase is associated with the particles due to hydration of the surface, or for some other reason, the "free" volume of continuous phase in which particles can move past one another when they are deflocculated is $(1-h_s\phi)$. Several viscosity equations have been proposed on this basis. They take a somewhat different form to Eq. (3.126).

$$\eta_{rel} = 1 + \frac{a\phi}{1-h_s\phi} \tag{3.127}$$

or

$$\ln \eta_{rel} = \frac{a\phi}{1-h_s\phi}.$$

Equation (3.127) reduces to Eq. (3.111) at low values of ϕ, and on expansion it shows a form similar to Eq. (3.115). The constant a has been interpreted in several ways, although it usually fitted the experimental data satisfactorily by taking it as 2·5. Robinson (1949, 1957) believed a to represent a coefficient of friction for dispersions, its precise value depending on particle shape and surface roughness. According to Mooney (1951) and Maron and his co-workers (1951, 1953), h_s represents the crowding effect which arises when particles of more than one size are packed together. In the simplest example, when only two different particle sizes are involved, h_s is a function of their size ratio. According to Vand (1948), h_s is the hydrodynamic interaction constant. Sweeney and Geckler (1954) found that h_s varied between 1·00 and 1·47, its value increasing as particle size decreased. Saunders (1961) observed a similar effect due to particle size in monodisperse latexes. In this case the particle size was reduced from $\sim 1\mu$ downwards to $\sim 0·1\mu$, and h_s increased from 1·118 to 1·357.

Brinkman (1952) wrote Eq. (3.115) as

$$\eta_{rel} = \frac{1}{(1-\phi)^a} \tag{3.128}$$

where a retains a value of 2·5 provided the globules move independently. When the flocculates are not broken down completely at high rates of shear, which may occur, for example, in some latex dispersions, a is larger than 2·5 (Gillespie, 1963).

Richardson (1933) restricted his attention to emulsions. By analogy with the influence of variable pressure on a material which obeys Hooke's law he calculated the "compressibility" of an emulsion whose disperse phase volume concentration is increased from ϕ to $\phi+\delta\phi$. From this he derived an expression for the relative viscosity of concentrated emulsions at any rate of shear.

$$\ln \eta_{rel} = K_R \phi \tag{3.129}$$

where K_R is a constant. Broughton and Squires (1938) found that this equation did not give values of η_{rel} which agreed with their experimental data, so they amended it in an empirical way to achieve more satisfactory agreement,

$$\ln \eta_{\text{rel}} = K_R \phi + Y_R \tag{3.130}$$

where Y_R is an additional constant. Neither Eq. (3.129) nor Eq. (3.130) has been used much by other workers in this field.

2. Viscosity

The particles in all dispersions can be regarded as having an infinitely high viscosity, so that any effect on dispersion viscosity due to this will remain constant. For emulsions the situation is quite different since the viscosity of the particles can vary over an extremely wide range. Several equations have been proposed to specifically account for this.

Taylor (1932) extended Eq. (3.111) to dilute suspensions of fluid particles. He assumed that the adsorbed layer of emulsifier around the particles would not hinder the transmission of tangential or normal stresses (Eq. (1.1)) from the continuous phase to the dispersed phase, and also that there was no slippage at the oil-water interface. The aforementioned stresses cause fluid circulation within the globules, and this reduces the distortion of the flow patterns around the particles. The magnitude of this effect depends on the ratio η_i/η_0,

$$\eta_{\text{rel}} = 1 + a \left(\frac{\eta_i + \frac{2}{5}\eta_0}{\eta_i + \eta_0} \right) \phi. \tag{3.131}$$

This equation differs from Eq. (3.111) by the factor

$$\left(\frac{\eta_i + \frac{2}{5}\eta_0}{\eta_i + \eta_0} \right),$$

which allows for the internal currents within the particles. When $\eta_i \gg \eta_0$, Eq. (3.131) reduces to Eq. (3.111), but under all other conditions η_{rel} is lower than for a dispersion of solid particles with the same value of ϕ. Leviton and Leighton (1936) extended Eq. (3.131) into a power series in ϕ, in accordance with a suggestion of von Smoluchowski (1916 b), so as to extend its range of validity to $\phi \approx 0.4$

$$\ln \eta_{\text{rel}} = a \left(\frac{\eta_i + \frac{2}{5}\eta_0}{\eta_i + \eta_0} \right) (\phi + \phi^{5/3} + \phi^{11/3}). \tag{3.132}$$

At low values of ϕ Eqs (3.131) and (3.132) are identical. The term $\phi^{5/3}$ corresponds to the next term in ϕ for a power series expansion, but $\phi^{11/3}$

represents an empirical addition to bring theoretical and experimental data into closer agreement.

Equation (3.131) has been applied to viscosity data for a wide range of dilute O/W emulsions (Nawab and Mason, 1958) which contained particles which were too small ($2 \cdot 5$–$4 \cdot 5\ \mu$) to deform significantly even at shear rates as high as 900 sec^{-1}. Some typical data are quoted in Table 3.3; also quoted are calculated values of the constant b derived by Eq. (3.115). The theoretical and experimental values of

$$a\left(\frac{\eta_i + \frac{2}{5}\eta_0}{\eta_i + \eta_0}\right)$$

agreed when the emulsifier layer around the oil particles did not inhibit the internal circulation of oil. When internal circulation was inhibited, the theoretical values were always lower than the experimental values. As η_i/η_0 increased, the emulsions which conformed to Eq. (3.131) showed increased values of

$$a\left(\frac{\eta_i + \frac{2}{5}\eta_0}{\eta_i + \eta_0}\right)$$

and of b.

Liquid flow around the particles in emulsions causes adsorbed emulsifier at the front of the particle to be swept to the rear, and this produces a surface pressure gradient (Linton and Sutherland, 1957). A gradient of a few dyne/cm^2 is sufficient to prevent fluid circulation within the particles, so that they behave like fluid drops with a high viscosity, or possibly even like solid spheres.

While the influence of η_i on the viscosity of dilute emulsions can be readily determined, its influence on the viscosity of concentrated emulsions is more difficult to analyse becomes of superimposed effects due to particle–particle interaction, flocculation, and possibly due to particle deformation also. Toms (1941) examined O/W emulsions which had been prepared with eleven different organic liquids as the internal phase, and with sodium laurate, potassium laurate, potassium myristate, sodium oleate, and potassium oleate as the emulsifier. He found that not only was there no correlation between η_i and η_{rel}, although the ratio η_i/η_0 increased from about $0 \cdot 3$ to 4, but that there was some form of interaction between the emulsifier and the oil phase. For example, in the oil–$0 \cdot 002$ M potassium laurate emulsions series the lowest η_{rel} was obtained when aniline was used as the dispersed phase, but the oil–$0 \cdot 002$ potassium oleate emulsions series using aniline as the dispersed phase produced the highest viscosity. A similar conclusion was reached by Shotton and White (1960) who found that the O/W emulsion with the highest η_{rel} was prepared from the oil with lowest η_i. No details are

TABLE 3.3

Viscosity data (Nawab and Mason, 1958) for dilute O/W emulsions at 25°C applied to Taylor's equation

Emulsion composition	η_i (centipoise)	η_0 (centipoise)	$\dfrac{\eta_i}{\eta_0}$	Range of ϕ	$a\left(\dfrac{\eta_i + 2/5\eta_0}{\eta_i + \eta_0}\right)$		b
					Experimental	Theoretical	
(a) Data agreeing with Taylor's theory							
Butyl benzoate in 25% aqueous sugar solution + 1% Tween 20	2·67	2·08	1·28	0·06−0·16	1·82	1·84	2·7
Butyl benzoate/water + 1% Tween 20	2·67	0·934	2·86	0·04−0·16	2·11	2·11	4·5
Butyl benzoate/water + 0·5% Tween 20	2·67	0·908	2·94	0·04−0·16	2·13	2·12	4·6
Castor oil solution in butyl benzoate/water + 1% Tween 20	4·89	0·934	5·23	0·07−0·16	2·23	2·26	6·8
(b) Data deviating from Taylor's theory							
Butyl benzoate solution of CCl$_4$ in 45% aqueous sugar solution + 1% Tween 20	1·99	3·79	0·53	0·06−0·16	2·01	1·52	7·5
Butyl benzoate solution of CCl$_4$ in 45% aqueous sugar solution + 0·2% Tween 20	1·91	3·77	0·51	0·06−0·16	1·99	1·50	6·3
Butyl benzoate solution of CCl$_4$ in 33% aqueous glycerine + 1% Tween 20	2·32	2·98	0·78	0·06−0·17	1·83	1·66	8·5

given in either publication of particle size but, judging from the repetitive homogenisation technique adopted, it is unlikely that the particles were large enough to deform significantly during shear.

The absence of any correlation between η_i and η_{rel} has been observed also in very concentrated non-Newtonian W/O emulsions (Sherman, 1955 b). Large variations in the ratio η_i/η_0 were achieved by adding 0–80% (wt/wt) glycerine to the dispersed phase, but in spite of this neither the viscosity (η_∞) at high rates of shear nor the extrapolated yield value showed any significant change (Table 3.4). When the aqueous phase contained 36·3%

TABLE 3.4

Influence of η_i on viscosity of W/O emulsions (Sherman, 1955 b)

Oil Phase composition (wt/wt)					89·02 Mineral oil ($\eta_{21} = 0.136$ poise) 8·29 Glyceryl monoricinoleate 2·69 Soya lecithin $\phi = 0.717$	
% Glycerine in water phase (wt/wt)	η_i (poise at 25°C)	η_0 (poise at 25°C)	η_i/η_0	η_∞ (poise at 25°C)	Extrapolated yield value	
0	0·0114	0·2475	0·046	3·10	2623	
6·5	0·0140	0·2475	0·057	3·10	2523	
13·0	0·0170	0·2475	0·069	3·16	2432	
19·5	0·0232	0·2475	0·094	3·10	2450	
25·0	0·0295	0·2475	0·119	3·14	2623	
70·0	0·2167	0·2475	0·875	3·06	2926	
80·0	0·5588	0·2475	2·257	3·00	2600	

propylene glycol, glycerine, or sorbitol syrup, the variation in η_i/η_0 was very much smaller, and, as was to be expected from the first series of tests, no significant differences were observed in η_{rel}. However, when a small amount of carbon black was added to the oil phase as an additional stabilizing agent there were distinct differences between the rheological properties of the three emulsions (Table 3.5). The emulsion containing propylene glycol in the aqueous phase, i.e. the aqueous phase with the highest η_i, gave the lowest η_∞ and the lowest extrapolated yield value. The explanation for this phenomenon probably rests in the observation that the specific absorption of the carbon black, i.e. degree of wetting by the aqueous phase, was lowest for the aqueous phase containing propylene glycol. Thus, the chemical constitution of the internal phase can affect stabiliser orientation at the oil–water interface even when η_i does not influence η_∞.

TABLE 3.5

Influence of η_i on viscosity of W/O emulsions containing carbon black
(Sherman, 1955 b)

Oil phase composition (wt/wt) $\begin{cases} 89\cdot52 \text{ Mineral oil } (\eta_{21} = 0\cdot126 \text{ poise}) \\ 1\cdot86 \text{ Blown rape oil } (\eta_{21} = 104\cdot4 \text{ poise}) \\ 8\cdot62 \text{ Soya lecithin} \end{cases}$
$\phi = 0\cdot661$

Additive to water phase (%wt/wt)	Additive to oil phase (%wt/wt)	η_i (poise at 25°C)	η_0 (poise at 25°C)	η_∞ (poise at 25°C)	Extra-polated yield value
36·30% Propylene glycol	—	0·0323	0·1437	2·10	1615
36·30% Glycerine	—	0·0275	0·1437	2·17	1514
36·30% Sorbitol syrup	—	0·0200	0·1437	2·18	1640
36·30% Propylene glycol	5·30% Carbon black	0·0323	0·1437	6·0	3632
36·30% Glycerine	5·30% Carbon black	0·0275	0·1437	8·9	5449
36·30% Sorbitol syrup	5·30% Carbon black	0·0200	0·1437	10·6	6659

3. Particle Size

Equation (3.111) does not include any term involving particle size. As already shown in Section 2.A.1 some of the expressions proposed for the viscosity of concentrated systems are extensions of this equation, and interaction between particles has been accounted for simply by incorporating higher powers of ϕ. Since viscosity can be changed merely by altering the particle size, while maintaining ϕ constant, such amendments to Eq. (3.111) are obviously not wholly satisfactory.

Much of the published literature makes no reference to the state of dispersion of the system studied. The situation is often little improved when this information is provided since the particle size, or mean particle size, may be anywhere within the limits $< 0\cdot1\,\mu$ to $> 100\,\mu$, and little significance has been attached to the actual values which are reported. For example, the validity of Eq. (3.115) has been studied by several workers using data in which the particle sizes varied from one extreme to the other, and the significance of these variations with respect to the values of the derived constants (Table 3.2) has not been appreciated.

Until comparatively recently there were few observations of any value regarding the influence of particle size. Of these two relate to emulsions, and another to dispersions of solid particles. In the latter case, Hauser and Le Beau (1939) found that the viscosity of bentonite sols increased as the particle size decreased. The two most important observations on emulsions appear at first sight to be contradictory. Leviton and Leighton (1936) observed that the viscosity of dilute, fluid, O/W emulsions did not change when particle size was reduced from $\sim 3{\cdot}0\,\mu$ to $0{\cdot}7\,\mu$. They suggested that particle size does not exert much effect on the viscosity of dilute emulsions as the particles are not packed closely together. No investigations were made with more concentrated emulsions to substantiate this view. Richardson (1950, 1953) restricted his attention to highly concentrated O/W emulsions. These emulsions were very viscous and exhibited non-Newtonian flow. Their viscosities were inversely related to their mean particle sizes, and the product of relative viscosity and mean particle size did not change provided the distribution of particle sizes around the mean value was not very wide. Much confusion has been created by Leviton and Leighton's, or Richardson's, observations being quoted to substantiate some particular experimental findings which were obtained with a different dispersed phase volume concentration than that on which the observations were originally made. It will be demonstrated later in this section that both views are correct within restricted regions of particle concentration.

The only equations which contain specific terms for the influence of particle size on viscosity were developed for dilute emulsions (Oldroyd, 1953, 1955; and Rajagopal, 1960). Oldroyd's equation also includes terms for η_i and the shear properties of the adsorbed emulsifier film around the particles. The latter factor will be discussed in detail in Section 3.D. When the emulsifier film is viscous, so that the particles can be deformed by shearing forces

$$\eta_{\mathrm{rel}} = 1 + a \left[\frac{\eta_i + \tfrac{2}{5}\eta_0 + \tfrac{2}{5}(2\eta_s + 3\eta_\beta)\,(1/r)}{\eta_i + \eta_0 + \tfrac{2}{5}(2\eta_s + 3\eta_\beta)\,(1/r)} \right] \phi \qquad (3.133)$$

where η_s and η_β are the shear viscosity and the area viscosity respectively of the emulsifier film, and the net effect is to increase η_i by $\tfrac{2}{5}(2\eta_s + 3\eta_\beta)\,(1/r)$.

Rajagopal (1960) proposed a relationship which was rather similar to Eq. (3.131), but in this case he introduced a coefficient (S_c) to allow for the possibility of slip at the interface between the particles and the continuous phase, presumably because of a lubrication effect by the emulsifier film.

$$\eta_{\mathrm{rel}} = 1 + a \left[\frac{\eta_i + \tfrac{2}{5}\eta_0 + (2\eta_0\,\eta_i/S_c)\,(1/r)}{\eta_i + \eta_0} \right] \phi \qquad (3.134)$$

When $\eta_i \to \infty$

$$\eta_{rel} = 1 + a \left[2 \cdot 5 + \frac{5\eta_0}{S_c} (1/r) \right] \phi \tag{3.135}$$

Some other viscosity equations for dispersions contain no specific term for particle size, but, instead, the values of certain constants depend on the particle size, and also on the particle size distribution. Sweeney and Geckler (1954), and Saunders (1961), were probably the first to establish a semi-quantitative relationship between the viscosity of dispersions and their particle size. In particular, Saunders (1961) showed that the value of h_s in Eq. (3.127) depended on particle size, and that it increased as particle size decreased. For particles within the range $0 \cdot 099\ \mu$–$0 \cdot 871\ \mu$ some influence due to particle size was observed when $< 0 \cdot 05$, so that Eq. (3.111) was no longer valid even within the concentration range for which its validity had previously been unquestioned, and with still smaller particles sizes of $0 \cdot 0075\ \mu$–$0 \cdot 0457\ \mu$, the influence of size becomes apparent at still lower values of ϕ (Greenberg et al., 1965). At a diameter of $0 \cdot 264\ \mu$ Eq. (3.111) could no longer be applied when $\phi > 0 \cdot 02$ (Cheng and Schachman, 1955), and with a diameter of $0 \cdot 088\mu$ η_{rel} showed a dependence on shear rate, i.e. the flow became non-Newtonian, at an unexpectedly low ϕ (Collins and Wayland, 1963).

Reverting now to emulsions, mean particle size exerts a large effect on the viscosity of pseudoplastic W/O and O/W emulsions over a wide range of ϕ (Sherman, 1960). Emulsions are rarely monodisperse, i.e. the particles do not all have the same size, so that the size is specified in terms of a mean size, or diameter, which is calculated in accordance with one of several well established equations. The particular equation selected for this calculation depends on the particular property of the dispersion, or emulsion, which is being studied. In the present discussion, mean particle size refers to the mean volume diameter, which is calculated according to

$$D_s = \left(\frac{n_1 D_1{}^3 + n_2 D_2{}^3 + n_3 D_3{}^3 \dots}{n_1 + n_2 + n_3 \dots} \right)^{\frac{1}{3}} = \left(\frac{\Sigma n_i D_i{}^3}{\Sigma n_i} \right)^{\frac{1}{3}} \tag{3.136}$$

where n_1, n_2, n_3, etc. refer to the number of particles per cc. which have diameters D_1, D_2, D_3, etc. When the mean size fell below about $2 \cdot 0\ \mu$, in emulsions in which the spread of sizes about the mean size was narrow, W/O emulsions showed a large curvilinear increase in η_∞ as mean particle size decreased for all values of ϕ which were employed (Fig. 3.7). In O/W emulsions the effect was less pronounced, and it appeared only when $\phi > 0 \cdot 5$ (Fig. 3.8). The difference in size dependence between the two types of emulsion is related to the differences in rheological properties of the inter-

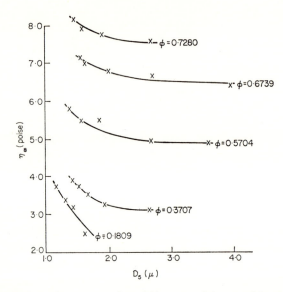

FIG. 3.7. Influence of D_s on η_∞ for sorbitan sesquioleate stabilised W/O emulsions (Sherman, 1960).

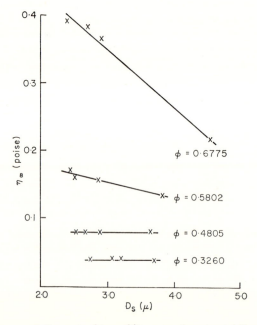

FIG. 3.8. Influence of D_s on η_∞ for sorbitan monolaurate stabilised O/W emulsions (Sherman, 1960)

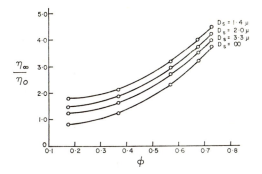

Fig. 3.9. Viscosity data for W/O emulsions stabilised with sorbitan sesquioleate (Sherman, 1965).

facial films produced by the emulsifiers in these emulsions. The W/O emulsifiers gave a film with a high surface rigidity, whereas the O/W emulsifier gave a film of low viscosity. If these data are replotted as $\eta_{rel} - \phi$ curves for selected values of D_s, where η_{rel} is now the ratio η_∞/η_0, it is readily apparent (Fig. 3.9) that any equation relating viscosity and volume concentration of dispersed phase must include a term for particle size, when the restrictions imposed in the development of Eq. (3.111) no longer apply.

At high rates of shear the particles in a dispersed system are deflocculated, and they are approximately equidistant from one another. Provided the particles behave as rigid spheres, and the particles in the previously mentioned W/O emulsions should fall within this category, the mean distance between the particles can be calculated (Sherman, 1960) from

$$a_m = D_s[(\phi_{max}/\phi)^{\frac{1}{3}} - 1] \tag{3.137}$$

with $\phi_{max} = 0.7403$ for monodisperse systems. When there is a wide range of particle sizes ϕ_{max} may be larger, because the small particles can then pack in the interstices between the larger particles. Provided the size distribution is narrow the value of 0.7403 can be adopted without unduly decreasing the accuracy of the calculation. When a_m was plotted against η_∞ (Fig. 3.10) for all the W/O emulsions prepared from a single emulsifier, a single exponential curve was obtained which covered all the values of ϕ. This form of plot for viscosity data has obvious advantages in that all relevant data are recorded on a single curve. Furthermore, it suggests an explanation for the influence of particle size on viscosity which is based upon the hydrodynamic interaction between particles. It also helps to explain the apparent anomalies in previously published literature.

The distance a_m in Eq. (3.137) is related to the distance L_s over which particles of finite size, on opposite sides of a hypothetical enclosure, interact in the presence of an intervening particle (Simha, 1952; Kynch, 1954) by

$$L_s = a_m + \frac{D_s}{2} \qquad (3.138)$$

and the hydrodynamic interaction coefficient I_H can be defined in an alternative way to Eq. (3.118),

$$I_H = \frac{[1-(D_s/2L_s)]}{D_s/2L_s} \qquad (3.139)$$

For $D_m/2L_s$ ratios between 0·5 and 1·0 this is equivalent to $2a_m/D_s$.

When a_m fell below $\sim 0.5\,\mu$ (Fig. 3.10) the increase in η_{rel} became progressively larger with further reduction in a_m, and when a_m did not exceed $0.1\,\mu$ the increase in η_{rel} was very pronounced. Equation (3.137) indicates that the critical value of $\sim 0.5\,\mu$ for a_m is reached with small particles at lower disperse phase concentrations than with large particles. If D_s does not exceed $2\,\mu$ a_m falls to $0.5\,\mu$ when ϕ is well below 0·5.

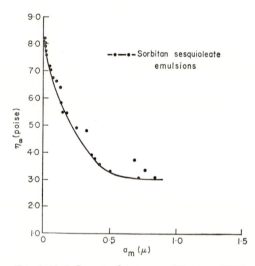

FIG. 3.10. Influence of a_m on η_∞ (Sherman, 1960).

A plot of η_{rel} against a_m for the O/W emulsions (Fig. 3.11) shows some anomalies as compared with Fig. (3.10). An exponential relationship is valid only up to $\phi = 0.634$, probably due to shear deformation of the larger size particles (Eq. 3.110), which for the O/W emulsions extended up to $\sim 16\,\mu$. When ϕ exceeded 0·634 the calculated values of a_m were larger than the theoretical values. The data for the O/W emulsions also conformed with

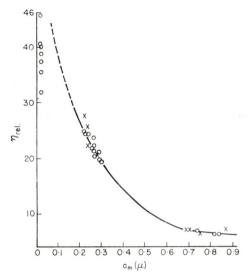

FIG. 3.11. The influence of a_m on η_{rel} for O/W emulsions (Sherman, 1963a). X freshly prepared emulsions; $\bigcirc$ aged emulsions.

$$\eta_\infty = x_s \left(\frac{1}{D_s}\right) + C_v \qquad (3.140)$$

where x_s is the gradient, and C_v is the intercept on the η_∞ axis. At $\phi = 0.678$, $C_v = 0$, so that the product $\eta_\infty D_s$ was constant, as observed by Richardson (1950, 1953), and Lawrence and Rothwell (1957). However, the importance of this observation should not be overemphasised as it did not hold for other values of ϕ.

Figure (3.10) and Fig. (3.11) show that $\ln \eta_{\text{rel}} = 0$, when a_m is very large, and that $\ln \eta_{\text{rel}}$ achieves a maximum value when a_m is infinitely small. Therefore

$$\ln \eta_{\text{rel}} = \ln \eta_\infty / \eta_0 \qquad \propto \alpha_v \left(\frac{1}{a_m}\right) \qquad (3.141)$$

where α_v is a constant which depends on D_s, since the maximum value of η_{rel} depends on D_s (Sherman, 1965). Equation (3.141) has been applied to published viscosity data for both Newtonian and non-Newtonian systems for which the particle size, or size distribution, was given in some detail. These data (Table 3.6) were for emulsions, latex dispersions, and dispersions of solid spheres. Several of the systems were monodisperse, and of the remainder, the majority had narrow size distributions. All data conformed with

$$\log \eta_{\text{rel}} = \alpha_v \left(\frac{1}{a_m}\right) - X_s \qquad (3.142)$$

where α_v is the slope of the line obtained when $\log \eta_{\text{rel}}$ is plotted against $1/a_m$, and X_s, the intercept on the $\log \eta_{\text{rel}}$ axis, always approximated to 0.15. A typical plot is shown in Fig. (3.12). Dilute systems, which exhibited Newtonian flow, also obeyed Eq. (3.142) as soon as ϕ reached such a value that Eq. (3.111) was no longer valid. Relevant data are shown in Table 3.6. When D_s exceeded 1μ

$$\alpha_v \approx 0.036 D_s^{2} \qquad (3.143)$$

but, below 1μ, α_v changes more drastically with changing D_s. This equation was not valid for heterodisperse systems. Presumably its form varies with the degree of inhomogeneity of the system, the dependence of α_v on D_s decreasing as the size distribution becomes broader. There is little evidence available at present to indicate the precise interrelationship.

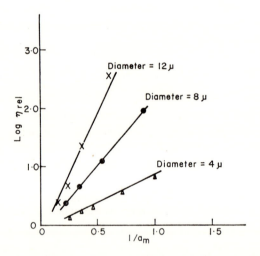

FIG. 3.12. Viscosity data for glass spheres in aqueous glycerol solution (Williams, 1953).

Thomas (1965) analysed a wide range of published viscosity data, and he found that when η_{rel} was plotted against ϕ there was $\pm 20\%$ spread about the mean η_{rel} at $\phi = 0.02$, and $\pm 75\%$ spread when $\phi = 0.5$. The sizes quoted ranged from 0.099μ to 435μ. He proposed that the influence of D_s and shear rate could be minimised by extrapolating the data to $D_s \to \infty$, and $D \to \infty$. Using a non-linear least squares procedure, a power series in ϕ was obtained resembling Eq. (3.115) which gave results similar to those predicted by Simha's (1952) theory (Eqs (3.117)–(3.123)). These observations confirm that

TABLE 3.6

Viscosity data applied to Eq. (3.142) (Sherman, 1965)

System	Particle size range (μ)	D_s (μ)	α_v	Optimum value of ϕ quoted	Reference
A. Emulsions					
1. W/O					
Emulsifying agent					
(a) Sorbitan mono-oleate	1·3—2·8	2·0	0·10	0·50	Sherman, 1965
(b) Sorbitan sesquioleate	1·0—2·5	1·75	0·15	0·57	Sherman, 1965
(c) Sorbitan trioleate	1·0—3·3	2·5	0·21	0·50	Sherman, 1965
(d) Polyoxyethylene sorbitan mono-oleate	2·7—3·2	3·0	0·38	0·40	Sherman, 1965
2. O/W					
Emulsifying agent					
(e) Sorbitan monolaurate	3·4—5·7	4·5	0·52	0·64	Sherman, 1965
(f) Asphalt/water emulsions		2·7	0·32	0·57	Eilers, 1941
(g) Latex	monodisperse	0·195	0·05	0·60	Maron et al., 1951
(h) Latex	monodisperse	0·1388	0·02	0·60	Maron and Ming-Fok, 1955
(i) Latex	monodisperse	0·11	0·03	0·52	Maron and Levy-Pascal, 1955
B. Dispersions of solid spherical particles					
(j) Glass spheres	3—4	3·5	1·0	0·37	Robinson, 1951
(k) Glass spheres	10—20	15	2·5	0·50	Robinson, 1949
(l) Glass spheres	20—30	25	4·5	0·45	Robinson, 1957
(m) Glass spheres	monodisperse	130	30	0·50	Vand, 1948
(n) Methyl methacrylate spheres	2—15	9·4	2·2	0·13	Eveson, 1959
(o) Methyl methacrylate spheres	15—30	21·9	5·2	0·20	Eveson, 1959
(p) Glass spheres	monodisperse	4	1	0·40	Williams, 1953
(q) Glass spheres	monodisperse	8	2·3	0·50	Williams, 1953

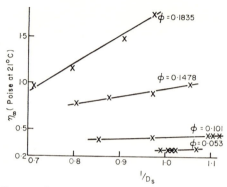

FIG. 3.13. The influence of D_s on the viscosity of O/W emulsions (Sherman, 1961). The crosses, reading from right to left, represent also homogenisation pressures of 1000; 1500; 2000; 2500; and 3000 lb/in.

the latter theory makes no allowance for the influence of particle size on viscosity, and consequently, it cannot be valid when particle size does not exceed a few microns.

Particle size analyses should be carried out as a routine when measuring viscosity because it does not follow, for example, that different emulsions will have identical mean sizes simply because they are prepared by the same technique. This applies particularly when a comparison is made between emulsions containing the same ingredients but having different concentrations of dispersed phase. For example ice cream emulsions (O/W) with ϕ ranging from 0·053 to 0·184 were homogenised at pressures between 1000 and 3000 lb/in^2 in increments of 500 lb/in^2 (Sherman, 1961). Graphs of the change in D_s with increasing homogenisation pressure (Fig. 3.13) indicated that higher homogenisation pressures had to be applied to the more concentrated emulsions to obtain the same D_s as in the dilute emulsions. The oil in the very dilute emulsions was very effectively dispersed at 500–1000 lb/in^2 since there was only a comparatively small further decrease in D_s at higher homogenisation pressures (Table 3.7). In the concentrated emulsions the lower

TABLE 3.7

Influence of homogenisation pressure and ϕ on dispersion (Sherman, 1961)

ϕ	Homogenisation pressure (psi) required to obtain D of 1·0μ	D_s following homogenisation at	
		1000 psi	3000 psi
0·0526	1500	1·01	0·94
0·1010	1600	1·17	0·84
0·1478	2200	1·24	0·95
0·1835	2800	1·41	1·03

homogenisation pressures gave a higher initial D_s, and higher pressures were required to reduce this effectively. At $3000 \, lb/in^2$ the variation in D_s for different values of ϕ was quite small.

Goulden and Phipps (1964) confirmed these observations for a wider range of ϕ. They also proposed an empirical relationship between D_s and homogenisation pressure (P_H) for dilute emulsions,

$$D_s = \left(\frac{P_0}{P_H}\right)^q \qquad (3.144)$$

where P_0 is the pressure required to produce unit globule size, and q is a factor which characterises the efficiency of the homogeniser. Both q and P_0 decreased with increasing ϕ. With concentrated emulsions the $\log D_s$–$\log P_H$ plot was no longer a straight line. Furthermore, the rate at which emulsion flowed through the homogeniser also influenced D_s.

4. *Particle Size Distribution*

It is very difficult to prepare monodisperse emulsions, so that the size distribution influences viscosity in addition to the mean particle size. This aspect of dispersed system rheology has not been studied to any degree, probably because in the past it was very difficult to control the size distribution. Now that monodisperse polymer dispersions are available it should be possible to study the influence of size distribution by mixing monodisperse systems of various particle sizes in different proportions.

Following Eq. (3.111) the viscosity of a dilute dispersion (a) is

$$\eta^a = \eta_0(1+a\phi^a) \qquad (3.145)$$

where η^a is the viscosity for a volume concentration ϕ^a of dispersed phase. If an additional small volume ϕ^b of material to be dispersed is now added the resultant viscosity (η^{a+b}) can be formulated in alternative ways. In the first, η^a now represents the viscosity of the new continuous phase, so that

$$\eta^{a+b} = \eta^a (1+a\phi^b) = \eta_0 [(1+a\phi^a)(1+a\phi^b)] \qquad (3.146)$$

According to the alternative way, η_0 remains the viscosity of the continuous phase and a volume concentration $(\phi^a+\phi^b)$ is dispersed in it,

$$\eta^{a+b} = \eta_0 [1+a(\phi^a+\phi^b)]. \qquad (3.147)$$

The difference between Eqs (3.146) and (3.147) is the term $a^2\phi^a\phi^b$, which will not be very significant provided ϕ^a and ϕ^b are small.

Eveson's (1959) study of the viscosity changes which occur when solid spheres of different sizes are mixed together follows the above line of thought.

He assumed that a dispersion of solid spheres containing only two different size ranges can be regarded as a suspension of the larger spheres in a continuous phase composed of the fluid phase plus smaller size of spheres. A further study was made on dispersions containing three different size ranges. In this case the viscosity (η^{a+b+c}) of the mixture was formulated in two ways

$$\eta_{\text{rel}}^{a+b+c} = \eta_{\text{rel}}^{a} \times \eta_{\text{rel}}^{b} \times \eta_{\text{rel}}^{c} \tag{3.148}$$

which for very dilute dispersions would reduce to

$$\eta^{a+b+c} = \eta_0 \left[(1+a\phi^a)(1+a\phi^b)(1+a\phi^c)\right] \tag{3.149}$$

where η_{rel}^{c}, η_{rel}^{b}, and η_{rel}^{a} are the relative viscosities of suspensions containing spheres of increasing size ranges with volume concentrations ϕ^a, ϕ^b, and ϕ^c in a common fluid phase. Alternatively, the final system can be regarded as a suspension of the largest spheres in a dispersion medium containing spheres of the two smaller sizes, i.e.

$$\eta_{\text{rel}}^{a+b+c} = \eta_{\text{rel}}^{a+b} \times \eta_{\text{rel}}^{c} \tag{3.150}$$

where η^{a+b} is the relative viscosity of a suspension containing a sphere concentration $(\phi^a + \phi^b)$. When the total volume concentration of spheres did not exceed 0·1, the experimentally derived values of $\eta_{\text{rel}}^{a+b+c}$ agreed satisfactorily with the values calculated by Eqs (3.149) and (3.150). At higher volume concentrations a discrepancy appeared which increased as the volume concentration increased. Equation (3.150) always gave values closer to experiment than Eq. (3.149)

In Eveson's (1959) experiments the spheres employed had rather large particle sizes. These sizes were certainly much larger than those found in most commercial emulsions and dispersions of interest, so that if the logic which led to Eqs (3.148)–(3.150) is to be applied to the latter systems some factor must be introduced to account for the particle sizes. One possible way to do this would be to make use of Eq. (3.127), subject to it being understood that h_s depends on D_s. Following the argument which gave Eq. (3.150), the viscosity of a dispersion containing two particle sizes is

$$\eta_{\text{rel}}^{a+b} = \frac{\eta^{a+b}}{\eta^a} = \exp\left(\frac{a\phi^b}{1-h_s^{b}\phi^b}\right) \tag{3.151}$$

where h_s^{b} is the value of h_s for the larger size spheres.

Since

$$\eta_{\text{rel}}^{a} = \frac{\eta^a}{\eta_0} = \exp\left(\frac{a\phi^a}{1-h_s^{a}\phi^a}\right) \tag{3.152}$$

$$\eta^{a+b} = \eta_0 \left[\exp\left(\frac{a\phi^a}{1-h_s^a\phi^a}\right) \exp\left(\frac{a\phi^b}{1-h_s^b\phi^b}\right) \right] .$$

For mixtures of three different particle sizes

$$\eta_{\text{rel}}^{a+b+c} = \frac{\eta^{a+b+c}}{\eta^{a+b}} = \exp\left(\frac{a\phi^c}{1-h_s^c\phi^c}\right) \qquad (3.153)$$

so that,

$$\eta^{a+b+c} = (\eta^{a+b})\left[\exp\left(\frac{a\phi^c}{1-h_s^c\phi^c}\right) \right]$$

$$= \eta_0 \left[\exp\left(\frac{a\phi^a}{1-h_s^a\phi^a}\right) \exp\left(\frac{a\phi^b}{1-h_s^b\phi^b}\right) \exp\left(\frac{a\phi^c}{1-h_s^c\phi^c}\right) \right]$$

For i different particle sizes

$$\eta^{a+b+c\ldots i} = \eta_0 \left[\exp\left(\frac{a\phi^a}{1-h_s^a\phi^a}\right) \exp\left(\frac{a\phi^b}{1-h_s^b\phi^b}\right) \exp\left(\frac{a\phi^c}{1-h_s^c\phi^c}\right) \ldots \right.$$

$$\left. \ldots \exp\left(\frac{a\phi^i}{1-h_s^i\phi^i}\right) \right] . \qquad (3.154)$$

The respective values of h_s for sizes a, b, c, ... i could be obtained, for example, from viscosity studies on model monodisperse polymer systems and applying Eq. (3.127). Relevant values of h_s for particle sizes below 1μ have already been published by Saunders (1961).

Collins and Wayland (1963) studied dispersions of spherical particles mixed with rod shaped particles. This study is particularly relevant to studies on emulsions containing added pigments. They assumed that the specific increase in viscosity ($\eta_{\text{sp}}^{\text{mix}}$) for the mixture was given by

$$\eta_{\text{sp}}^{\text{mix}} = \eta_{\text{sp}}^{\text{spheres}} + \eta_{\text{sp}}^{\text{rods}} + \Delta\eta_{\text{sp}} \qquad (3.155)$$

where $\eta_{\text{sp}}^{\text{spheres}}$ and $\eta_{\text{sp}}^{\text{rods}}$ were the respective specific increases in viscosity for suspensions of rods and spheres separately, and that $\Delta\eta_{\text{sp}}$ which arises from the interaction between the two types of particles, is bilinear in both their concentrations. The continuous phase was the same for both suspensions of spheres and of rods, both volume concentrations were very low, and

$$\Delta\eta_{\text{sp}} = f(g)\,\phi^{\text{spheres}}\phi^{\text{rods}} \qquad (3.156)$$

where $f(g)$ is a function of the geometric parameters of the two types of particles. Experimentally, it was found that $\Delta\eta_{sp}$ increased linearly with the product $\phi^{spheres}\phi^{rods}$, and this was considered to be due to the spheres preventing the rods from becoming oriented during the shear induced flow. Presumably, if this is correct, and both solid phases had been spheres the $\Delta\eta_{sp}$ term in Eq. (3.155) would disappear and it would then reduce to Eq. (3.147).

One interesting point which arises from Eveson's (1959) studies, and also those of Eveson *et al.* (1951), is that their approach predicts a minimum in the relative viscosity curve for mixtures of two different particle sizes at approximately equal proportions of the two sizes. The presence of a minimum was confirmed experimentally, but for many dispersions it did not appear at the predicted point. Chong (1964) suggested that the smaller particles enter the voids between the larger particles wherever possible so "... that the fine spheres act like ball bearings between large spheres. If the particle size ratio of small sphere to large sphere is less than 1/10, the ball bearing action appears to cease gradually and the small spheres seem to behave like a fluid toward the large spheres." The particle settling studies of Fidleris and Whitmore (1961) were quoted in support of this view. The latter workers found that with size ratios of small spheres/large spheres of $\frac{1}{3}$ to 1/100 the large spheres encountered the same resistance to motion when passing through a suspension of the small spheres as they did when settling through a fluid with the same density and viscosity.

Binary mixtures of protein thickened polystyrene latexes with particle sizes of 880 Å and 8140 Å have been found to exhibit pronounced hysteresis in their rate of shear-shear stress behavior (Saunders, 1967). The area of the hysteresis loop was at its maximum in mixtures which contained 30-40% of the larger particle size. This corresponds approximately to the concentration of smaller particles which would be required for complete coverage of the surfaces of the larger particles. It can be shown quite readily that the probability of collision between small particles and large particles is greater than the probability of collision between small particles, so that the large particles behave as nuclei around which the small particles collect (Sherman, 1967b). On the basis of this theory a minimum should still appear in the relative viscosity curve for mixtures of two different particle sizes, but its position will now shift to lower concentrations of the smaller particle size as the size ratio decreases, because fewer small particles will be required for surface coverage of the larger particles.

5. *Particle Shape*

The particles in O/W emulsions deform after flocculation, and this allows them to pack together more closely than in dispersions of undeformed spherical particles. If the particles in the flocculates eventually assume a rhomboidal dodecahedral configuration (Lissant, 1966) they may occupy as much as 94% of the total flocculate volume. As a result, the viscosity of such a system at very low rates of shear will be lower than that of a flocculate containing the same volume concentration of undeformed spheres. At very high rates of shear the particles may deform even more if their diameters are large, so that eventually they may become ellipsoids.

The viscosity of non-spherical particles depends on their orientation with respect to the direction of flow. This orientation is modified by two effects (Frisch and Simha, 1956). The first is particle rotation due to the shear applied to the fluid phase, and the second, which opposes the effect due to the first, is rotational Brownian motion of the particles. Jeffery (1922) calculated the rotational velocity of an ellipsoid in a flow field which was devoid of the perturbations produced by Brownian motion, and from this he calculated the viscosity arising through the dissipation of energy. The viscosity was found to be greatly influenced by the initial orientation of the particles. Boeder (1932), Kuhn (1932), Peterlin (1938), and Kuhn and Kuhn (1951) found that particles were oriented at an angle of $\sim 45°$ to the direction of flow at low rates of shear, but with increasing rate of shear this angle decreased until eventually, at very high rates of shear, the particles were oriented in the direction of flow.

Little study has been made of the viscosity of concentrated dispersions of ellipsoidal particles. For dilute systems Eq. (3.111) was modified (Kuhn and Kuhn, 1945) into three different forms, the form to be used in a particular case depending on the axial ratio (p_a) of the particle, the latter being the ratio of the major axis (L_A) to the minor axis (B_A).

When, $0 < p_a < 1$

$$\eta_{rel} = 1 + a\phi + \frac{32}{15\pi}\left(\frac{1}{p_a} - 1\right)\phi - 0.628\left[\frac{[(1/p_a) - 1]}{[(1/p_a) - 0.75]}\right]\phi \quad (3.157)$$

When $1 < p_a < 15$

$$\eta_{rel} = 1 + a\phi + 0.4075\,(p_a - 1)^{1.508}\phi \quad (3.158)$$

When $p_a > 15$

$$\eta_{rel} = 1 + 1.6\phi + \frac{(p_a)^2}{5}\left[\frac{1}{3(\ln 2p_a - 1.5)} + \frac{1}{\ln 2p_a - 0.5}\right]\phi \quad (3.159)$$

Simha (1940, 1945), on the other hand, found that when $p_a > 5$

$$\eta_{rel} = 1 + \frac{14}{15}\phi + \frac{(p_a)^2}{5}\left[\frac{1}{3(\ln 2p_a - \lambda_a)} + \frac{1}{\ln 2p_a - \lambda_a + 1}\right]\phi \quad (3.160)$$

with $\lambda_a = 1\cdot5$ for an ellipsoid, and $\lambda_a = 1\cdot8$ for a cylindrical rod. Equation (3.160) is similar in form to Eq. (3.159). When $p_a = 1$ all the above equations reduce to Eq. (3.111).

Equations (3.157)–(3.160) are of the form

$$\eta_{\text{rel}} = 1 + c_1\,\phi + c_2 f(p_a)\phi + \dots \tag{3.161}$$

or,

$$\frac{\eta_{\text{sp}}}{\phi} = [\eta] + C_1[\eta]^2\phi + C_2[\eta]^3\phi^2 + \dots$$

Studies by various workers indicate that for rigid rods, or dumbells, $C_1 = 0\cdot73$–$0\cdot77$. For spherical particles reported value of C_1 range from as low as $0\cdot745$ to $2\cdot26$.

Brodnyan (1959) extended this treatment to concentrated dispersions of ellipsoidal particles by adopting Mooney's (1951) approach, in which the crowding effect due to the introduction of more particles is considered. The final equation was

$$\eta_{\text{rel}} = \exp\left[\frac{a\phi + 0\cdot399\,(p_a-1)^{1.48}\phi}{1 - h_s\,\phi}\right] \tag{3.162}$$

which resembles Eq. (3.127). The function $c_2 f(p_a)$ in Eq. (3.161) was assumed to be of the form $a_b(p_a-1)^{b_b}$, and from the experimental data the values of a_b and b_b were found to be $\sim 0\cdot4$ and $\sim 1\cdot5$ respectively. These values closely agreed with the theoretical calculations of Kuhn and Kuhn (1945). When $p_a = 1$, then $h_s = 1\cdot35$ in closely packed systems of spheres, according to Mooney (1951). Brodnyan (1959) calculated that $h_s = 1\cdot91$ for ellipsoids with a large value of p_a, so if these two assumptions are correct then h_s increases rapidly, and asymptotically, as p_a increases.

Eirich et al. (1936) found that suspensions of glass and silk rods obeyed the relationship

$$\eta_{\text{rel}} = 1 + aF_A\,\phi + 8F_A^2\phi^2 + 40F_A^3\phi^3 \tag{3.163}$$

where F_A is a factor which depends on the axial ratio but not on the size of the rods, or on their concentration.

Little is known about the viscosity relationships for other, less well defined, shapes of particles, but significant differences can be anticipated particularly at very low rates of shear since the attraction potential between flocculated particles varies with shape (Vold, 1954), and also with their distance of separation. This point is discussed in Section 4 of this chapter.

Lewis et al. (1935) found that Eq. (3.111) was obeyed by dilute suspensions of clay particles, with the form of small flat plates, when their dimensions were approximately the same in all directions. This was confirmed by Matsui

(1948), but in other cases, where the particles were rod shaped (Hong Kong Kaolin), Eq. (3.111) was not valid. Irregular shaped particles of carborundum ($\sim 0 \cdot 2$–$0 \cdot 3 \, \mu$) gave values of $2 \cdot 5$–$2 \cdot 9$ for the constant a in Eq. (3.115), and a mean value of $19 \cdot 2$ for the constant b. These values were attributed to turbulence in the continuous phase fluid which was produced by the irregularities in the surfaces of the particles (Duclaux and Sachs, 1931). When the particle surfaces are very irregular the dispersion behaves as if it has an effective volume which exceeds the theoretical volume concentration of particles by the volume of fluid which is immobilised within the indentations. Rough methyl methacrylate spheres with mean sizes ranging from $38 \, \mu$ to 279μ (Ward and Whitmore, 1950) showed viscosities which suggested that the layer of fluid immobilised within the indentations was, on average, about $6 \, \mu$ thick. This value, which was confirmed by sedimentation tests, and also by micro-photographs, is too large for the fluid to be held to the surfaces of the particles by any form of chemical binding. Provided the surface irregularities are small with respect to the size of the particles, they behave like smooth spheres.

B. Continuous Phase

The one point on which all viscosity equations for dispersions and emulsions agree is that their viscosity is directly proportional to the viscosity of the continuous phase. The latter, as represented by η_0, is the viscosity of the entire continuous phase, and not merely the viscosity of the basic fluid to which the emulsifier, hydrocolloids, etc. are added. In dispersed systems η_0 is slightly, but usually insignificantly, reduced by the adsorption of emulsifier or surfactant at the oil-water or particle-water interface. Thus, η_0 can in most cases be safely assumed to correspond to the viscosity of the continuous phase in bulk. However, this assumption may be incorrect in the light of some recent studies which suggest that the viscosity of a thin film of liquid (100–250Å) is very much higher than the viscosity of the same liquid in bulk. In aqueous films these discrepancies are attributed to electrical charge effects (Derjaguin and Samygin, 1951; Derjaguin and Titijevskaya, 1959). Similar effects are now claimed for non-polar films (Fuks, 1958; Karasova and Derjaguin, 1959). Applying these observations to concentrated dispersed systems in which the particles are separated by very thin films of continuous phase, it is possible that now the viscosity of bulk continuous phase is no longer representative of η_0. The very high viscosity of such emulsions may be partially due to an abnormally high value of η_0, in addition to the increased hydrodynamic interaction between the particles. Conversely, the initial rapid decrease in viscosity when shear is applied to a flocculated emulsion may be due in part to a decrease in η_0 as the particles move further apart.

When the continuous phase is non-Newtonian in its flow characteristics, this behavior will be reflected in the properties of the dispersion or emulsion even when the volume concentration of dispersed phase is very low.

C. Emulsifying Agent

The influence of the emulsifier on fluid circulation within the particles of emulsions, and hence on viscosity, was discussed in Section 3.A.2. In addition, the chemical constitution of the emulsifier influences viscosity (Wilson and Parkes, 1936; Broughton and Squires, 1938; Sumner, 1940). Emulsions prepared with the same volume concentration of dispersed phase in all cases, and with the same continuous phase, can show different viscosities when different emulsifiers are used (Sherman, 1955a). Table 3.8 illustrates this point. Six emulsions gave relative viscosities which were reasonably close together ($\eta_{rel} = 13\cdot9-15\cdot2$), but the remaining emulsions had relative viscosities which were either much lower or much higher than this.

TABLE 3.8

Influence of chemical constitution of emulsifying agent on emulsion viscosity
(Sherman, 1955a)

Emulsifying agent	ϕ	η_0	η_∞	$\dfrac{\eta_\infty}{\eta_0}$	Extrapolated yield value
Soya lecithin	0·688	0·168	1·10	6·55	807
Sorbitan sesquioleate	0·687	0·176	2·60	14·77	1766
Sorbitan mono-oleate	0·687	0·176	2·67	15·14	1456
Sorbitan trioleate	0·686	0·169	3·22	19·03	1922
Polyethylene glycol (200)* Mono-oleate	0·686	0·151	2·10	13·91	1214
Diglycol monolaurate	0·686	0·144	2·14	14·86	1221
Propylene glycol Mono-oleate	0·686	0·149	2·20	14·76	1614
Blown soya bean oil	0·686	0·184	2·72	14·78	2339
Glyceryl polyricinoleate	0·686	0·358	4·21	11·76	2825
Polyethylene glycol Polyricinoleate	0·686	0·939	3·90	4·15	2371

*Refers to the molecular weight of the polyethylene glycol chain

Part of the influence exerted by the emulsifier on viscosity may arise from its influence on mean particle size, and particle size distribution. One of the ways of accounting for the latter effect is to plot the data in the form ln η_{rel}

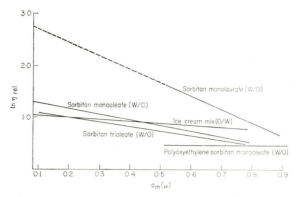

FIG. 3.14. Influence of the chemical nature of the emulsifier on emulsion viscosity.

against a_m, where $\eta_{rel} = \eta_\infty/\eta_0$. Figure 3.14 illustrates plots of this form for several emulsion systems. The O/W emulsions stabilised with sorbitan monolaurate showed abnormal effects at small values of a_m due to particle distortion, so that the dotted line extrapolation indicates only theoretical values of $\ln \eta_{rel}$ when $a_m < 0.5$.

The chemical nature of the emulsifier will be important also in low rate of shear studies on flocculated systems. For systems with either polar or non-polar continuous phases it affects the distance separating the particles. In the former, the emulsifier influences the height of the potential energy barrier (V_{max}) to closer approach of the particles and the location of the secondary minimum in the $V-H_0$ curve, while in the latter the particles are separated by a distance which approximates to twice the length of the hydrocarbon chain of the emulsifier.

Emulsifier concentration also influences emulsion viscosity, the effect becoming more pronounced as dispersed phase concentration increases. Few comparable studies have been made on dispersions. Analysis of published literature suggests that the effect can be represented qualitatively by an empirical equation which closely resembles Eq. (3.130) with K_R replaced by emulsifier concentration multiplied by another constant (Sherman, 1959). This rise in viscosity due to the emulsifier concentration has often been attributed to increased adsorption of emulsifier around the particles, thus leading to an increase in the effective dimensions of the particles. While this explanation may be valid in certain instances, e.g. when proteins are used, for many other emulsifiers it is highly improbable that the adsorbed layer of emulsifier around the particles ever exceeds a thickness of one molecule. When a condensed monomolecular layer has been established there is no further adsorption at the particle surface, and any excess emulsifier molecules remain in the continuous phase where they associate to form micelles. These

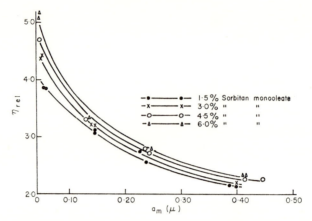

FIG. 3.15. Sorbitan mono-oleate stabilised water-in-liquid paraffin emulsions: the influence of emulsifier concentration on η_{rel} (Sherman, 1963b).

micelles immobilise continuous phase fluid within themselves, in much the same way as soap micelles do in aqueous media. In bulk solutions of surface active agents this is reflected in the non-linearity of the viscosity-concentration curve, but in emulsions there is an additional effect. Immobilisation of continuous phase alters the volume ratio of dispersed phase to "free" continuous phase, thereby causing an increase in the viscosity. The degree of fluid immobilisation increases as the excess of emulsifier increases.

Figure (3.15) illustrates how the viscosity of concentrated W/O emulsions

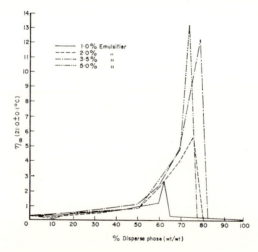

FIG. 3.16. Variation of reciprocal mobility with concentration of dispersed phase (Sherman, 1950a).

($\phi = 0.37$–0.68) rises when the concentration of the emulsifier, sorbitan mono-oleate, increases from 1.5% to 6.0% (Sherman, 1963b). The data are plotted as η_{rel} (η_∞/η_0) against a_m in order to eliminate the influence of variable mean particle size. From these curves the effect of fluid immobilization on η_{rel} has been calculated for three values of a_m (Table 3.9). For all three values of a_m there is a significant change in η_{rel} as the concentration of emulsifier increases. As might be expected, the influence of emulsifier concentration increases as a_m decreases.

TABLE 3.9

Influence of emulsifying agent concentration on viscosity
(Sherman, 1963b) of W/O emulsions ($D_s = 1.504\mu$)

Concentration of sorbitan mono-oleate wt/wt	As multiple of C.M.C.	a_m (μ)	$\dfrac{\eta_\infty}{\eta_0}$
	1		2·99
1·5	88·2		3·07
3·0	182·8	0·15	3·16
4·5	284·6		3·25
6·0	294·4		3·36
	1		2·48
1·5	72·9		2·55
3·0	150·1	0·25	2·63
4·5	232·1		2·71
6·0	319·7		2·80
	1		2·20
1·5	64·1		2·25
3·0	131·8	0·35	2·31
4·5	203·0		2·38
6·0	277·5		2·45

Emulsifier concentration has another indirect influence on emulsion viscosity through its effect on the inversion point of emulsions (Sherman, 1950a; Becher, 1958). W/O emulsions stabilised by non-ionic emulsifiers showed an approximately linear increase in η_∞ as ϕ rose to ~ 0.5 ($D_s = 2.0\,\mu$). When ϕ exceeded 0.5 in emulsions which contained 2.0–5.0% emulsifying agent, or ~ 0.6 when only 1.0% emulsifying agent was present, η_∞ increased more rapidly (Fig. 3.16). The maximum value of η_∞ was reached when

$\phi = 0.75-0.80$ for all emulsions apart from those which contained only 1.0% emulsifying agent. For the latter the maximum in η_∞ was reached when $\phi = 0.625$. Very slight increases in ϕ beyond these values caused inversion to O/W emulsions, and this was accompanied by a drastic reduction in η_∞ to little more than the viscosity of the aqueous phase (Sherman, 1950a).

The solubility of the emulsifier in the continuous phase influences the stability to inversion. For example, W/O emulsions stabilised with mannide mono-oleate, mannitan mono-oleate, and sorbitan sesquioleate, showed no change in η_∞ as the pH of the aqueous phase was increased through the range 3.0-9.0. When the pH exceeded 9.0 all the emulsions inverted to O/W emulsions, and this was associated with a drastic fall in η_∞ (Sherman, 1950b). At pH 9.0 the emulsifiers were completely soluble in the aqueous phase, with possible conversion to fatty acid soaps, whereas at pH 7.0 they were dispersible but not soluble in the aqueous phase.

D. Rheological Properties of the Adsorbed Film of Emulsifier

Reference has already been made to the influence of the rheological properties of the adsorbed emulsifier film on deformation of particles in emulsions. In dispersions these film properties would not be of great importance, except possibly in very concentrated dispersions. As yet no method has been devised for studying the rheological properties of these films, *in situ* in emulsions. The standard procedure is to study films which are developed at a flat, extended, interface between bulk water and oil phases (Criddle, 1960). There is some doubt whether the results of such investigations can be applied to the behavior of emulsifier films in sheared emulsions, because in the latter the films are far from flat, and the pronounced curvature may exert an important, but hitherto unidentified, effect.

Equation (3.133) indicates the relevance of some rheological properties of emulsifier films to the viscosity of dilute emulsions. Singularly little effort appears to have been made to experimentally assess the validity of this equation. The quantity $(2\eta_s + 3\eta_\beta)$ was calculated for several dilute O/W emulsions (Nawab and Mason, 1958), and it was found to lie within the range 0.92×10^{-4} to 0.014×10^{-4} g.sec^{-1}. These values resemble those for η_s which were obtained with emulsifier films spread at an air-water interface (Joly, 1956), but as the data obtained for the emulsions relate to an oil-water interface the two sets of data cannot really be compared.

Shearing forces cause molecules of the emulsifier film to be displaced from their equilibrium position, and also molecules of the oil and water phases which are in the immediate vicinity. The stress which is developed depends on the molecular rearrangements. This phenomenon has been treated theoretically in different ways by Joly (1954, 1956) and Oldroyd (1953, 1955).

Joly's (1954, 1956) approach to the problem was based on absolute reaction rate theory (Ewell and Eyring, 1937; Moore and Eyring, 1938)

$$\sinh \frac{\eta_s D_f \sigma}{2kT} = \frac{D_f h \exp (\Delta F/kT)}{2kT} \qquad (3.164)$$

where σ is the molecular area, D_f is the velocity gradient in the film, and ΔF is the free energy of activation per molecule.

When $D_s f_s \sigma \ll 2kT$

Eq. (3.164) reduces to

$$\eta_s = \left(\frac{h}{\sigma}\right) \exp \left(\frac{\Delta F}{kT}\right) \qquad (3.165)$$

and this relationship is valid when the emulsifier film shows Newtonian flow.

The experimental value of σ is a mean value for a mixture of molecules in stable neighboring states

$$\sigma = (1-x_p) \sigma_i + x_p \sigma_{i-1} \qquad (3.166)$$

where x_p is the proportion of the total number of molecules/cm^2 in the state $i-1$, the molecules occupying a finite number of stable equilibrium states $1, 2, ... i-1, i$, with molecular areas $\sigma_1, \sigma_2, ... \sigma_i$. ΔF is defined in a similar way to σ,

$$\Delta F = (1-x_p) \Delta F_i + x_p \Delta F_{i-1} \qquad (3.167)$$

where ΔF_{i-1} and ΔF_i are the activation energies for the $i-1$ and i states, respectively. Two terms are incorporated in ΔF_i, one for the energy required to make a hole in the film into which an activated adjacent molecule can move, and the other term is for the energy required by a molecule to pass from an equilibrium position to a free adjacent position.

The measured surface viscosity (η_z) includes contributions for the molecules of the oil and water phases adjacent to the emulsifier film which move with it during shear if there is no slippage. Consequently,

$$\Delta F = \Delta F_c + \Delta F_w + \Delta F_{\text{oil}} \qquad (3.168)$$

where ΔF_c, ΔF_w, and ΔF_{oil} are the free energies of activation of the emulsifier molecules, the water molecules, and the oil molecules respectively.

Combining Eqs (3.165) and (3.168)

$$\eta_z = \eta_s \exp \left[-\left(\frac{\Delta F_w}{kT} + \frac{\Delta F_{\text{oil}}}{kT}\right)\right] \qquad (3.169)$$

which suggests that films which have a high viscosity will show non-Newtonian behavior.

Oldroyd (1955) pointed out that when the emulsifier molecules in the film are relatively far apart, as for example when repulsion forces operate, the work required to shear the film is much less than the work necessary to alter its area. The interfacial tension at any point, resulting from displacement of molecules, is then a simple function of the local surface dilatation, or the rate at which it changes. When the molecules in the emulsifier film are packed closely together, so that they are bound together by hydrogen bonding, much more work is required to shear the film at constant area. A perturbation treatment similar to that of Fröhlich and Sack (1946) led to Eq. (3.133) for very dilute emulsions in which the emulsifier film around the particles is fluid, and to Eq. (3.111) when the film is elastic. The two equations give similar results when the particles are small. This conclusion also follows from Eq. (3.110) when particle size is small.

E. Electroviscous Effect

When very dilute dispersions or emulsions which contain electrically charged particles are sheared, the symmetry of the electrical double layer around each particle is distorted. The interaction between ions in the electrical double layer and the electrical charge on the particle surfaces are affected, and this leads to an extra dissipation of energy and an increased viscosity (Conway and Dobry-Duclaux, 1960).

von Smoluchowski (1916) amended Eq. (3.111) for spherical rigid particles to account for this first electroviscous effect

$$\eta_{sp} = a\phi \left[1 + \frac{1}{\eta_0 \, \kappa r^2} \left(\frac{\varepsilon_d \, \zeta}{2\pi} \right)^2 \right] \tag{3.170}$$

where κ approximates to the specific conductivity. Thus, the increase in η_{sp} due to the electrical charge is

$$\frac{a\phi}{\eta_0 \, \kappa r^2} \left(\frac{\varepsilon_d \, \zeta}{2\pi} \right)^2 .$$

Equation (3.170) indicates that the first electroviscous effect will exert a significant effect only when r is not greater than a few hundred Angstroms, in contrast to von Smoluchowski's own assumption that the thickness of the double layer is small compared with particle size. This contradiction disappears only when the ionic strength is greater than about 0·01.

Booth (1950) extended Eq. (3.170) in the form of a power series of particle charge. He also introduced terms for the thickness of the diffuse double layer $(1/\chi)$, ion concentration (n_i), and the valency (Z_i) of the ions,

$$\eta_{sp} = a\phi \left[1 + q_b \left(\frac{\varepsilon_d \zeta}{kT} \right) (1 + \chi r^2) \, Z(\chi r) \right] \qquad (3.171)$$

where

$$q_b = \frac{\varepsilon kT \sum_{1}^{i} n_i z_i \rho_i}{\eta_0 \, e_c^{\,2} \sum_{1}^{i} n_i z_i^{\,2}}$$

and ρ_i is the frictional coefficient of an ion i in the continuous phase, e_c is the elementary charge, $Z(\chi r)$ is a complex function of χr such that $(1 + \chi r^2)$ $Z(\chi r)$ decreases as χr increases.

The double layer thickness is large when (χr) is small, and

$$Z(\chi r) = \frac{1}{200\pi(\chi r)} + \frac{11(\chi r)}{3200\pi}$$

so that the electroviscous effect contributes significantly to viscosity, and the contribution increases proportionately with $(1/\chi r)$.

When the double layer thickness is small, i.e. (χr) is large

$$Z(\chi r) = \tfrac{1}{2}\pi(\chi r)^4$$

and the electroviscous effect is very small, because the double layer is not distorted to any extent during particle flow.

Booth's (1950) treatment predicts a lower contribution to η_{sp} than von Smoluchowski's (1916) treatment. This agrees with most experimental findings. Studies with sulfonated polystyrene latex at ionic strengths between 10^{-3} and 10^{-4} indicated, however, that whereas Eq. (3.170) gave theoretical values which were higher than those derived experimentally, Eq. (3.171) gave values which were too low (Chan and Goring, 1966).

Street (1958) modified Eq. (3.171) by considering the relative motion between the continuous and dispersed phases during flow

$$\eta_{sp} = a\phi \left[1 + \frac{\phi}{2\eta_0 \, \kappa r^2} \left(\frac{\varepsilon \zeta}{2\pi} \right) (1 + \chi r^2) \right] \qquad (3.172)$$

In more concentrated emulsions the distance between particles is small so that their double layers may overlap, and the viscosity increases due to mutual repulsion. This effect, which is known as the second electroviscous effect, was first reported by Harmsen $et\ al.$ (1953). Its magnitude is directly proportional to ϕ^2. At constant ϕ it increases as the ionic strength decreases because the electrical double layer will increase, thus raising the probability of double layer interaction. Schaller and Humphrey (1966) observed the secondary electroviscous effect in monodisperse polystyrene latices with sizes 0.088μ, 0.557μ, and 1.305μ dispersed in media of various ionic strengths, the effect being most pronounced for the smallest size.

Because it is difficult to produce monodisperse particles of the desired size in emulsions there has been little study of electroviscous effects for these systems. van der Waarden (1954) determined the viscosities of a series of O/W emulsions which had mean particle sizes not exceeding 0.205μ. The maximum concentration of emulsifier used was unusually large since it formed about 12% of the total weight. At the higher emulsifier concentrations viscosities calculated by Eq. (3.111) were not in agreement with experimental values (Table 3.10). The discrepancy was much larger than suggested by either Eq. (3.170) or (3.171), so that distortion of the diffuse double layer around the particles could not be the cause. The strongly ionised emulsifier which was adsorbed on the particle surfaces was considered to produce a high electric field strength of 10^5–10^6 V/cm, to which a layer of water molecules was strongly bound. The thickness of the water layer, as indicated by the apparent increase Δr in particle radius, was

TABLE 3.10

Electroviscous effect in O/W emulsions (van der Waarden, 1954)

Emulsifier in oil phase (gm/gm)	ϕ	Globule size (μ)	η_{rel}	a	Δ_r (μ)
0·05	0·0565		1·17		
	0·112		1·40		
	0·222	0·2050	2·12	2·6 − 2·7	0·0014 − 0·0028
	0·328		3·52		
0·10	0·0559		1·21		
	0·111		1·51		
	0·219	0·1038	2·56	3·0 − 3·1	0·0031 − 0·0037
	0·325		5·64		
0·17	0·0549		1·25		
	0·109		1·68		
	0·216	0·0586	3·94	3·4 − 3·5	0·0032 − 0·0035
	0·321		22·15		
0·35	0·0531		1·35		
	0·106		1·96		
	0·210	0·0276	6·50	4·8 − 5·0	0·0033 − 0·0036
	0·313		92·8		

0.0014–0.0037μ, reaching an approximately steady value at the higher emulsifier concentrations. Mukerjee (1957) pointed out that contrary to normal practice, van der Waarden introduced the emulsifier into the oil phase. If allowance was made for diffusion of emulsifier across the interface and into the aqueous phase when emulsification took place (by inversion of the initially prepared W/O emulsion), then Eq. (3.171) was approximately followed.

van der Waarden's observations can be explained in two other ways. First, as already pointed out, Eq. (3.111) is not valid for the very small particle sizes employed so that the discrepancies between the calculated and experimental data are to be expected. Second, although each of his emulsions was reasonably monodisperse, they did not all have the same particle size. In fact, sizes between 0.0276μ and 0.205μ were used, and in this range slight changes in particle size have a profound effect on viscosity. If the data are replotted as η_{rel} against $1/a_m$ a series of straight lines are obtained, their gradients (α_v) decreasing as the particle size decreases. When α_v is plotted against particle size (Fig. 3.17) a relationship is derived which closely resembles that obtained with Saunder's (1961) data. In the latter case no electroviscous effect was reported, so that it is quite possible that it was also absent from van der Waarden's emulsions.

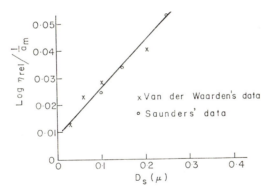

FIG. 3.17. Correction of van der Waarden's data for particle size variation before assessing the influence of the eletroviscous effect (Sherman, 1965).

The ζ potential of some W/O emulsions may be as high as $100\,\text{mV}$ so that a first electroviscous effect might be anticipated. However, several emulsions ($\phi = 0.03$–0.33) which contained different emulsifying agents, and which had ζ potentials of 15–$100\,\text{mV}$, all gave approximately the same value of a when Eq. (3.111) was applied to their viscosity data (Albers, 1957). The particle sizes exceeded 1μ in these emulsions. A first electroviscous effect of about

1% was indicated by Eq. (3.171), thus confirming that the effect is small in systems with a low dielectric constant.

The electrical double layer is several microns thick in W/O emulsions, so that a secondary electroviscous effect might be expected in the more concentrated systems, but, as the double layer is very diffuse, the increase in viscosity due to this effect would also be very small.

When traces of electrolyte, which originated from the polymerisation initiator and the surface active agent employed to stabilise the system, were removed from latex, the viscosity at low rates of shear increased quite appreciably and the flow became non-Newtonian (Brodnyan and Kelly, 1965). Removal of the electrolyte led to an increase in the thickness of the diffuse double layer, so that under conditions approaching those of the stationary state a higher concentration of the continuous phase was held between the flocculated particles. The net effect was an increase in the effective volume concentration of dispersed phase, since the flocculates flowed as single units at low rates of shear. On the addition of electrolyte the situation was reversed, and the viscosity decreased as the electrolyte concentration increased, until eventually it reached a minimum value. This was accompanied by a change from non-Newtonian to Newtonian flow. Sodium lauryl sulphate was far less effective than sodium chloride in reducing the viscosity, e.g. 1.71×10^{-5} mole sodium lauryl sulphate/gm latex reduced the viscosity, as measured at a rate of shear of $1 \sec^{-1}$, from 505 poise to 425 poise, but an identical concentration of sodium chloride reduced the viscosity to 0.367 poise.

F. Hydrocolloids, Pigments, and Crystals

When hydrocolloids are dissolved in an aqueous continuous medium they increase the viscosity of the latter to an extent which depends on the concentration employed, and on the molecular weight. The solution may exhibit non-Newtonian flow in shear. If the concentration of added hydrocolloid is large a gel-like consistency may be developed. A distinction should be drawn between non-Newtonian behavior which arises from the dispersed phase and non-Newtonian behavior due to the properties of the continuous phase. Some hydrocolloids, e.g. gum acacia, stabilise emulsions, in the absence of any other surface active material, by adsorbing at the oil-water interface to form a hydrated rigid film. Viscosity data for dilute emulsions prepared with gum acacia (Shotton and White, 1963) suggested an apparent increase in the volume concentration of dispersed phase ranging from 19%, when a mixture of n-heptane and carbon tetrachloride was used as the oil medium, to 35% when n-heptane was used by itself. Since any effect due to the size of the oil drops would be small in dilute emulsions, it was concluded that the thickness of the gum acacia film around the drops would have to be $0·14 \mu$ to $0·16 \mu$ to give this level of apparent increase in dispersed phase concentration.

Finely divided pigments also migrate to the oil-water interface in emulsions where they form a protective layer around the particles. Hydrous oxides, e.g. the hydrated forms of vanadium pentoxide, ferric oxide, and alumina, are all surface active. Apart from any increase in the viscosity of freshly prepared emulsions which results from their use, additional increases may appear during storage due to progressive hydration of the oxide layer. Eventually a gelatinous layer may develop around the particles. Concentrated W/O emulsions showed this phenomenon at ambient temperature when alumina was incorporated in the water phase (Sherman, 1955c). When propylene glycol was added to the water phase in concentrations up to 20% these changes were retarded to an extent which increased with increasing propylene glycol concentration. At concentrations in excess of 20% gel layer formation was completely inhibited. Several other polyalcohols had the same effect.

The influence of non-surface active pigments, or crystals, on the viscosity of emulsions and dispersions can probably be treated along the lines of Eqs (3.146) to (3.156) if the particles are spherical; otherwise, some modifications will be necessary, as indicated by Eqs (3.157) to (3.160), depending on particle shape.

4. RHEOLOGICAL PROPERTIES OF DISPERSED SYSTEMS AT VERY LOW SHEARING STRESSES

In Section 3.A.3 of this chapter strong evidence was provided for the profound influence of particle size on the flow properties of dispersed systems over a wide range of shear rates. Under such shear conditions the interlinked structure which develops between particles following flocculation in the static system is broken down to varying degrees. There has been no systematic study, however, of the influence of particle size on the rheological properties of dispersed systems under such conditions of shear that minimal damage is caused to the interlinked network of flocculated particles. This is probably due to the difficulty of selecting a suitable model system for study.

Both dispersions of glyceryl tristearate in groundnut oil or paraffin oil (van den Tempel, 1961) and carbon black in Nujol (Payne, 1964) showed viscoelastic behavior at low shearing stresses in creep compliance-time tests. An attempt was made to interpret the first series of data (van den Tempel, 1961) by a simple network model in which the particles form long chains which are distributed in three dimensions, with the shearing stress acting in any one plane being transmitted along these chains from one particle to the next. This model closely resembles those used by Goodeve (1939) and Casson (1959) and more recently by Cross (1965) to explain the influence of higher shearing forces on the viscosity of non-Newtonian dispersed systems. If the dispersion contains N particles per cm^3 with diameter D_s, then the volume concentration of particles (ϕ) is given by

$$\phi = \frac{\pi N D_s^3}{\phi} \tag{3.173}$$

and, the total number of particles in any one plane is $N/3$, or $2\phi/\pi D_s^3$. When the distance separating flocculated particles is small compared with particle size, the total chain length in a plane is $2\phi/\pi D_s^2$.

Van den Tempel (1961) calculated the attraction between the anisometric glyceride particles on the basis of an equation for V_A which was intermediate between that for spherical particles (Eq. 3.99) and that for cubes with each side of length D_s($V_A = -AD_s^2/12\pi H_0^2$), so that

$$V_A = -\frac{AD_s^{1.5}}{12H_0^{1.5}} \tag{3.174}$$

The force of attraction (F_v) between adjacent particles in a chain is derived by differentiating Eq. (3.174)

$$F_v = -\frac{AD_s^{1.5}}{8H_0^{2.5}} \tag{3.175}$$

and the stress (p) transmitted across 1 cm^2 of the chain in any one plane is given by

$$p = \frac{2\phi}{\pi D^2} \times \left(-\frac{AD_s^{1.5}}{8H_0^{2.5}}\right) = -\frac{A\phi}{4\pi D_s^{0.5}H_0^{2.5}} \tag{3.176}$$

The elastic shear modulus (G_s) of the three dimensional network in the stationary state is derived from

$$G_s = \frac{D_s}{3}\left(\frac{dp}{dH_0}\right) = \frac{5A\phi D_s^{0.5}}{24\pi H_0^{3.5}} \tag{3.177}$$

The most striking fact about Eq. (3.177) is that it suggests that the size of the anisometric particles has little influence on the shear modulus. Furthermore, if a similar treatment to that given in Eqs (3.174)–(3.177) is applied to Eq. (3.97), so as to develop an equation for the shear modulus of dispersions of spherical particles, no particle size term is found. Experimental data do not support these conclusions. Nederveen (1963) observed discrepancies between theoretical values of the elastic modulus, calculated on the basis of a linear chain model, and experimentally derived values. In particular, he found large anomalies with dispersions containing very small particles, where the elastic modulus appeared to be proportional to ϕ^8 rather than to ϕ (Eq. 3.177). It is possible, however, that this was due to particle size effects which are not covered by Eq. (3.177), since Nederveen (1963) concludes that "Further understanding of the mechanical behavior of suspensions could be gained by the study of model systems with systematic and well known variations ... in the shape and size of the particles." Payne (1964) found that for any one type of carbon black the elastic modulus decreased as the particle size increased, but no quantitative relationship was established between the two variables.

On the strength of Weymann's (1965) argument that a state of minimum potential energy is achieved by a long chain polymer when it assumes a coiled configuration, the author has developed a relationship which indicates a relationship between the elastic modulus and particle diameter (Sherman, 1968). No particular geometry was assumed other than that the particles are distributed uniformly throughout the fluid continuous medium, so that flocculation is non-localised. This implies the presence of a reasonably large concentration of particles. Following from the number of contact points between unit volume of a particle and its immediate neighbors it can be shown that

$$G_s = \frac{\phi(1 + 1 \cdot 828 v_f)A}{36\pi D_s{}^3 H_0{}^3} \qquad (3.178)$$

where v_f is the volume fraction of fluid held in the voids between the particles, and it will therefore depend on the packing geometry. Equation (3.178) indicates a linear relationship between G_s and the ratio $\phi/D_s{}^3$; this has been found to hold satisfactorily with a range of data for W/O emulsions.

The dependence of rheological properties on particle size, and of course, on particle size distribution also, have been confirmed by low stress studies on emulsions (Sherman, 1967b). These emulsions, which were of both the O/W and W/O type, showed viscoelastic behavior at stresses from 2–5 dynes/cm^2 in creep compliance-time studies (Chapter 1, Section C.1). Following preparation, the mean particle size of the emulsions increased with time due to coalescence of the particles, and this was accompanied by a decrease in the magnitude of all the viscoelastic parameters. The changes in all the parameters were particularly significant during the first two to three days, this being the time during which the particles at the lower end of the size range coalesced and disappeared. A plot of the logarithm of any parameter against mean particle size yielded a straight line with a negative gradient, the gradients varying from one parameter to another.

5. Rheological Changes in Dispersed Systems During Aging

Following preparation of a dispersion the particles come together and, provided they are present in sufficient number, they will eventually form an interlinked structure. This network traps fluid within the voids between the particles, so that the network behaves at low shear as if it had a volume which is larger than that of its constituent particles. Thus, it is to be expected that for dispersions of solid particles in fluid media the viscosity at low rates of shear should gradually increase with time to a maximum, as the complex structure develops, in accordance with the theory outlined in Section 2.C of this chapter. Similarly, in accordance with Section 4, the viscoelastic behavior at very low stresses should become more pronounced.

Emulsions show a more complex pattern of behavior, because following contact between the particles they do not retain their separate identities indefinitely. Sooner or later they merge or coalesce together, with the net result that the mean size and size distribution increase with time, the rate determining factor being the rate of coalescence. Coalescence represents the last stage in the three-stage process of coagulation, viz. flocculation, drainage of the thin films of fluid separating the particles, and coalescence. If the rate of coalescence is slow compared to the rate of flocculation then any rheological parameter which is measured should increase with time. Alternatively, if the rate of coalescence is high compared with the rate of flocculation the magnitude of the parameter should decrease with time. Detailed studies have been made with emulsions whose behavior falls between these two extremes (Sherman, 1963, 1967).

The particles in concentrated W/O emulsions flocculate very rapidly, but nevertheless, the viscosities of these emulsions, as measured over a wide range of shear rates, decreased with time (Sherman, 1963, 1967a). At rates of shear which are high enough to ensure complete deflocculation the decrease in viscosity with time was directly related to the rate of increase in mean particle size. The interrelationship could be defined by Eq. (3.142). The rate of change in mean particle size can be calculated from

$$N_t = N_0 \exp(-Ct) \qquad (3.179)$$

where N_0 and N_t are the number of particles per cm^3 at zero time and time t respectively, and C is the rate of particle coalescence, so that

$$\ln D_{s(t)} = \ln D_{s(0)} + \frac{Ct}{3} \qquad (3.180)$$

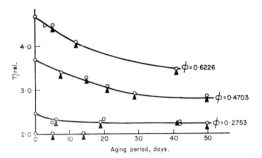

FIG. 3.18(a). W/O emulsions—comparison of experimental and calculated changes in η_{rel} on aging: ○, experimental data; ▲, calculated data (Sherman, 1963a).

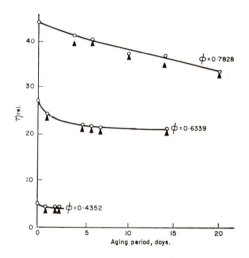

FIG. 3.18(b). O/W emulsions—comparison of experimental and calculated changes in η_{rel} on aging: ○, experimental data; ▲, calculated data. (Sherman, 1963a).

where $D_{s(0)}$ and $D_{s(t)}$ are the mean particle diameters at zero time and time t respectively. Thus, by determining the rate of particle coalescence C the change in viscosity with time can be calculated using Eqs (3.180) and (3.142). Experimental and calculated data were in close agreement (Fig. 3.18).

At rates of shear between 0·13 and 10·77 secs^{-1} deflocculation was only partial, and under such conditions the relationship between viscosity and particle size had to be modified to

$$\log \eta_{\text{rel}} = \alpha_v \left[\frac{1}{D_s[\sqrt[3]{(\phi_{\max}/f_s\,\phi)-1}]} \right] - 0 \cdot 15 \qquad (3.181)$$

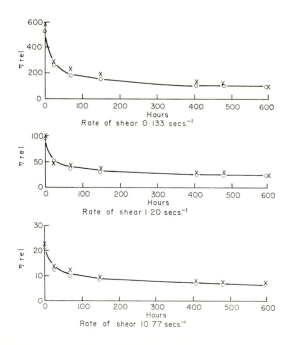

FIG. 3.19. Comparison of experimental and calculated changes in η_{rel} on aging (Sherman, 1967a). O—O, experimental data. X—X, theoretical data.

where f_s, the swelling factor in Eq. (3.181) was exponentially related to the rate of shear by

$$f_s = \exp\left(0.28 D^{-0.15}\right) \tag{3.182}$$

This relationship was valid over 8 decades from 10^{-5} to 10^3 secs^{-1}. Figure (3.19) illustrates the satisfactory agreement between experimental data and data derived by Eqs (3.180) and (3.181). Values of $\phi_a (= f_s \phi)$ were obtained from viscosity-dispersed phase volume concentration plots, as described in Section 2.C, and the corresponding values of f_s were calculated using Eq. (3.108). Table (3.11) summarises values of ϕ_a and f_s which were calculated from viscosity data obtained at 1.33 secs^{-1} for a W/O emulsion with $\phi = 0.47$, and which was allowed to stand at ambient temperature for about $2\frac{1}{2}$ weeks. Also included are data derived from Eq. (3.82) to show how incorrect conclusions may be reached if the wrong viscosity equation is used. As derived from Eq. (3.181) the values of ϕ_a and f_s remain constant at about 0.63 and 1.34 respectively irrespective of the change in mean particle size

TABLE 3.11

Changes in aggregate structure in W/O emulsions when aged (Sherman, 1967b)

Aging time (hours)	D_s (μ)	ϕ_a Equation (3·82)	ϕ_a Equation (3·142)	f_s Equation (3·82)	f_s Equation (3·142)
0	1·25	0·71	0·64	1·51	1·36
24	2·09	0·70	0·63	1·49	1·34
44	2·29	0·68	0·63	1·45	1·34
68	2·46	0·67	0·63	1·43	1·34
189	2·97	0·66	0·66	1·40	1·34
405	3·27	0·64	0·63	1·36	1·34

This suggests that the particles in the flocculates behave as if they are packed together in a dodecahedral geometry, with the particles constituting 74·04% of the volume.

At a lower rate of shear of 0·33 sec^{-1} the value of f_s was 1·4, and at 10^{-5}–10^{-6} sec^{-1} the value was not less than 1·6, so that flocculation in the absence of shear must lead to a fairly open packing geometry.

The lowest rates of shear quoted were obtained by using a constant shear stress of 4·9 dyne/cm^2 (Chapter 2, Section 2.E). Creep compliance-time studies at this stress on W/O emulsions (Sherman, 1967b), in which the particles were non-deformable spheres, indicated viscoelastic behavior in accordance with Section C.1 of Chapter 1. The viscoelastic parameters found for freshly prepared emulsions, which contained a significant proportion of particles with diameters of 0·5μ or less, were an instantaneous elastic modulus (E_0), elastic moduli E_1 and E_2 and viscosities η_1 and η_2 associated with the retarded elasticity, and a Newtonian viscosity η_N. Because of the inherent instability of the very small particles (Sherman, 1967c) they disappeared very rapidly due to mutual coalescence, or coalescence with larger particles. This occurred within a few hours after preparing the emulsions, and because of the very large influence of these small particles on the rheological parameters the latter decreased very sharply during this time. Subsequently the rate of decrease in the magnitude of all parameters was very much slower. Figure (3.20) gives some typical creep compliance-time response curves for samples of a 50% (wt/wt) W/O emulsion at various times after preparation. Analysis of such data for emulsions containing 65·0, 50·0, and 30·0% (wt/wt) dispersed phase concentration gave the results shown in Table (3.12). After the disappearance of the very small particles only one elastic modulus (E_1) and a single viscosity (η_1) could be derived from the retarded elastic compliance region of the creep compliance-time curves.

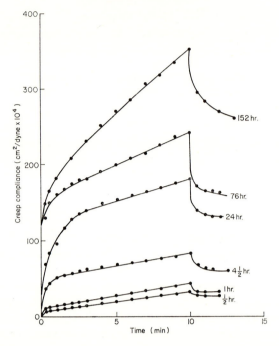

FIG. 3.20. Creep compliance of 50% (wt/wt.) W/O emulsions after different aging times (Sherman, 1967b).

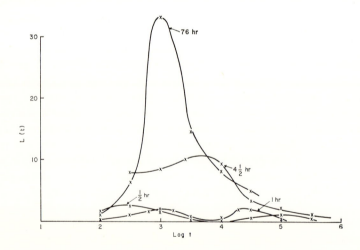

FIG. 3.21. Retardation spectra of aged 50% (wt/wt) W/O emulsions (Sherman, 1967b).

Following Eqs (1.36) and (1.37) the retardation spectra were derived (Fig. 3.21). The retardation spectra for the emulsion containing 50% (wt/wt) dispersed phase volume concentration, for example, indicated that the retardation times are spread over a wide range of times rather than being specific times, as suggested by the data in Table 3.12. After $\frac{1}{2}$ h and 1 h aging there appeared to be two peaks in the respective spectra, but these were not sharply defined so that there is a rather diffuse spread of time dependent retardation mechanisms. At $4\frac{1}{2}$ h there was only one peak, which was rather better defined, and it became even more sharply defined with increasing aging time. At higher values of t the function $L(t)$ approaches zero as a state of steady flow is reached. Emulsions with 30% (wt/wt) dispersed phase showed similar behavior to the 50% (wt/wt) emulsions except that at any given time the creep compliance was greater for the former. Their retardation spectra also showed two peaks at $\frac{1}{2}$ h and at 1 h after the emulsions had been prepared, but emulsions containing 65% (wt/wt) dispersed phase showed only one peak at these aging times. All the emulsions, irrespective of the dispersed phase content, retained little elasticity after a few days aging, but both η_1 and η_N remained relatively high.

Concentrated O/W emulsions (50%–75% wt/wt dispersed phase) stabilised with ionic surface active materials also showed viscoelastic behavior at very low stress (Sherman, 1968b). In this case it was necessary to use a lower stress (2 dyne/cm^2) than for the W/O emulsions, but in spite of this all the rheological parameters were smaller than for the latter. Another striking difference was that all the parameters did not decrease with aging time, as had been found for the W/O emulsions. Instead, they actually increased for several days to optimum values (Table 3.13), after which they decreased steadily with aging time. Viscosity data obtained over a wide range of shear rates ($0 \cdot 004632 \, \text{sec}^{-1}$ to $190 \cdot 3 \, \text{sec}^{-1}$) showed a similar trend. Flocculation of particles into the secondary minimum of the potential energy-distance of separation curve proceeds readily due to the absence of a potential energy barrier (Section B.7). Microphotographs indicated that the flocculated oil particles were not spherical, and that each region of their surfaces which faced another particle was flattened. Thus, the overall structure resembled that of a polyhedral foam with the gaseous phase now replaced by oil, and the water was held as thin lamellae between the distorted oil particles.

With increasing numbers of particles packing into the flocculates, each particle will become surrounded by a larger number of other particles and will undergo distortion at an increasing number of points on its surface. If the particles eventually assume a rhomboidal dodecahedral configuration they can form well in excess of 74·04% of the total flocculate volume with a minimum of surface and angular distortion (Lissant, 1966). A heterogeneous particle size distribution promotes tighter packing than in a monodisperse

TABLE 3.12

Changes in the rheological parameters of W/O emulsions when aged (Sherman, 1967a)

% disperse phase (wt/wt)	Aging period (h)	E_0 (dynes/cm²)	E_1 (dynes/cm²)	E_2 (dynes/cm²)	η_1 (Poise)	η_2 (Poise)	η_N (Poise)	D_s (μ)	% visible globules with diameter less than 0.5μ
65·0	0·5	8350	8330	—	3.55×10^7	—	6.29×10^5	1·22	32·0
	1	3570	7690	—	2.31×10^7	—	—	1·40	23·0
	4	2000	1390	—	1.06×10^7	—	2.95×10^5	2·08	1·0
	24	560	1080	—	2.90×10^6	—	1.84×10^5	2·88	0
	32	340	510	—	1.70×10^4	—	1.76×10^5	2·96	0
	75	140	280	—	1.70×10^4	—	1.18×10^5	3·03	0
50·0	0·5	2270	5030	24300	3.27×10^5	4.0×10^5	2.43×10^5	1·41	42
	1	1210	4250	10000	1.45×10^5	1.0×10^5	1.90×10^5	1·81	23
	4·5	325	455	—	1.90×10^4	—	—	2·17	1·0
	24	185	—	—	8.50×10^3	—	1.13×10^5	2·89	0
	76	64	240	—	7.10×10^3	—	6.10×10^4	3·44	0
	152	73	207	—	1.37×10^3	—	3.80×10^4	4·02	0
30·0	0·5	346	869	2707	5.17×10^4	1.72×10^4	8.75×10^4	1·48	9·0
	1	165	429	—	1.01×10^4	—	7.23×10^4	1·50	13·0
	2	164	189	—	—	—	2.21×10^4	2·40	23·0
	4	83	58	—	1.86×10^4	—	1.94×10^4	2·83	3·0
	24	66	56	—	1.13×10^4	—	3.00×10^3	3·36	0·5
	30	36	27	—	5.5×10^3	—	1.31×10^3	4·06	0

TABLE 3.13

Rheological parameters calculated from creep compliance—time data
(Sherman, 1968b)

Aging period (days)	E_0 (dynes/cm^2)	E_1 (dynes/cm^2)	E_2 (dynes/cm^2)	η_1 (poise)	η_2 (poise)	η_N (poise)
0·16	251	73	192	3888	1,960	21 400
1	392	221	630	13 950	5,790	29 600
2	690	1300	2500	74 500	14,500	302 000
3	830	1270	—	67 000	—	763 000
6	860	1430	—	69 400	—	991 000
7	970	1673	—	84 900	—	1 020 000
14	630	920	—	42 500	—	92 100
28	580	620	—	29 850	—	42 100
41	583	632	—	25 010	—	56 000

system. In the freshly prepared O/W emulsions particle coalescence proceeds at a slower rate than flocculation due to there not being many very small particles, so that the latter process exerts the greater influence on the rheological parameters initially. The flocculates grow to their maximum size for several days, with the development of a maximum number of interlinkages, and maximum distortion of the enclosed particles. Particle coalescence, which reduces the number of particles, then becomes the rate controlling factor and gradually reduces the volume (v_f) of continuous phase which is held in the lamellae between particles, in accordance with

$$v_f = \frac{\phi_{a(D)} - \phi}{K_{(D)}D} \qquad (3.183)$$

where $\phi_{a(D)}$ is the value of ϕ_a at any rate of shear D, and K_D is a constant defining the dependence of v_f on the rate of shear.

Within the range of shear rates $0·004632 \sec^{-1} - 21·15 \sec^{-1}$, i.e. over nearly 4 decades, all the O/W emulsions gave approximately straight lines in a double logarithmic plot of viscosity against rate of shear. Hence they followed a power law relationship as given by Eq. (3.1). In order to avoid the criticisms levelled against this type of equation on dimensional grounds (Reiner, 1960), although it has been suggested recently that these may not be justified if it holds over several decades of shear rate (Scott Blair and Prentice, 1966; Scott Blair, 1967), the amended version (Eq. 3.3) was subsequently used. The plots for any series of emulsions prepared with a

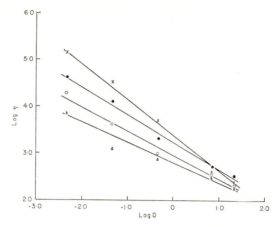

FIG. 3.22. Double logarithmic plot of viscosity data against rate of shear (Sherman, 1968). ●—●, aged 2 days; X—X, aged 6 days; ○—○, aged 14 days; △—△, aged 28 days.

single emulsifying agent showed that a maximum gradient, and hence the maximum value of n_s, was obtained at the same aging time as that at which both a maximum viscosity and maximum values of the rheological parameters, as derived from the creep compliance-time curves, were observed.

BIBLIOGRAPHY

Albers, W. (1957). Doctoral dissertation, University of Utrecht.
Albers W. and Overbeek, J. Th. G. (1959). *J. Colloid Sci.* **14**, 510.
Albers W. and Overbeek, J. Th. G. (1960). *J. Colloid Sci.* **15**, 489.
Becher P. (1958). *J. Soc. Cosmetic Chemists* **9**, 141
Boeder P. (1932). *Z. Physik.* **75**, 258.
Booth, F. (1950). *Proc. R. Soc.* A.**203**, 533.
Brinkman, H. C. (1952). *J. chem. Phys.* **20**, 571.
Brodnyan, J. G. (1959). *Trans. Soc. Rheology* **3**, 61.
Brodnyan, J. G. and Kelly, E. L. (1965). *J. Colloid Sci.* **20**, 7.
Broughton, J. and Squires, L. (1938). *J. Phys. Chem. Ithaca* **42**, 253.
Casson, N. (1959). in "Rheology of Disperse Systems" (C. C. Mill ed.) pp. 84-104 Pergamon Press, London.
Chan, F. S. and Goring, D. A. I. (1966). *J. Colloid Interface Sci.* **22**, 371.
Cheng, P. Y. and Schachman, H. K. (1955). *J. Polym. Sci.* **16**, 19.
Chong, J. S. (1964). Doctoral dissertation, University of Utah.
Collins, D. J. and Wayland, H. (1963). *Trans. Soc. Rheology* **7**, 275.
Conway, B. E. and Dobry-Duclaux, A. (1960). in "Rheology. Theory and Applications" (F. Eirich, ed.) Vol. 3, pp. 83-120, Academic Press, New York.
Criddle, D. W. (1960). in "Rheology. Theory and Applications" (F. Eirich, ed.) Vol. 3, pp. 429-442. Academic Press, New York and London.
Cross, M. M. (1965). *J. Colloid Sci.* **20**, 417.

Davies, J. T. and Rideal, E. K. (1963). "Interfacial Phenomena" 2nd edn. p. 29, Academic Press, New York and London.

de Bruijn, H. (1942). *Rec. Trav. Chim. Pays-Bas Belg.* **61**, 263.

de Bruijn, H. (1948). Proc. 1st Intern. Congr. Rheology II, 95.

Derjaguin, B. V. (1934). *Kolloidzeitschrift* **69**, 155.

Derjaguin, B. V. (1939). *Trans. Faraday Soc.* **36**, 203

Derjaguin, B. V. (1940). *Acta phys.-chim. URSS*, **10**, 333.

Derjaguin, B. V. and Samygin, M. H. (1951). *Discuss. Faraday Soc.* **18**, 24.

Derjaguin, B. V. and Titijevskaya, A. S. (1957). Proc. 2nd Intern. Conf. Surface Activity, **1**, 211.

de Vries, A. J. (1963). in "Rheology of Emulsions" (P. Sherman, ed.) pp. 43-58, Pergamon Press, London.

de Waele, A. (1925). *Kolloidzeitschrift* **36**, 332.

Duclaux, J. and Sachs, D. (1931). *J. chim. phys.* **28**, 511.

Eilers, H. (1941). *Kolloidzeitschrift* **97**, 313.

Einstein, A. (1905). *Annln. Phys.* **17**, 549.

Einstein, A. (1906). *Annln. Phys.* **19**, 289, 371.

Einstein, A. (1911). *Annln. Phys.* **24**, 591.

Eirich, F. R., Margaretha, H. and Bunzl, M. (1936). *Kolloidzeitschrift* **75**, 20.

Eveson, G. F. (1959). "Rheology of Disperse Systems" (C. C. Mill, ed.) pp. 61-80, Pergamon Press, London.

Eveson, G. F., Ward, S. G. and Whitmore, R. L. (1951). *Disc. Faraday Soc.* **11**, 11.

Ewell, R. H. and Eyring, H. (1937). *J. chem. Phys.* **5**, 726.

Fidleris, V. and Whitmore, R. L. (1961). *Rheol. Acta* **1**, 573.

Frankel, N. A. and Acrivos, A. (1967). *Chem. Eng. Sci.* **22**, 847.

Frisch, H. L. and Simha, R. (1956). in "Rheology. Theory and Applications" (F. Eirich, ed.) Vol. 1, pp. 525-613, Academic Press, New York.

Fröhlich, H. and Sack, R. (1946). *Proc. R. Soc. Ser. A* **185**, 415.

Fuks, G. I. (1958). *Kolloid Zh.* **20**, 705.

Gabrysh, A. F., Eyring, H., Shimizu, M. and Asay, J. (1963). *J. appl. Phys.* **34**, 261.

Gillespie, T. (1960a). *J. Colloid Sci.* 15, 219.

Gillespie, T. (1960b). *J. Polym. Sci.* **46**, 383.

Gillespie, T. (1963a). *J. Colloid Sci.* **18**, 32.

Gillespie, T. (1963b). in "Rheology of Emulsions" (P. Sherman, ed.) pp. 115-124, Pergamon Press, London.

Glasstone, S., Laidler, K. J. and Eyring, H. (1941) "The Theory of Rate Processes", McGraw Hill, New York.

Goodeve, C. F. (1939). *Trans. Faraday Soc.* **35**, 342.

Goulden, J. D. S. and Phipps, L. W. (1964). *J. Dairy Res.* **31**, 195.

Greenberg, S. A., Jarnutowski, R. and Chang, T. N. (1965). *J. Colloid Sci.* **20**, 20.

Guth, E. and Simha, R. (1936). *Kolloidzeitschrift* **74**, 266.

Harmsen, G. J., van Schooten, J. and Overbeek, J. Th. G. (1953). *J. Colloid Sci.* **8**, 64.

Hatschek, E. (1911). *Kolloidzeitschrift* **8**, 34.

Hauser, E. A. and Le Beau, D. S. (1939). *Kolloidzeitschrift* **86**, 105

Herschel, W. H. and Bulkley, R. (1926). *Kolloidzeitschrift* **39**, 291.

Higginbotham, G. H., Oliver, D. R. and Ward, S. G. (1958). *Brit, J. appl. Phys.* **9**, 372.

Houwink, R. (1958). "Elasticity, Plasticity, and Structure of Matter" Dover Publications, New York.

Jeffery, G. B. (1922). *Proc. R. Soc.* **A102**, 161.

Joly, M. (1954). Proc. 2nd Intern. Congr. Rheology, pp. 365-370, Butterworths, London.

Joly, M. (1956). *J. Colloid Sci.* **11**, 519.

Karasova, V. V. and Derjaguin, B. V. (1959). *Zh. fiz. Khim.* **33**, 100.

Kim, W. K., Hirai, N., Ree, T. and Eyring, H. (1960). *J. appl. Phys.* **31**, 358.

Kuhn, W. (1932). *Z. physik. Chem.* **A161**, 1.

Kuhn, W. and Kuhn, H. (1945). *Helv. Chim. Acta* **28**, 97.

Kynch, G. J. (1954). *Brit. J. appl. Phys. Suppl.* No. 3, S5.

Kynch, G. J. (1956). *Proc. R. Soc.* **A237**, 90.

Lawrence, A. S. C. and Rothwell, E. (1957). Proc. 2nd Intern. Congr. Surface Activity, **1**, 499.

Leviton, A. and Leighton, A. (1936). *J. phys. Chem., Ithaca* **40**, 71.

Lewis, W. K., Squires, L. and Thompson, W. I. (1935). *Trans. Am. Inst. Mining Met. Engnrs.* **114**, 38.

Linton, M. and Sutherland, K. (1957). Proc. 2nd Intern. Congr. Surface Activity, **1**, 494.

Lissant, K. J. (1966). *J. Colloid Interface Sci.* **22**, 462.

Manley, R. St. J. and Mason, S. G. (1954). *Can. J. Chem.* **32**, 763.

Maron, S. H. and Ming Fok, S. (1955). *J. Colloid Sci.* **10**, 482.

Maron, S. H. and Madow, B. P. (1953). *J. Colloid Sci.* **8**, 130.

Maron, S. H., Madow, B. P. and Kreiger, I. M. (1951). *J. Colloid Sci.* **6**, 584.

Maron, S. H. and Levy Pascal, A. (1955). *J. Colloid Sci.* **10**, 494.

Maron, S. H. and Pierce, P. E. (1956). *J. Colloid Sci.* **11**, 80.

Maron, S. H. and Sisko, A. W. (1957). *J. Colloid Sci.* **12**, 99.

Matsui, T. (1948). *J. Japan Ceram. Assoc.* **56**, 127.

Metzner, A. B. and Reed, J. C. (1955). *A.I.Ch.E. Journal* **1**, 434.

Mooney, M. (1946). *J. Colloid Sci.* **1**, 195.

Mooney, M. (1951). *J. Colloid Sci.* **6**, 162.

Moore, W. J. and Eyring, H. (1938). *J. chem. Phys.* **6**, 391.

Mukerjee, P. (1957). *J. Colloid Sci.* **12**, 267.

Nawab, M. A. and Mason, S. G. (1958). *Trans. Faraday Soc.* **54**, 1712.

Nederveen, C. J. (1963). *J. Colloid Sci.* **18**, 276.

Oldroyd, J. C. (1953). *Proc. R. Soc.* **A218**, 122.

Oldroyd, J. C. (1955). *Proc. R. Soc.* **A232**, 567.

Ostwald, W. (1925). *Kolloidzeitschrift* **36**, 99, 157, 248.

Payne, A. R. (1964). *J. Colloid Sci.* **19**, 744.

Peterlin, A. (1938). *Z. Physik.* **111**, 232.

Rajagopal, E. S. (1960) *Z. physik. Chem.* **23**, 342.

Ree, T. and Eyring, H. (1955). *J. appl. Phys.* **26**, 793.

Reiner, M. (1961). "Deformation, Strain, and Flow" 2nd edn., H. K. Lewis, London.

Richardson, E. G. (1933). *Kolloidzeitschrift* **65**, 32.

Richardson, E. G. (1950). *J. Colloid Sci.* **5**, 404.

Richardson, E. G. (1953). *J. Colloid Sci.* **8**, 367.

Robinson, J. (1949). *J. phys. Colloid Chem.* **53**, 1042.

Robinson, J. (1951). *J. Phys. Colloid Chem.* **55**, 455.

Robinson, J. (1957). *Trans. Soc. Rheol.* **1**, 15.

Roscoe, R. (1952). *Brit. J. appl. Phys.* **3**, 267.

Saito, H. (1950). *J. phys. Soc. Japan* **5**, 4.

Saunders, F. L. (1961). *J. Colloid Sci.* **16**, 13.
Saunders, F. L. (1967). *J. Colloid Interface Sci.* **23**. 230.
Schaller, E. J. and Humphrey, A. E. (1966). *J. Colloid Interface Sci.* **22**, 573.
Schenkel, J. H. and Kitchener, J. 1960. *Trans. Faraday Soc.* **56**, 161.
Scott, J. R. (1931). *I.R.I. Trans.* **7**, 2.
Scott Blair, G. W. (1949). "A Survey of General and Applied Rheology" Pitman, London.
Scott Blair, G. W. (1965). *Rheol. Acta* **4**, 53, 152.
Scott Blair, G. W. (1966). *Rheol. Acta* **5**, 184.
Scott Blair, G. W. (1967). *Rheol. Acta* **6**, 201.
Scott Blair, G. W. and Prentice, J. H. (1966). *Cah. Groupe franc. Rheol.* **1**, 75.
Shangraw, R., Grim, W. and Mattocks, A. M. (1961). *Trans. Soc. Rheol.* **5**, 247.
Sherman, P. (1950a). *J. Soc. chem. Ind. London* **69**, S70.
Sherman, P. (1950b). *J. Soc. chem. Ind. London* **69**, S74.
Sherman, P. (1955a). *J. Colloid Sci.* **10**, 63.
Sherman, P. (1955b). *Kolloidzeitschrift* **141**, 6.
Sherman, P. (1955c). *Mfg. Chem.* **26**, 306.
Sherman, P. (1959). *Kolloidzeitschrift* **165**, 156.
Sherman, P. (1960). Proc. 3rd Intern. Congr. Surface Activity II, 596.
Sherman, P. (1961). *Fd. Technol.* **15**, 394.
Sherman, P. (1963a). *J. phys. Chem. Ithaca* **67**, 2531.
Sherman, P. (1963b). in "Rheology of Emulsions" (P. Sherman, ed.) pp. 77-90, Pergamon Press, London.
Sherman, P. (1965). Proc. 4th Intern. Congr. Rheology 3, 605. Interscience, New York.
Sherman, P. (1967a). *J. Colloid Interface Sci.* **24**, 107.
Sherman, P. (1967b). *J. Colloid Interface Sci.* **24**, 97.
Sherman, P. (1968a). Paper presented at 5th Intern. Congr. Rheology, Kyoto, Japan, October.
Sherman, P. (1968b). *J. Colloid Interface Sci.* **27**, 282.
Shotton, E. and White, R. F. (1960). *J. Pharm. Pharmacol.* **12**, 108T.
Shotton, E. and White, R. F. (1963). in "Rheology of Emulsions" (P. Sherman, ed.) pp. 59-71, Pergamon Press, London.
Sibree, J. O. (1930). *Trans. Faraday Soc.* **26**, 26.
Sibree, J. O. (1931). *Trans. Faraday Soc.* **27**, 161.
Simha, R. (1940). *J. Phys. Chem.* **44**, 25.
Simha, R. (1945). *J. Chem. Phys.* **13**, 188.
Simha, R. (1952). *J. appl. Phys.* **23**, 1020.
Simha, R. and Somcynsky, T. (1965). *J. Colloid Sci.* **20**, 278.
Srivastava, S. N. (1965). *Indian J. Chem.* **3**, 376.
Street, N. (1958). *J. Colloid Sci.* **13**, 288.
Sumner, C. G. (1940). *Trans. Faraday Soc.* **36**, 372.
Sweeney, R. H. and Geckler, R. D. (1954). *J. appl. Phys.* **25**, 1135.
Taylor, G. I. (1932). *Proc. R. Soc.* **A138**, 41.
Taylor, G. I. (1934). *Proc. R. Soc.* **A146**, 501.
Thomas, D. G. (1964). *A.I.Ch.E. Journal* **10**, 517.
Thomas, D. G. (1965). *J. Colloid Sci.* **20**, 267.
Toms, B. A. (1941). *J.C.S.* 542.
Tuorila, P. (1927). *Kolloidchem. Beihefte* **24**, 1.
van den Tempel, M. (1961). *J. Colloid Sci.* **16**, 284.

van den Tempel, M. (1963). in "Rheology of Emulsions" (P. Sherman, ed.) pp. 1-14, Pergamon Press, London.

van der Waarden, M. (1954). *J. Colloid Sci.* **9**, 215.

Vand, V. (1948). *J. phys. Colloid Chem.* **52**, 277, 300.

Verwey, E. J. W. and Overbeek, J. Th. G. (1948). "Theory of the Stability of Lyophobic Colloids" Elsevier, Amsterdam.

Vold, M. J. (1954). *J. Colloid Sci.* **9**, 451.

Vold, M. J. (1961). *J. Colloid Sci.* **16**, 1.

Vold, M. J. (1963). *J. Colloid Sci.* **18**, 684.

von Smoluchowski, M. (1916a). *Phys. Z.* **17**, 557, 585.

von Smoluchowski, M. (1916b). *Phys. Z.* **18**, 190.

von Smoluchowski, M. (1917). *Z. physik. Chem.* **92**, 129.

Van Wazer, J. R., Lyons, J. W., Kim, K. Y. and Colwell, R. E. (1963). "Viscosity and Flow Measurement" Interscience, New York

Weymann, A. D. (1965). Proc. 4th. Intern. Congr. Rheology **3**, 573. Interscience, New York.

Ward, S. G. and Whitmore, R. L. (1950). *Brit. J. appl. Phys.* **1**, 325.

Williams, P. S. (1953). *J. appl. Chem.* **3**, 120.

Williamson, R. V. (1929). *Ind. Engng. Chem.* **21**, 1108.

Wilson, G. L. and Parkes, J. (1936). *Quart. Jl. Pharm. Pharmac.* **9**, 188.

Yerazanis, S., Bartlett, J. W. and Nissan, A. H. (1962). *Nature* **195**, 33.

CHAPTER 4

Rheological Properties of Foodstuffs

1. Fats, Margarine, and Butter 185
2. Ice Cream 198
3. Chocolate 207
4. Cake, Bread, and Biscuits 215
5. Cheese 229
6. Fruit and Vegetables 235
7. Dough 252
8. Meat 274
9. Hydrocolloids 288
10. Milk and Cream 305

Bibliography 316

In this chapter, as in Chapters 5–6, discussion will be centred around the generalised application of the principles and techniques outlined in Chapters 1–3 to certain materials of industrial interest. Where other relevant techniques are available, which have restricted application to a smaller number of materials, they will be discussed in the appropriate section.

Food materials are far too numerous for it to be practical to discuss them all. The categories which will be discussed are fats, margarine, and butter; ice cream; chocolate; cake, bread, and biscuits; cheese; fruit and vegetables; dough; meat; and milk and cream; in the order enumerated so as to embrace a wide range of consistencies.

1. Fats, Margarine, and Butter

A large number of instruments have been employed to derive both quantitative and qualitative information about the consistency of fats, margarine, and butter.

Their creep compliance-time response at low shearing stresses has been studied by torsion of a hollow cylinder of the sample (van den Tempel, 1961), with a parallel plate viscoelastometer (Shama and Sherman, 1968), and by compression between parallel plates (Sone, 1961; Sone *et al.*, 1962). Dynamic rheological moduli have been determined also, using the principles of free torsional vibrations and bending vibrations (Nederveen, 1963), and with a vibrating plate viscometer (Sone, 1961, 1962).

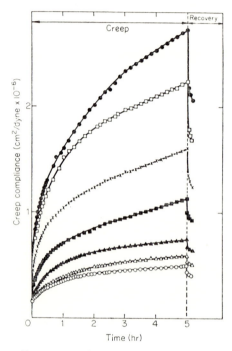

FIG. 4.1. Creep compliance curves for margarine stored at 15° (Shama and Sherman, 1968). Shearing stress 1843 dyne/cm². •—• 1 day old; □—□ 2 days old; x—x 3 days old; ■—■ 7 days old; ▲—▲ 17 days old; △—△ 22 days old; o—o 35 days old.

The stresses applied in the torsion tests were much higher (36 400 and 27 300 dyne/cm²) than those employed in the tests using the parallel plate viscoelastometer (1 843 dyne/cm²). Figure 4.1 shows some typical curves for margarine obtained at 15°C with the latter equipment at different times after manufacture. At low shearing stress margarine shows linear viscoelastic behavior, but at higher shearing stress the creep behavior is non-linear, the compliance increasing as the shear stress increases. The parameters derived from these curves using Eqs (1.26)–(1.34) are shown in Table 4.1. Each creep curve in Fig. 4.1 requires a minimum of ten parameters to define its shape. The value of E_0 is initially $\sim 10^6$ dyne/cm², and it increases rapidly for about 10 days, after which it moves asymptotically to a steady value of $\sim 10^7$ dyne/cm². The changes in the other parameters follow a similar trend

TABLE 4.1

Decrease in hardness of margarine on kneading (after Haighton, 1965)

Sample	$p_0(u)$ (g/cm²)	$p_0(w)$ after working			W_d after working		
		1×	2×	3×	1×	2×	3×
Margarine I (15°C)	1090	195	165	140	82·1	84·8	87·1
Margarine II (15°C)	760	245	230	—	67·8	69·8	—
Special purpose shortening I (20°C)	2300	1300	1190	910	43·4	48·2	60·3
Special purpose shortening II (20°C)	2250	1120	1060	970	50·3	52·9	56·9

but they take somewhat longer to reach steady optimum values. The retardation spectrum $L(t)$ was derived by Eq. (1.37), and it shows (Fig. 4.2) that the number of retardation mechanisms involved in creep compliance does not alter as the margarines age, nor does the range of time elements alter. However, the contribution from the lower end of the time scale becomes progressively smaller with aging time, and the initial distribution of time elements alters to a much flatter distribution. This indicates the onset

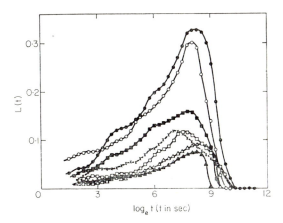

FIG. 4.2. Influence of storage time on the retardation spectrum of margarine (Shama and Sherman, 1968). •—• 1 day old; o—o 2 days old; ■—■ 3 days old; x—x 7 days old; □—□ 17 days old; △—△ 22 days old; ▲—▲ 35 days old.

of a more even distribution of time elements, although the peak in each curve still remains in the vicinity of the larger time elements.

Sone (1961, 1962) observed similar changes in the parameters for butter which was stored at 20°C after it had been worked for 30 min on a roller mill. He assumed that the increase in viscosity during storage was proportional to the decrease in specific volume so that he could treat the data by Avrami's (1939, 1940, 1941) theory. This theory deals with the kinetics of phase changes during crystal growth around randomly distributed nuclei in terms of the change in specific volume. If specific volume is replaced by η_N it becomes

$$1 - \frac{\eta_{N(0)} - \eta_{N(t)}}{\eta_{N(0)} - \eta_{N(\infty)}} = \exp\left(-k_A \, t^{n_A}\right) \tag{4.1}$$

where $\eta_{N(0)}$, $\eta_{N(t)}$, and $\eta_{N(\infty)}$ are the values of η_N initially, at time t, and after infinite time respectively, k_A is a constant which is proportional to the mechanism of nucleation and the rate of crystal growth, and n_A is a constant whose value lies between 1–4 depending on the nucleation mechanism. Eq. (4.1) can be rewritten as

$$\frac{\eta_{N(\infty)} - \eta_{N(t)}}{\eta_{N(\infty)} - \eta_{N(0)}} = \exp\left(-k_A \, t^{n_A}\right)$$

so that

$$\ln \ln \left(\frac{\eta_{N(\infty)} - \eta_{N(t)}}{\eta_{N(\infty)} - \eta_{N(0)}} \right) = -(n_A \ln t + \ln k_A)$$

If this approach is valid, then a plot of the left hand side of the equation against $\ln t$ should give a straight line with a gradient equivalent to n_A and an intercept equal to $\ln k_A$. Figure 4.3 shows such a plot for viscosity changes during storage at 20°C. The gradient is approximately 1·2, which suggests that the crystals grow in a needle-like form. Similar tests at -5°C gave a gradient of 2·0, which still indicates needle-like growth.

An unworked fat consists of an interlocking network of fat crystals. According to Soltøft (1947) "... if the rate of crystallisation is widely different in different directions, so that the crystals become elongated, an open reticular structure must be formed during the solidifying. The fat will therefore be capable of sustaining quite considerable stresses during elastic deformation, until the yield value is reached. As soon as the yield value is exceeded the reticular structure breaks down, the elastic capacity becomes substantially smaller, and the greater part of the stress is converted into movement so that the fat shows a very strong movement as soon as flow begins". When crystallisation is accompanied by mechanical agitation the fat continues to harden with age, the rate and extent depending on such factors as the storage temperature, solid fat content, fat crystal size, and past thermal history (De Man, 1963). In the absence of agitation during crystallisation the fat does not continue to harden on storage. This suggests that the continued hardening during storage is not due to crystallisation or polymorphism, i.e. differences in the internal molecular structure of the crystals, but that it originates from a thixotropic rearrangement of the crystals into a three dimensional scaffolding structure (De Man and Wood, 1959).

By analogy with the equations of motion for polymers subjected to a shearing stress (Tobolsky and Eyring, 1943) a more fundamental theory has been proposed for the rheological behavior of fats (Van den Tempel, 1961; Haighton, 1965). It is suggested that two distinct types of bonds are involved. Those bonds of the structural network which are not ruptured during a test

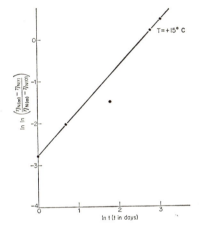

FIG. 4.3. Analysis of the change in viscosity of margarine during storage at 15°C by Avrami's theory.

at low shearing stress are known as primary bonds, while those bonds which break and may be re-established during the test period are called secondary bonds. The latter originate from weak London–van der Waals' attraction forces between neighboring particles. The primary bonds may develop from relatively strong London–van der Waals' forces of attraction between suitably oriented particles during shear by mechanical entanglements of different regions of the structural network, or during precipitation of crystalline material from liquid fat when cooled. If the stress applied to a fat is large enough for some of the primary bonds to break, in addition to rupture of the secondary bonds, then they take much longer than the test period to be re-established. When a large number of primary bonds is present the fat is hard and brittle.

The sub-division of bond strengths into primary and secondary categories appears to be a rather arbitrary concept, for in practice there will be a spectrum of bond strengths (Section I.C.I., Chapter 1), i.e. a less sharply defined gradation of bond strengths.

Although the cone penetrometer (Section 3.B, Chapter 2) does not provide quantitative information about consistency it can be used (Haighton, 1965) to derive qualitatively useful data about the types of bonds which are destroyed during work softening of a fatty material. In addition, an index can be calculated from penetration data for worked and non-worked butter and margarine which shows a significant correlation with the spreadability characteristics of these two materials. Following Eq. (2.76) let $p_{0(u)}$ and $p_{0(w)}$ represent the yield value before and after working respectively, then the decrease in hardness (W_d) expressed in terms of the original hardness is

$$W_d = \frac{p_{0(u)} - p_{0(w)}}{p_{0(u)}} \times 100\% = \left[1 - \frac{(d_{p(u)})^{1.6}}{d_{p(w)}}\right] \times 100\% \qquad (4.2)$$

where $d_{p(u)}$ and $d_{p(w)}$ represent the depths of penetration by the cone before and after working respectively.

The samples are kneaded in a cylindrical tube fitted with a plunger to which is attached a perforated plate with 24 holes of diameter 0·06 in. The plunger is moved up and down the tube many times. Table 4.1 shows data which were obtained with some typical fat systems following repeated working. It is seen that while much of the network structure is destroyed by working, it is not destroyed completely. Most of the structural change occurred during the first working. Subsequent workings produced relatively small changes.

When the worked samples were stored at a constant temperature for some time they slowly began to harden again. Penetrometer tests indicated that hardness increased approximately logarithmically with time. "This solidification is due to the slow renewal of the crystal lattice of secondary bonds. At the beginning there are loose crystal chains slowly forming ramifications which thus form a secondary network, but later on primary bonds are slowly formed by melting and precipitation of insoluble glycerides reinforcing the lattice" (Haighton, 1965).

For both butter and margarine the spreadability index is defined by

$$S_i = p_{0(u)} - 0.75\,[p_{0(u)} - p_{0(w)}] \qquad (4.3)$$

where the factor 0·75 derives from a comparison of penetrometer data with panel assessment of spreadability. Butter shows a lower work softening (50–55%) than margarine (70–75%), and because of this test panels will rate margarine as more easily spreadable even when the two materials have the same initial hardness.

Vasić and De Man (1968) interpreted cone penetrometer data, derived with a flat nosed cone, in terms of a hardness index (H_p) which is expressed as the ratio of the force required to make the indentation to the area of the impression

$$H_p = \frac{mg}{A} = \frac{mg \times 10^3}{\left[d_p \pi \frac{\tan a_p}{\cos a_p} \left(d_p + \frac{2r}{\tan a_p} \right) + r^2 \pi \right] 10^4} \tag{4.4}$$

where mg is the total weight of the moving parts, and r is the radius of the flat surface of the cone tip (in units of 1/10 mm).

The decrease in hardness following working is given by

$$W_d = \frac{H_{p(0)} - H_{p(w)}}{H_{p(0)}} \times 100\% \tag{4.5}$$

where $H_{p(0)}$ and $H_{p(w)}$ represent the values of H_p before and after working respectively.

TABLE 4.2

Hardness and work softening of conventionally and continuously made butters (Vasić and De Man, 1968)

Type of Butter	No. holes in perforated plate used for working samples	5°C		15°C	
		hardness (kg/cm²)	% decrease in hardness when worked	hardness (kg/cm²)	% decrease in hardness when worked
Conventional method of manufacture	control	0·411	—	0·109	—
	1	0·212	48·4	0·083	23·8
	19	0·207	49·6	0·082	24·8
	43	0·202	50·9	0·078	28·4
Continuously made	control	0·895	—	0·181	—
	1	0·271	69·7	0·117	35·4
	19	0·264	70·5	0·115	36·5
	43	0·257	71·3	0·111	38·7

Table 4.2 shows hardness data at 5°C for butters, made in the conventional way and by a continuous process, both before and after working. The two butters had the same chemical composition. The essential difference between the two manufacturing processes is that in the conventional process the butter is subjected to mechanical treatment after the fat has crystallised,

whereas in the continuous process most of the mechanical treatment occurs before the fat has crystallised. The numbers in the "Sample treatment" column refer to the number of holes in the plate through which the samples were extruded during working. It is readily apparent that the hardness following working is not affected significantly by the number of holes in the extrusion plate. Continuously made butter has a much higher initial hardness than conventionally made butter, and it undergoes a much larger structural alteration when it is worked. At 15°C the differences between the two types of butter are much less marked.

When the two types of butter were stored at 5°C for 15 days following working their hardness increased slightly with storage time (Table 4.3), but this increase is not significant compared with the previous loss in hardness when worked. Vasić and De Man (1968) suggest that continuously made butter contains more primary bonds than butter manufactured in the conventional manner because the fat crystals are larger and more irregular in shape. Since bonds between large fat crystals are more likely to be permanently ruptured than the bonds between small fat crystals, the continuously made butter undergoes a greater degree of work-softening.

The rheological properties of fats, margarine, and butter have been further studied by extrusion through a capillary tube (Soltøft, 1947; Scherr and Witnauer, 1967), and through a narrow orifice (Prentice, 1952, 1954, 1956). In these types of test the sample is subjected to much higher shearing forces than when it is tested by the cone penetrometer, and therefore it undergoes a greater degree of structural damage than in the latter test.

Soltøft (1947) measured the consistency of hardened fats with a variable pressure capillary viscometer which is essentially a modification of the original Bingham (1922) instrument. The fat was introduced into a bronze tube (diameter 1·2 in, length 5·8 in), which was then connected up at one end to a long capillary tube of suitable diameter. Compressed air was introduced at the other end of the sample tube through an inlet tube, and this produced movement of a plunger at the rear of the fat sample so that a pressure was applied to the fat and it was extruded through the capillary. Air pressure was not applied directly to the fat surface because this would have made the fat run out through the tube thus producing a funnel in the rear of the fat. Eventually the air would have forced its way right through the funnel until it reached the capillary tube. With a plunger positioned at the back of the sample the rear surface of the latter was usually plane and a steady flow of fat through the capillary was maintained. The whole equipment was thermostatted at the required temperature.

Capillaries of metal and glass were used ranging in diameter from approximately 0·04 in to 0.2 in, the length in all cases being approximately 4 in. Thus the length/diameter ratios were between 1/100 and 1/20.

TABLE 4.3

Hardness and residual hardness of conventional and continuously made butters at different times (at 5°C) after working (Vasić and De Man, 1968).

Type of Butter	No. holes in perforated plate used for working samples	hardness (kg/cm²) No. days after mechanical treatment			residual hardness (%) No. days after mechanical treatment		
		1	8	15	1	8	15
Conventional method of manufacture	control	0·411	0·425	0·439	100	100	100
	1	0·212	0·227	0·251	51·6	53·4	57·2
	19	0·207	0·222	0·239	50·4	52·2	54·5
	43	0·202	0·222	0·239	49·2	52·2	54·5
Continuously made	control	0·895	0·995	0·995	100	100	100
	1	0·271	0·350	0·397	30·3	35·2	38·9
	19	0·264	0·340	0·385	29·5	34·2	38·7
	43	0·257	0·330	0·373	28·7	33·2	37·5

H

The principle of operation was to determine the weight of fat which was extruded through a given capillary tube in a fixed period of time using a particular air pressure. The procedure was then repeated several times using different air pressures. In this way a relationship could be established between pressure and rate of flow through the capillary which, by applying the appropriate corrections, could then be converted into a stress-rate of shear curve. Air pressures within the range 1–3 kg/cm^2 were normally used, but in exceptional cases where very hard fats were tested pressures as high as 15–20 kg/cm^2 were required.

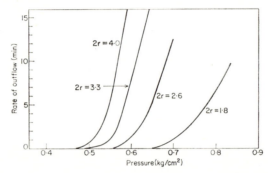

FIG. 4.4. Correction curves for different outlet diameters (Soltøft, 1947).

As with the FIRA-NIRD extruder (Section 3E, Chapter 2) a significant part of the applied pressure is used to overcome the frictional resistance between the sample and the cylinder walls. It was found, by replacing the capillary with an orifice of the same diameter, that about one third of the total pressure was used up in this way. Some correction curves are shown in Fig. 4.4.

The shear stress and rate of shear operating during passage of the fat through the capillary were calculated using the same equations (Table 2.3) as for fluid flow, which is questionable, to say the least, in the case of hard fats. Figure 4.5 shows some typical curves for a series of mixtures of two grades of hardened Groundnut oil which had melting points of 30°C and 40°C respectively. The tests were made at 24°C. Figure 4.6 shows similar data for butter and various grades of margarine. In all cases flow did not commence until a certain shearing stress was achieved. Then, the rate of shear increased rapidly with increasing shearing stress over a limited range of the latter until the curve became linear, i.e. the ratio of the two variable parameters was constant. The curves, therefore, resemble those for non-Newtonian, plastic, flow behavior. Since the flow curves included contributions from both plug flow and normal flow, Soltøft (1947) extrapolated the initial rectilinear portion over the whole experimental pressure range and then subtracted these values from the original curve in order to derive the true flow curves. These curves conformed to the Buckingham–Reiner equation (Eq. 2.9). The consistency of each sample was thus defined by two parameters, viz. the yield value and the apparent viscosity.

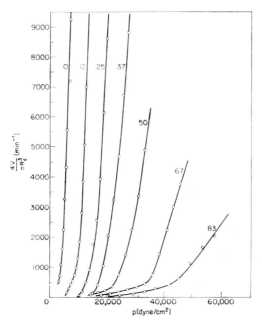

FIG. 4.5. Flow curves for mixtures of two grades of hardened Groundnut oil (Soltøft, 1947).

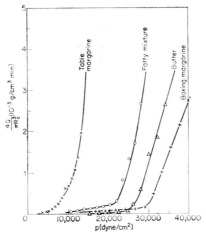

FIG. 4.6. Flow curves for butter and margarine (Soltøft, 1947).

Since the consistency of fatty materials depends greatly on the mechanical and thermal treatment, Soltøft (1947) applied a standardised preliminary treatment to each sample. The fat was melted, and then solidified by immersion in iced water. Each sample was stored for 3 days at the test temperature. It was then extruded through a narrow capillary, presumably to soften it by breaking down some of the crystalline structure so that flow would begin at a lower shearing stress than in non-softened samples.

Scherr and Witnauer (1967) used an extrusion capillary viscometer to study the consistency of lard. In this case the barrel containing the sample was made from low carbon, lead-bearing, screw-stock steel bar. The plunger was made from steel bar, and to it was attached a Teflon plug which acted as a piston. The capillaries, which were attached to the sample barrel by a special threading arrangement which did not require any gasketing material to give a good seal, were made from stainless steel hypodermic needle tubing. Pressure was applied to the sample in this case by using the cross head of an Instron Tensile Tester Model TT–B, which provided a load-weighing capacity ranging from 0–1000 lb, in conjunction with a specially designed compression bridge. Data obtained on lard at 23·4°C indicated that the shearing stress-rate of shear relationship could be defined by an empirical power law (Eq. (2.1)) with a value of n_s which deviated sharply from unity, and suggested pronounced non-Newtonian behavior. There was no indication of time-dependent viscous effects as the lard flowed through the capillary.

Prentice (1952, 1954, 1956) assessed the consistency of compound fat, margarine, and butter, using a wide range of instruments (including a cone penetrometer, a sectilometer, and a FIRA-NIRD extruder) and he compared the data with judgements of spreadability and firmness which were made by a panel of people who were experienced in handling these materials. In order to judge spreadability the panel members were required to spread the samples on a slice of bread, and to describe the ease of spreading by one of the following terms—very difficult, difficult, fairly difficult, moderate, fairly easy, easy, very easy. An average score was then calculated on the basis of the responses. Firmness was assessed by placing a small piece of the sample on the tip of a knife, and then squeezing it between the blade of the knife and a pastry board. Statistical analysis of the physical and sensory data indicated that the FIRA-NIRD extruder data gave the highest correlation coefficients with the mean scores for spreadability and firmness, with the sectilometer following close behind. The correlation coefficients for the cone penetrometer data were not quite so good as for the other two instruments.

These observations were confirmed by Naudet and Sambuc (1959) who studied 12 different margarines at 12·5°C, 17·5°C, and 22·5°C, using a needle penetrometer, a cone penetrometer, a sectilometer, and a FIRA-NIRD extruder. The correlation coefficients which they derived are shown in Table 4.4. It appears that the correlation between cone penetrometer data and firmness is somewhat lower than that between cone penetrometer data and spreadability.

A partial explanation, at least, for the poorer correlation of cone penetrometer data with the panel assessments could be that this instrument comes into contact with a much smaller volume of sample than does either the sectilometer or the FIRA-NIRD extruder. Margarine, fats, etc. are far from being homogeneous materials so that a much wider spread of test results is obtained with the cone penetrometer, and the calculated mean value is correspondingly less meaningful.

TABLE 4.4

Correlation of instrumental and sensory assessments of spreadibility and firmness of margarine (Naudet and Sambuc, 1959)

Instrument	Correlation coefficient	
	spreadibility	firmness
needle penetrometer	0·854	−0·844
cone penetrometer	0·789	−0·775
FIRA-NIRD extruder	−0·903	0·907
sectilometer	−0·867	0·853

The ease with which butter can be spread on bread has been studied (Huebner and Thomsen, 1957) using a modification of Mohr and Haesing's (1949) technique which measured the force required to draw a knife blade across the surface of a rectangular block of butter. A block of butter was placed in a stainless steel box, and its upper surface was levelled out. A knife blade, beveled at an angle of 45° to a height of $\frac{1}{16}$ in, was then pulled through the sample by means of a constant speed motor. The resistance encountered by the knife blade as it moved forward was partially registered by a torque meter, which had been calibrated previously with a range of known weights. All butter samples were tempered for 48 h at a mean temperature of 13·1°C before testing. Spreadability data obtained in this way agreed closely with consumer panel assessment of the same property. As would be expected the instrumental technique proved more sensitive to small variations in spreadability.

A highly significant statistical correlation was obtained also between instrumental measurements of spreadability and hardness, the latter property being assessed using a sectilometer. The regression equation was

$$\text{resistance to spreading (g)} = 90 \cdot 9 + 2 \cdot 63 \times \text{hardness (g)}$$

with a standard deviation of 43·4 g, which gave confidence limits at the 5%
significance level of ± 86·7 g. It was concluded that " . . . while the most
important factor affecting spreadability is hardness, certain other factors
may exert minor influences".

2. ICE CREAM

The ice cream mix from which the frozen product is prepared consists
essentially of a dilute fat-in-water emulsion (Table 4.5). The fat is either
butter fat or a vegetable fat, and the aqueous phase contains milk or skim
milk or milk powder, sugar, and a vegetable gum or gelatine or alginate as
stabiliser. Milk protein provides the emulsifying agent, and another commercial emulsifying agent such as glycerol monostearate is usually added
also at a low concentration. The function of the stabiliser is to assist the
frozen ice cream retain its shape as it melts. Following emulsification and
homogenisation the ice cream mix is frozen at a low temperature, usually
−6°C to −7°C, in a freezer barrel while being subjected to violent mechanical
agitation. Simultaneously air is introduced into the freezing mix. Eventually
the frozen product is extruded through a nozzle as a continuous ribbon
which is automatically cut by a wire frame into rectangular blocks.

TABLE 4.5

Typical ice cream formulation (Sherman, 1961)

Phase	ingredient	% composition (wt/wt)
Oil	Hardened vegetable fat	10·0
	Molecularly distilled, soap-free,	
	glycerol monostearate	0·18
Aqueous	Water	60·87
	Spray dried milk powder	13·10
	Sugar	15·65
	Vegetable gum	0·20

Ice cream mix exhibits an anomalously high plastic viscosity (Sherman,
1961), which suggests that the fat droplets behave as if their hydrodynamic
diameter is very much larger than the actual diameter observed under the
microscope. For example, if the viscosity data are analysed by the Guth–
Simha (1936) amended version of the Einstein (1906) equation then the
viscosity (η_∞) at infinitely high rate of shear is

$$\eta_\infty = \eta_0 \left[1 + 2\cdot5\,(X_a \phi) + 14\cdot1\,(X_A \phi)^2 \right] \qquad (4.6)$$

where X_a is the apparent increase in the volume fraction of fat. Calculation indicates that the droplets behave as if they are covered by a layer of about $0 \cdot 3 \, \mu$ thickness. This apparent thickness is very much greater than the thickness of an adsorbed monomolecular layer of protein around the fat droplets, so that the layer may be several molecules thick. Even if the viscosity data are analysed by Eq. (2.127) the anomaly still remains. It is conceivable, therefore, that the various layers within the adsorbed protein film have undergone different degrees of denaturation and coagulation depending on their distance from the oil-water interface. Only the layers which are in closest proximity to the oil-water interface would undergo complete denaturation. Those layers which are furthest away from the interface will retain some capacity to bind water from the aqueous phase.

Frozen ice cream can be regarded as a solid foam, with air forming somewhat more than 50% of the total volume. Microscopic examination of frozen sections indicates (Sherman, 1965) that the lamellae between the air cells range between $30-300 \, \mu$ in thickness, which is very much thicker than the lamellae in detergent foams. The air cells and the ice crystals range between $50-150 \, \mu$ diameter, but the fat crystals are very much smaller and do not exceed a few microns in diameter. Up to about 75% of the water in the aqueous phase is converted into ice during the freezing of ice cream mix. Thus the residual, unfrozen, aqueous phase will consist of a concentrated, highly viscous, solution of sugar and stabiliser.

The rheological properties of frozen ice cream have been studied with the parallel plate viscoelastometer (Shaw, 1963; Shama and Sherman, 1966) described in Section 2.1 of Chapter 2. At low shearing stresses up to 4000 dyne/cm^2 the creep compliance with time at $-11°C$ indicates linear viscoelastic behavior. Figure 4.7 shows typical data for ice creams with an air volume content of 55% which contained 0–10% fat. When the ice cream contained less than 10% fat, the weight of the total aqueous phase was proportionately larger. Detailed analysis of the creep compliance data indicated that the minimum number of components required for the spring-dashpot mechanical model (Section 1.C.3, Chapter 1) is eight, viz. one spring to define the instantaneous elastic compliance, 2 Kelvin–Voigt units in series to define the retarded elastic compliance, and a dashpot to represent the Newtonian viscosity. Equation (1.30) can now be applied in the form

$$J(t) = J_0 + J_1[1 - \exp(-t/\tau_1)] + J_2[1 - \exp(-t/\tau_2)] + \frac{t}{\eta_N} \qquad (4.7)$$

where $\tau_1 \, (= J_1 \eta_1)$ and $\tau_2 \, (= J_2 \eta_2)$ are the two retardation times.

Since the microscopical structure of frozen ice cream is reasonably well understood it has been possible to relate the parameters calculated from the creep compliance data with different parts of this structure (Shama and

Sherman, 1966). Apart from studying the influence of fat content on the magnitude of these rheological parameters, the influence of air content and test temperature were also investigated. Air contents of 1%, 25%, and 55% were employed, and test temperatures of $-11°C$, $-13°C$, and $-15°C$. Varying the test temperature altered the amount of ice in the sample. Figures 4.8, 4.9, and 4.10 show the influence of the three variables on some of the creep parameters. The latter are all several orders of magnitude lower than for pure ice, which has, for example $E_0 \sim 10^{11}$ dyne/cm^2, $E_1 \sim 10^{11}$ dyne/cm^2, $\eta_1 \sim 10^{13}$ poise, and $\eta_N \sim 10^{14}$ poise (Jellinek and Brill, 1956). If frozen ice cream can be regarded as an ice crystal network

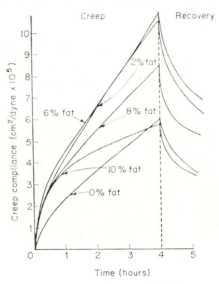

Fig. 4.7. Creep compliance curves for ice creams with different fat contents (Shama and Sherman, 1966).

which has been modified primarily by introducing fat and air, then air content causes by far the greater change in the magnitude of the original ice crystal parameters.

The Newtonian viscosity can be represented by

$$\eta_N = A_b \exp (E/RT) \tag{4.8}$$

where A_b is a constant, and E is the activation energy for plastic flow. Within the temperature range $-11°C$ to $-15°C$, where the transformation rate of water to ice in ice cream is low, $E = 5\,200$ cal/mole for a 10% fat ice cream containing 55% air by volume. The corresponding value for pure ice is $\sim 16\,100$ cal/mole (Jellinek and Brill, 1956).

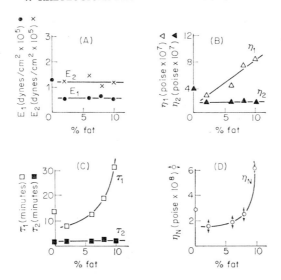

FIG 4.8. Variation of the viscoelastic parameters of ice cream with fat content (Shama and Sherman, 1966).

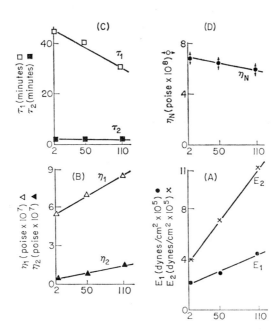

FIG. 4.9. Variation of viscoelastic parameters of ice cream with overrun (Shama and Sherman, 1966).

As the fat content was increased (Fig. 4.8) E_1, E_2, and η_2 did not change significantly, whereas η_1 increased. When the fat content exceeded 7–8 per cent, η_N showed a very marked increase. Increasing the air content from 1 per cent to 55 per cent (Fig. 4.9) actually produced a slight decrease in η_N, whereas E_1, E_2, η_1, and η_2 all increased. Lowering the temperature from $-11°C$ to $-15°C$ produced little change in E_1 and η_N, but E_2, η_1, and η_2 increased quite appreciably.

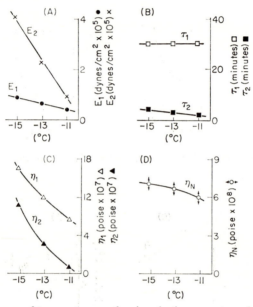

FIG. 4.10. Influence of temperature on the viscoelastic parameters of ice cream (Shama and Sherman, 1966).

From the relative effects of fat content, freezer temperature, i.e. the proportion of water frozen into ice, and air content, on the rheological parameters the principal factors influencing the consistency of frozen ice cream were derived. For example, the retarded elastic moduli E_1 and E_2 are probably associated with the air cell–protein–water interface and the weak gel structure constituted by the unfrozen water plus the sugar, stabiliser etc. E_2 is much more influenced by air content than E_1, so E_1 may be associated with the retarded elastic properties of the weak gel whereas E_2 is associated with the air cell–protein–water interface. (Because of its surface active properties the milk protein should be adsorbed at the air–water interface in addition to the fat–water interface, so that the air cells should be coated with a protein film also.) The viscosity η_1 associated with retarded elasticity is influenced to a greater degree by fat content than by overrun, so that this parameter is probably associated with the onset of flow following bond rupture in the fat crystal network. Since the ice crystals are separated by relatively large distances of several microns there is little chance of bonds being formed between them. The main interactions must be between the fat crystals located in the spaces between the ice crystals and air cells, i.e. in the fluid which has not been frozen, and possibly also between fat crystals and ice crystals. The value of η_2 is affected by overrun but not by fat content, which suggests a viscous damping effect due to the air cell structure and the surrounding weak gel. Increased interaction between fat crystals during viscous flow explains the influence of increasing fat content on η_N. E_0 decreased as the fat content or air content increased. The ice crystal contribution to E_0 must be of primary importance therefore, since the ice crystal content decreased as the fat content increased.

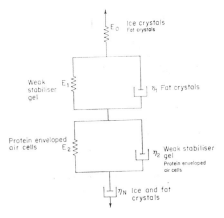

FIG. 4.11 Six element model for frozen ice cream showing rheological associations with structural components (Shama and Sherman, 1966).

Figure 4.11 shows the interrelationships between the rheological parameters and the structural elements of frozen ice cream in terms of a mechanical spring-dashpot model. The structural elements which appear to exert the major and minor effects on the rheological parameters are indicated by large and small print respectively. These observations support the view (Sherman, 1965) that frozen ice cream has an aerated ice crystal structure whose consistency is modified by a superimposed fat crystal network. The response of this structure to a shearing stress depends on the strength of the lamallae between the air cells.

If the model given in Fig. 4.11 is substantially correct then it should be possible to predict the corresponding model, with its associated parameters, for melted ice cream, since the structural changes which occur during the freezing of ice cream mix and during any subsequent melting are reasonably well known. Furthermore, since sensory assessment of texture on the palate depends partly on the rheological properties of melted ice cream, it should be possible to determine which elements in this mechanical model are responsible for good texture by examining a range of good and poor texture samples.

When ice cream mix is subjected to mechanical agitation during freezing the fat droplets begin to coagulate. This produces a reduction in the number of droplets per unit volume, and a corresponding increase in droplet size (Sherman, 1965). Coagulation is gradually retarded, and finally stopped, by the simultaneous conversion of water into ice and a sharp rise in the viscosity of the unfrozen fluid as a result of progressive concentration of the sugar plus stabiliser solution. However, the freezing temperature plus the

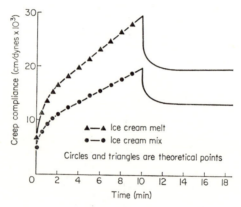

FIG. 4.12. Creep compliance v. time for ice cream mix and melted ice cream (Sherman, 1966).

violent agitation will have damaged many protein membranes, so that when ice cream melts out at room temperature or in the mouth at some later date, the ice crystals melt away and coagulation will continue. Therefore, melted ice cream should have a weaker structure than the original ice cream mix from which the ice cream was made.

Creep compliance–time studies at a shearing stress of 4·9 dyne/cm² in a coaxial cylinder viscometer (Section 2.E, Chapter 2) confirm this belief (Sherman, 1966). Both ice cream mix and melted ice cream show viscoelastic behavior (Fig. 4.12), but there is very little instantaneous or retarded elasticity. The main contribution to the overall compliance is due to Newtonian flow. Consequently, there is no substantial recovery when the shearing stress is removed. At any given time following the application of stress, melted ice cream shows a larger creep compliance than the ice cream mix, i.e. the former has a weaker structure. Table 4.6 shows the parameters, and their values, derived from the analysis of the creep compliance-time curves. Some typical values for frozen ice cream are also given for comparison. The tests on melted ice cream were made at 20°C because this is approximately the temperature which frozen ice cream reaches within a very short time of being placed on the palate.

TABLE 4.6

Rheological parameters for ice cream mix and melted ice cream
(Sherman, 1966)

Rheological parameter	Sample		
	mix (20°C)	melted ice cream (20°C)	frozen ice cream (−11°C)
E_0(dyne/cm^2)	200	143	$10^5 - 10^6$
E_1(dyne/cm^2)	301	156	45 000
E_2(dyne/cm^2)	1430	—	112 000
η_1(poise)	13 400	5,200	8.5×10^7
η_2(poise)	8300	—	1.6×10^7
τ_1(sec)	44.8	33.4	1866
τ_2(sec)	5.8	—	144
η_N	5.4×10^4	3.7×10^4	6.0×10^8

Both ice cream mix and melted ice cream have very much weaker structures than frozen ice cream (Table 4.6). Six parameters define the creep behavior of ice cream mix, as were found for frozen ice cream (Fig 4.11), but only four parameters are required for melted ice cream. The equation defining the creep compliance-time curve for ice cream mix therefore will be identical to Eq. (4.7), while the equation for melted ice cream will be

$$J(t) = J_0 + J_1 [1 - \exp(-t/\tau_1)] + \frac{t}{\eta_N} \qquad (4.9)$$

When ice cream melts the ice crystals disappear, the fat crystals melt, and the air content is lost, so that fat coagulation can now proceed once again. On the basis of Fig. 4.11 this means that both E_0 and η_N will drop substantially. The water provided by melting of ice crystals will dilute the viscous solution of sugar and stabiliser, and the latter will now contain approximately four times the amount of water which it contained in frozen ice cream. This dilution effect should produce a sharp drop in E_1. The viscosity η_2 will show an even greater reduction in value partly for the same reason, and partly because it depends to some extent on the now non-existent air cell structure. E_2 should disappear completely as it depends entirely on the air cell structure. The viscosity η_1 will decrease because the attraction forces between flocculated fat droplets in an aqueous medium are weaker than in a predominantly solid ice medium. The 4-component mechanical model for melted ice cream therefore follows logically from the 6-component model for frozen ice cream. Using the same line of reasoning one would anticipate that a 4-component model could represent the rheological behavior of ice cream mix. The experimental data (Table 4.6) offer no corroboration for this assumption. Nevertheless, the parameters E_2 and η_2 obviously cannot be allocated to the same structural elements as in frozen ice cream. The essential difference between ice cream mix and melted ice cream is that the fat droplets are larger in the latter due to partial coagulation. In addition, ice cream mix contains fat droplets which are $0 \cdot 5 \mu$ in diameter, or even smaller. It is these very small droplets which contribute an additional Kelvin–Voigt unit to the mechanical model (Sherman, 1967a). The potential energy curves for these very small droplets will have a peak which is no greater than $\sim 2kT$ so that they are able to flocculate in the primary minimum at a distance of separation which does not exceed a few Angstroms. The larger droplets flocculate in the secondary minimum. It can be shown (Sherman, 1967b) that the small droplets are less likely to have a compact layer of adsorbed protein molecules around their surfaces than are the larger droplets, and this will facilitate their flocculation in the primary minimum.

A coaxial cylinder viscometer has been used to compare the rheological properties of good and poor texture ice creams (Whitehead and Sherman, 1967). All ice cream samples were prepared in accordance with the formulation shown in Table 4.5. The only difference in manufacturing procedure between the two categories of ice cream was that the poor texture batches were frozen at a higher temperature than the good texture batches. The rheological data (Table 4.7) suggest that all the parameters are higher for good texture ice creams than for poor texture ice creams, but the differences are most noticeable in the values of E_0 and E_1. The poor texture samples have values of E_0, E_1, η_1, and η_N which are similar to those for ice creams prepared at the normal freezing temperature but containing less than 8% fat (Table 4.8). These results indicate the importance of structure development due to the fat droplets in the sensory assessment of texture. For a smooth texture on the palate the fat droplets should not be too large, otherwise a "buttery" sensation is experienced (Frandsen and Arbuckle, 1961). This implies that the fat droplets should not undergo too much coagulation, especially as the rheological properties are directly influenced by mean droplet size and size distribution (Sherman, 1967a). As the mean droplet size and the droplet size distribution increase each parameter decreases almost exponentially in value.

TABLE 4.7

Rheological parameters of good and poor texture ice creams after melting
(Whitehead and Sherman, 1967)

Texture quality	Sample No.	E_0 (dyne/cm^2)	E_1 (dyne/cm^2)	η_1 (poise)	η_N (poise)
	1	143	156	5.2×10^3	3.7×10^4
good	2	716	278	6.6×10^3	5.6×10^4
	3	793	299	13.3×10^3	8.6×10^4
	4	62	33	2.6×10^3	2.3×10^4
poor	5	21	14	0.7×10^3	1.1×10^4
	6	42	25	1.9×10^3	1.0×10^4

TABLE 4.8

Influence of fat content on the rheological properties of melted ice cream
(Whitehead and Sherman, 1967)

% Fat (wt/wt)	E_0 (dyne/cm^2)	E_1 (dyne/cm^2)	η_1 (poise)	η_N (poise)
0	22.2	25.0	594	6.46×10^3
2	20.8	43.5	905	9.7×10^3
6	41.7	43.5	1.24×10^3	1.14×10^4
8	179	87.0	3.13×10^3	4.54×10^4
10	416	278	6.62×10^3	5.64×10^4

3. CHOCOLATE

Chocolate is primarily a dispersion of non-fatty particles of sugar and cocoa in a continuous phase of solid cocoa butter (Aylward, 1960). The non-fatty particles usually range between 10–100 μ in diameter, while the cocoa butter melts completely at about 34°C. The only fat present in plain chocolate is cocoa butter, which is the natural fat of the cocoa bean, whereas in milk chocolate butter fat and the non-fatty solids of milk are dissolved in the cocoa butter. Cocoa butter, which has a complex chemical composition, contains six triglycerides (Aylward, 1960) but the two which are present in the largest amount are 2 oleo-palmitostearin ($\sim 57\%$) and 2-oleodistearin ($\sim 22\%$).

A much greater interest has been shown in the rheological properties of molten chocolate than in the rheological properties of the hard, non-molten, chocolate. In the latter case there do not appear to have been any serious quantitative studies, although the author has found (unpublished observations) that the parallel plate viscoelastometer is admirably suited to this type of study using shearing stresses of 15 000–20 000 dyne/cm^2 at temperatures of 20°C. Detailed qualitative studies have been made using an adaptation of the Brinell hardness test and a cone penetrometer (Guice *et al.*, 1959; Lovegren *et al.*, 1958).

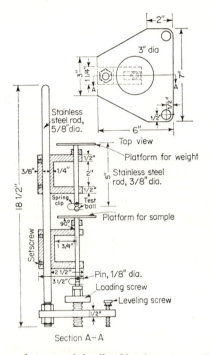

FIG 4.13. Dimensions and structural details of hardness tester. (Lovegren *et al.*, 1958).

It can be shown (Rosenberg, 1954) that when a sphere of given diameter under a given force penetrates an elastic isotropic material the depth of the indentation depends on the elastic modulus of the material provided the strain is small. The Brinell hardness test makes use of this principle. In the particular version (Fig. 4.13) used to measure the hardness of chocolate Lovegren *et al.*, 1958; Guice *et al.*, 1959) a steel ball was held in a conical recess by means of a spring clip. The recess was at the end of a stainless steel shaft, which was attached to the platform which was loaded with a weight when the test began. The total weight of the moving parts, i.e. platform, shaft, pin etc., plus the steel ball was 200 g, and this was the minimum weight which could be applied to the sample. The sample was placed on another platform which was located initially at some distance below the shaft which held the steel ball. Levelling screws mounted on the base plate of the instrument were used for aligning the steel shafts which supported the steel ball and the sample platform. This platform was then moved upwards by a loading screw until the steel ball rested on the sample. A single-row radial bearing, containing 10 small ball bearings to provide a friction-free connection, and mounted on a steel nut, was located between the loading screw and the $\frac{3}{8}$ in steel ball at the bottom of the shaft supporting the sample platform. This bearing ensured that no circular motion was imparted to the sample.

Test balls of carbon or stainless steel were used. For hard fats such as chocolate or cocoa butter a steel ball of $\frac{3}{8}$ in diameter was used. With softer fats diameters of $\frac{1}{8}$ in, $\frac{3}{16}$ in, $\frac{1}{4}$ in, and $\frac{1}{2}$ in could be used.

An impression was made in the sample by turning the loading screw until the weight of the ball, platform, shaft, pin, etc. was wholly supported by the sample. The sample was subjected to this weight, plus any additional weight that proved necessary by loading the upper platform, for 1 min exactly. Then the loading screw was reversed and the sample platform was lowered. A cathetometer, or a magnifying glass with a built-in scale, was used to measure two diameters of the impression at right angles, and the average value was used to calculate the hardness index (H_p), where

$$H_p = \frac{mg \times 100}{\frac{1}{2}\pi D_b \left[D_b - (D_b^2 - d_i^2)^{\frac{1}{2}}\right]} \qquad (4.10)$$

and mg and D_b are the weight of the steel ball and its diameter (mm) respectively, and d_i is the diameter (mm) of the impression. The denominator of Eq. (4.10) represents the curved area of the impression in terms of the diameters of the ball and the impression. The only requirements for this kind of test to proceed satisfactorily are that the surface of the sample should be perfectly horizontal throughout the course of the test, the thickness of the sample at any point where an impression is made should be at least ten times the depth of the impression, and that the distance of the centre of the impression from the edge of the sample or from the edge of another impression is at least two and a half times the diameter of the impression. Using the $\frac{3}{8}$ in steel ball the applied weight should be such that $d_i/D_b = 0.15-0.45$.

H_p depends not only on the variables quoted in Eq. (4.10), but also upon the time for which the steel ball is applied to the sample, the thermal history of the sample, and the test temperature. The influence on the hardness of cocoa butter of test weight, D_b, and the time for which the steel ball is applied to the sample are shown in Table 4.9. Provided the test was carried out for at least 1 min H_p was not influenced to any degree by the test time. When using this test time the influence of applied weight and steel ball diameter on H_p were smaller than that of test times which were shorter than 1 min.

Table 4.10 shows the influence on H_p when cocoa butter was subjected to different thermal treatments. It is apparent that the hardness can be drastically reduced by appropriate thermal treatment. H_p decreases linearly with increasing test temperature up to a temperature of about 26°C (Fig. 4.14)

TABLE 4.9

Influence of test time, applied load, and ball diameter on hardness measurements for cocoa butter (Lovegren *et al.*, 1958)

Ball diameter (in)	Weight applied (kg)	Duration of test (sec)	d_i/D_b	H_p
	0·2	10	0·252	17·4
3/16	0·2	60	0·294	12·7
	0·2	120	0·294	12·7
	0·2	60	0·228	11·9
	0·4	10	0·268	17·3
1/4	0·4	60	0·315	12·4
	0·7	60	0·441	10·8
	0·2	60	0·168	9·8
	0·5	10	0·210	15·8
	0·5	30	0·220	14·3
3/8	0·5	60	0·262	10·0
	0·5	120	0·262	10·0
	1·2	60	0·409	9·6
	0·7	60	0·228	10·3
1/2	1·2	60	0·307	9·9
	1·7	60	0·346	10·8

after which the rate of decrease in H_p with further increase in temperature up to approximately 30°C is somewhat slower. As the temperature rises from about 12°C to 20°C H_p is halved, so that accurate control of test temperature when determining H_p is extremely important. Once again it is found that thermal history influences the rate of decrease of H_p with increasing temperature. Since the solid/liquid ratio does not alter very much between 12°C–20°C the drop in H_p must be due to softening of the fat crystals and not to partial melting.

Partially hydrogenated oils, which contain glycerides differing in fatty acid composition and configuration from those found in cocoa butter, have been added to chocolate to reduce its sensitivity to temperatures greater than $27.8°C$. Hydrogenated cottonseed oil increases H_p only slightly (Guice *et al.*, 1959). When 0.3% lecithin, based on the weight of fat, is added to a chocolate formulation containing 10% completely hydrogenated fat H_p falls

TABLE 4.10

Influence of thermal history on the hardness of cocoa butter
(Lovegren *et al.*, 1958)

Thermal treatment	Weight applied (kg)	d_i/D_b	H_p†
Bars tested as received from manufacturer	0·7	0·181	29·8
	1·2	0·229	31·7
Cocoa butter melted by heating to 38°C, melt seeded on solidification, tempered for 5 h at 27°C, cooled for 10 min at 5°C to release from mold	0·7	0·181	29·8
	1·2	0·232	30·8
Cocoa butter melted by heating to 15°C, solidified, and held for 5 h at 5°C	0·2	0·134	15·6
	0·7	0·252	15·2
	1·2	0·331	15·0

† At 15°C, using ⅜ in steel ball, and load applied for 1 min

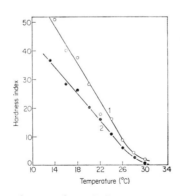

FIG. 4.14. Hardness curves for cocoa butter B (Lovegren *et al.*, 1958). (1) bars moulded by manufacturer and stored at room temperature (24–28°C) for several months before testing; and (2) test samples obtained by melting some of the bars, seeding the melt as it resolidified, and tempering the resolidified cocoa butter for 24 h. Tests made with a ⅜-in ball and load applied for 1 min.

$\sim 25\%$ at room temperature. The main disadvantage to the use of hydro-genated oils in chocolate is that they may increase the melting and softening range of the cocoa butter. The hardness of chocolate has also been studied, with particular reference to the influence of hydrogenated oils during stor-age at 36°C for periods up to 72 h, using a cone penetrometer.

Since chocolate is essentially a dispersed system when in the molten state several attempts have been made to apply the equations for non-Newtonian flow to it, but the published literature is very contradictory. For example, Campbell (1940) found that the flow properties of several types of molten chocolate conformed with Williamson's theory of pseudoplasticity (Section 2.B.1., Chapter 3). More recently, Steiner (1959) studied the flow properties of three different chocolate samples with five different types of rotational viscometer which included a Haake "Rotovisko", a Portable Ferranti, and a Drage-Epprecht viscometer. All the viscometers gave shear stress–rate of shear curves which were in good agreement in the sense that at rates of shear exceeding about 40 sec^{-1}, which corresponded to a stress of about 600 dyne/cm^2, the plots became straight lines. Extrapolation of these straight lines, however, gave three different yield values, viz. ~ 350 dyne/cm^2, ~ 230 dyne/cm^2, and ~ 110 dyne/cm^2.

Steiner (1958) adapted Casson's equation for non-Newtonian flow (Eq. (2.65)) to interpret chocolate flow data. As pointed out in Section 2.B.3, Chapter 3, Casson assumed that the particles in flocculated dispersions assume the configuration of rigid cylindrical rods, and Steiner presents no argument or experimental evidence to justify the selection of this model for molten chocolate. The relevance of Eq. (2.65) to the rheological properties of various dispersions was tested with a cone-plate viscometer. When using a coaxial cylinder viscometer the rate of shear is not constant across the gap, so Steiner amended Casson's equation to allow for the shear gradient present in his studies. This led to an equation which was valid for rates of shear greater than 1 sec^{-1}

$$[1+(R_1/R_2)]\,(D_N)^{\frac{1}{2}} = \frac{1}{K_c}[1+(R_1/R_2)\,p^{\frac{1}{2}}-2K_0] \qquad (4.11)$$

or

$$(D_N)^{\frac{1}{2}} = \frac{1}{K_c}\left[p^{\frac{1}{2}} - \frac{2K_0}{[1+(R_1/R_2)]}\right]$$

where D_N is the Newtonian rate of shear at the inner cylinder $[\,= 2\omega_a R_2{}^2/(R_2{}^2-R_1{}^2)]$, $K_c{}^2$ is the plastic viscosity, and $K_0{}^2$ is the yield stress. Thus, by plotting $[1+(R_1/R_2)]\,(D_N)^{\frac{1}{2}}$ against $[1+(R_1/R_2)]\,p^{\frac{1}{2}}$ a straight line should be obtained with gradient $1/K_c$ and an intercept on the latter axis equal to $2K_0$.

Steiner (1958) found that when rate of shear-stress data for several molten chocolates, which had been derived originally with the Haake Rotovisko and Portable Ferranti over ranges of shear rates equal to $1 \cdot 3$–$216 \cdot 0 \sec^{-1}$ and $8 \cdot 7$–$125 \cdot 0 \sec^{-1}$ respectively, were replotted in accordance with Eq. (4.11) straight lines were obtained whereas the original plots had been curvilinear. Furthermore, the discrepancies previously observed between viscosity data derived with two different rotors of the Haake Rotovisko now disappeared. All chocolate samples now showed slightly lower viscosities and considerably lower yield values than when Bingham body behavior was assumed.

Heimann and Fincke (1962) studied the flow properties of many samples of molten chocolate at $37 \cdot 8°C$ using rates of shear between 2 and $75 \sec^{-1}$. This temperature is above the temperature corresponding to the melting point of the most stable (β) crystal form of cocoa butter. These workers found that plain chocolate conformed approximately with Eq. (4.11), but that this equation was not followed by many milk chocolate samples. All samples, irrespective of type, gave a linear plot of $(p^2)^{\frac{1}{3}}$ against $(D^2)^{\frac{1}{3}}$, and based on this observation it was found that all the flow data conformed with the equation proposed by Heinz (1959) for paints. This latter equation is also basically a modification of Eq. (2.65), viz.

$$(p^2)^{\frac{1}{3}} = K_0 + K_c (D^2)^{\frac{1}{3}}$$

or

$$D = \left(\frac{p^{1/m} - K_0}{K_c} \right)^m \tag{4.12}$$

where m is a constant. When $m = 1$ the equation reduces to that for Bingham body flow, and when $m = 2$ the equation reduces to Eq. (2.65), so that basically it has a very general form. It has the disadvantage, however, that it contains three unknown variables, viz. K_0, K_c, and m. For the milk chocolate samples m was $1 \cdot 5$, and for the plain chocolate samples m was 2. Heinz (1959) found that for some types of chocolate the precise value of m depended on the dispersion techniques employed to prepare the samples.

A flow equation has been proposed (Harbard, 1956) specifically for chocolate which relates the viscosity of molten chocolate to the cocoa butter content. Assuming that part of the cocoa butter is associated with the surface of the solid particles the "free" volume concept (Eq. (3.127)) was applied, and resulted in

$$\frac{\eta_\infty}{\eta_0} = \left[1 - \frac{\phi}{(1 - v_d)} \right]^{-a} \tag{4.13}$$

where v_d is the fraction of voids between the solid particles after they have been packed tightly together by centrifuging the molten chocolate at high speed, and the constant a varied in this case between $2 \cdot 1$ and $2 \cdot 6$.

Equation (4.13) can also be expressed in another way if ϕ is small

$$\ln \frac{\eta_\infty}{\eta_0} = -k \ln \left[1 - \frac{\phi}{(1-v_d)} \right] = k \left(\frac{\phi}{1-v_d} \right)$$

so that, when the dispersed phase has a wide range of particle sizes, i.e. $v_d \to 0$

$$\frac{\eta_\infty}{\eta_0} = \exp\left(+k\phi\right)$$

which is the Arrhenius equation.

TABLE 4.11

Apparent viscosity of cocoa butter as a function of holding temperature and time (Sterling and Wuhrmann, 1960)

Holding temperature (°C)	Apparent viscosity (cP) after the following holding time (h)					
	1	3	5	7	9	13
30	72·2	72·7	74·2	74·9	75·4	75·4
28	86·4	88·3	88·8	89·5	89·5†	91·0
27	98·6	102·0	102·5	103·0	103·4†	106·1
26	111·4	112·6†	113·0	114·7	117·1	121·7
25	124·1	126·5†	126·5	131·5	143·3	148·6
24	138·0†	139·9	140·2	153·4	164·6	209·3

† First appearance of crystals (polarising microscope)

Purified cocoa butter has a slight non-Newtonian viscosity (Sterling and Wuhrmann, 1960), the precise value depending on the test temperature, e.g. $\eta_\infty = 0·73$ poise at 30°C and $1·524$ poise at 24°C. The non-Newtonian flow behavior becomes more marked the lower the temperature at which the test is made. If the cocoa butter is stored at various temperatures between 24°C and 30°C, after first melting at 100°C, the apparent viscosity increases with holding time (Table 4.11), the rate of increase being much larger at the lower temperatures. This is due to the rapid onset of fat crystallisation at the lower temperatures, followed by aggregation of the fat crystals into a network. When cocoa butter is stored at the still higher temperature of 60°C for 270 days and it is then cooled down to 0°C it shows many gel-like properties and is plastic, whereas samples stored at lower temperatures (17·5°C–45°C) are brittle at 0°C (Sterling et al., 1960).

The temperature dependence of molten chocolate has been defined (Fincke and Heinz, 1956) by the general formula

$$\eta = a_h \exp\left(\frac{b_h}{T - c_h}\right) \qquad (4.14)$$

where a_h, b_h, and c_h are constants. Since c_h is very small, the equation becomes

$$\ln \eta = \ln a_h + \frac{b_h}{T}.$$

Thus, a plot of $\ln \eta$ as ordinate against $1/T$ as abscissa gives a straight line of gradient b_h with an intercept $\ln a_h$. Heimann and Fincke (1960) found that in the temperature range $28°$–$75°$C $\ln a_h = -12.5$, and that b_h was related to the viscosity by the following empirical relationship

$$b_h = (3.9 + 0.3 \ln \eta) \, 10^3 \qquad (4.15)$$

when the viscosity was calculated by Eq. (4.12).

If cocoa butter or chocolate is to have a good texture it is obvious from what has been previously stated that it must be subjected to careful temperature control before it hardens. Vaeck (1951, 1952, 1955) has identified four polymorphic forms of cocoa butter, viz. a very unstable (γ) form with a low melting point (18°C), an α form with a melting point of $23.5°$C, β' form with a melting point of 28°C, and a β form with a melting point of $34.7°$C. Pure triglycerides have a more pronounced stiffening effect in the β' form than in the β form because the latter form has a more compact crystal habit. Mixed triglycerides, such as cocoa butter, have a greater stiffening effect than a fat composed mainly of a single fatty acid presumably because they assume the β form less readily. Aylward (1960) has pointed out that the temperature control prior to hardening should be so adjusted that it meets the two requirements of giving a high proportion of crystals of the stable (β) form plus a well distributed amount of the smoother, more glossy, less-stable (α and β') forms.

4. Cakes, Bread, and Biscuits

The consistencies of bread and cake have been assessed qualitatively using penetrometers, compressimeters, the Bread and Cereal T.N.O. Panimeter, a rotary cutter, and a Tenderometer which simulates the masticatory process.

Quantitative measurements have been made at low shearing stresses with the parallel plate viscoelastometer (Shama and Sherman, 1968). Figure 4.15 shows creep compliance-time curves for one-day-old Madeira cake, the recipe for which is given in Table 4.12, at three different low shearing stresses. It is readily apparent that the cake shows non-linear viscoelastic behavior, the creep compliance at any given time following the application of stress increasing with increasing shearing stress. In this respect the rheological

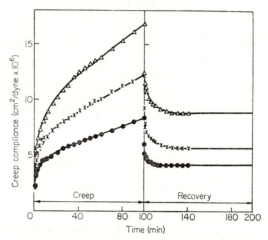

FIG. 4.15. Creep curves at various stresses for one-day-old Madeira cake (Shama and Sherman, 1968). △—△ stress 7750 dyne/cm²; ×—× stress 5900 dyne/cm²; •—• stress 4000 dyne/cm².

behavior of cake resembles the behavior of both bread dough and bread. The data in Fig. 4.12 indicate that the creep compliance is approximately proportional to the square of the applied stress over the range of shearing stresses which were investigated.

TABLE 4.12

Recipe for Madeira Cake

Ingredient	Content (parts by weight)
Margarine	310
Sugar	362
Egg (whole)	362
Flour	413
Water	103

At a shearing stress of 1106 dyne/cm² the creep compliance of Madeira cake is defined, in terms of Eq. (1.30), by

$$J(t) = J_0 + J_1 [1 - \exp(-t/\tau_1)]$$

$$+ J_2 [1 - \exp(-t/\tau_2)] + J_3 [1 - \exp(-t/\tau_3)] + \frac{t}{\eta_N} \quad (4.16)$$

TABLE 4.13

Influence of staling period on the rheological properties of Madeira cake; shearing stress 1106 dyne/cm²
(Shama and Sherman, 1968)

staling period	E_0 (dyne/cm²)	E_1 (dyne/cm²)	E_2 (dyne/cm²)	E_3 (dyne/cm²)	η_1 (poise)	η_2 (poise)	η_3 (poise)	η_N (poise)
1 h	1.23×10^5	1.83×10^5	2.51×10^5	3.61×10^5	1.44×10^8	2.71×10^7	5.62×10^6	3.00×10^9
2 h	1.84×10^5	2.59×10^5	4.15×10^5	6.21×10^5	2.60×10^8	8.63×10^7	16.42×10^6	5.33×10^9
3 h	2.51×10^5	3.05×10^5	5.62×10^5	8.80×10^5	3.58×10^8	8.40×10^7	21.50×10^6	6.54×10^9
4 h	6.89×10^5	11.90×10^5	21.74×10^5	20.62×10^5	13.22×10^8	46.30×10^7	52.57×10^6	10.30×10^9
24 h	8.85×10^5	12.81×10^5	22.22×10^5	25.71×10^5	16.10×10^8	30.20×10^7	52.90×10^6	12.00×10^9
48 h	12.79×10^5	13.70×10^5	38.46×10^5	46.95×10^5	13.23×10^8	54.70×10^7	101.3×10^6	14.40×10^9
96 h	12.99×10^5	18.42×10^5	30.68×10^5	40.82×10^5	19.83×10^8	54.80×10^7	113.4×10^6	15.10×10^9

At this level of shearing stress the creep behavior is still linear. When cake hardens during staling under normal ambient conditions, all the rheological parameters derived by Eq. (4.16) alter quite appreciably (Table 4.13). All eight parameters increase with storage time, but the major part of the change occurs during the first day. During the latter period the creep compliance is non-linear at shearing stresses above 2000–3000 dyne/cm², but with continued storage the creep compliance becomes progressively more linear. Cake which had been stored for seven days showed linear behavior for shearing stresses up to 8000 dyne/cm². This observation is extremely interesting because the compressibility of freshly baked bread under stress is believed to be due mainly to plastic flow, and that during staling the contribution due to flow decreases and elasticity assumes greater importance (Bice and Geddes, 1949). Staling of cake cannot be due to a similar transformation because the data in Table 4.13 indicate that all the viscosity parameters increase quite appreciably in addition to the increases in the elastic parameters. The big difference in fat content between cake (Table 4.13) and bread (Table 4.14) may also have great significance.

TABLE 4.14

Recipe for bread (Cornford *et al.*, 1964)

Ingredient	Content
flour, baker's grade	280 lb
yeast	4 lb
salt	4½ lb
calcium propionate	9 oz
water	14–17 gals
	depending on grade of flour
hydrogenated vegetable shortening	0 or 1·99 lb

The consistencies of cakes made with different fats, so as to give different crumb strengths, have been examined with the Panimeter (Section 3.F, Chapter 2) and rotary cutter (Section 3.G, Chapter 2) one day after baking. Because of the non-linear behavior of the cakes Panimeter tests were made in two ways by using either a constant stress or a constant strain. In the former series the compression was determined using loads of 400 g on 3 cm diameter samples and 750 g on 4 cm diameter samples; in the constant strain tests the load required to give a compression of 10% was determined. Comparative data for the Panimeter and rotary cutter tests are given in Table 4.15. With increasing crumb strength the % compression under a fixed load tended to decrease, but the correlation was less satisfactory for the 4 cm

TABLE 4.15

Panimeter and rotary cutter for sandwich cakes (Robson, 1966)

		Panimeter data						Rotary cutter (area under graph) (– arbitrary units)
		constant stress				constant strain		
		3 cm diameter samples 400 g load		4cm diameter samples 750 g load		force (g) required for 10% compression		
Sample	Known crumb strength	% compression	% recovery	% compression	% recovery	maximum	minimum	
A	weakest	38	14	43	23	350	250	0·308
B		37	21	51	20	385	290	0·358
C		26	32	47	28	490	340	0·321
D		13	62	32	45	405	280	0·369
E		18	32	33	28	620	390	0·357
F	toughest	17	38	33	43	670	450	0·348

diameter samples than for the 3 cm samples, probably because the former samples suffered much greater damage to crumb structure. Tests at constant strain, so that all samples underwent the same degree of structural change, showed the same trend as for the 3 cm diameter samples under a load of 400 g. The tests with the rotary cutter were far less satisfactory, and showed no definite relationship with crumb strength. However, when tests were made on a series of Madeira cakes which had still greater crumb strengths (Table 4.16) the rotary cutter data showed good agreement with increasing crumb strength, but curiously the constant stress data using applied loads of 500 g, 400 g, and 300 g now showed less correlation with crumb strength than in the previous tests.

TABLE 4.16

Panimeter and rotary cutter data for madeira cakes (Robson, 1966)

Sample	known crumb strength	Panimeter data at constant stress; % compression at following loads			Rotary cutter (area under graph–arbitrary units)
		500 g	400 g	300 g	
A	weakest	25	21	20	0·34
B		23	15	7	0·38
C		22	18	8	0·42
D		23	16	10	0·51
E		18	8·5	6·5	0·51
F	toughest	16	9	3·5	0·53

Because of the heavy stress involved in tests with a Panimeter the sample suffers much greater structural alteration than in tests with the parallel plate viscoelastometer. Nevertheless, in terms of overall deformation a plot of the creep compliances of five different sponge cakes after 100 min using a shearing stress of 3900 dyne/cm^2 against per cent compression in the Panimeter for an applied load of 300 g (Shama and Sherman, 1968) yielded an approximately straight line (Fig. 4.16). Because of the non linear behavior of fresh

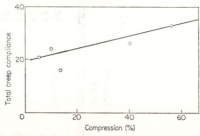

FIG. 4.16. Correlation of total creep compliance with total % compression (Panimeter) for sponge cakes (Shama and Sherman, 1968).

cakes too much significance should not be attached to Fig. 4.16 because the linear relationship will vary with the test conditions for both instruments.

The original version of the compressimeter (Platt, 1930) consisted essentially of a balance with one of its pans located directly over a plunger, which had the form of a rod with a larger diameter solid core at the other end, and an adjustable platform which supported the sample. A suitable weight (often 100 g) was placed on this pan and a chain of the same weight on the other pan. The sample was introduced onto the platform, which was then raised so that the surface of the sample just touched the base of the plunger. The chain was removed gently from its pan, so that the weight acted on the sample. Compression during 1 min was recorded by the indicator needle and scale of the balance. Most of the compression occurred during the first 15 sec. The degree of elasticity was determined from the degree of recovery by the plunger when the load on the sample was removed by gently replacing the chain, and also from the way in which it obeyed Hooke's law when different weights were used. This latter test was believed to simulate the conditions under which the baker assesses "springiness".

The addition of a weight to one of the balance pans may cause some damage to the surface of the sample by introducing a jarring effect. This may be eliminated by hanging the weight on a hook which, in turn, is suspended by a rubber band (Platt and Powers, 1940). Stretching of the latter absorbs the jarring motion.

In a more sophisticated version of the compressimeter (Cornford, 1963) the sample was compressed between circular brass plates (Fig. 4.17). The lower plate was stationary, while the upper one was attached by a vertical metal rod to a platform on which a weight was placed when the stress was to be applied to the sample. The platform was attached, in turn, to a metal framework which was linked by a thread over a pulley to a weight which counterbalanced the weight of the upper plate and platform. The pulley was mounted on a shaft which carried a counterbalanced pointer to record the compression. In operation one determined the weight required to compress samples which were 1 cm thick and of either 3·2 cm or 2·5 cm diameter. This particular version of the compressimeter covered a range of 30–500 kilodyne/cm^2 for 3·2 cm diameter samples (from 14 oz and 28 oz bread loaves) or 40–750 kilodyne/cm^2 for 2·5 cm diameter samples (from 4 oz bread loaves). It was found to operate efficiently with bread or sponges, but it was not suitable for testing cakes with a readily fractured crumb, or even very fresh bread.

Platt and Powers (1940) studied the factors which affect the quality of bread crumb other than staling. For the latter effect flour strength appears to be of primary importance, the rate of staling in terms of loss in compressibility being a non-linear function of the protein content of the flour

(Steller and Bailey, 1938). Compressibility of freshly prepared bread increases with increasing shortening content, but the addition of small amounts of sugar has little effect. As would be anticipated the baking time has a big effect on compressibility. Maximum compressibility appears to be associated with optimum fermentation.

As bread stales there are changes in the proportion of soluble starch, in the enzyme susceptibility of the starch, and in the X-ray diffraction pattern of the starch (Bechtel, 1955, 1961). These changes suggest that starch exerts the principal influence on staling. Changes in the compressibility of bread

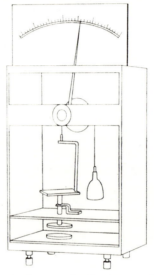

Fig. 4.17. The compressimeter for assessing the texture of bread and cakes (Cornford, 1963).

crumb with time have been employed to study the kinetics of staling (Cornford *et al.*, 1964). It was assumed that if the elastic modulus E is linearly related to the extent of starch crystallisation then Avrami's (1939, 1940, 1941) theory can be applied in a form resembling Eq. (4.1) but with viscosity replaced by elastic modulus

$$\frac{E_\infty - E_t}{E_\infty - E_0} = \exp\left(-k_A t^{n_A}\right) = \exp\left(-t^{n_A}/t_x\right) \qquad (4.17)$$

where E_∞, E_t, and E_0 are the respective elastic moduli after infinite time, after a time t, and at $t = 0$.

The elastic modulus was calculated as the stress/strain ratio from readings of the weight required to compress cylindrical samples 1 cm thick $\times$ 3·2 cm diameter to half their original thickness in 1 min (Cornford, 1963). Alternatively, for soft crumbs, the modulus was calculated from penetration data using a 90° cone, after a penetration time of 1 min. The modulus was derived in this case after calibrating the instrument with bread samples which had also been examined with the compressimeter. The observed relationship was

$$\frac{m}{d_p^{\,2}} = 1\cdot53E + 0\cdot00260\,E^2 \qquad (4.18)$$

where E is quoted in kilodyne/cm^2.

Crumb modulus data were plotted as log $(E_\infty - E_t)$ against t and fitted the following relationship

$$\log(E_\infty - E_t) = \log(E_\infty - E_0) - (t\log e)/t_x \qquad (4.19)$$

or

$$E_t = E_0 + (E_\infty - E_0)\,[1 - \exp(-t/t_x)]$$

Bread stored at 21°C showed an increase in modulus with age. Eventually after about 8 days the modulus approached a limiting value (E_∞) of 210 kilodyne/cm^2, with $t_x = 3\cdot22$ days. Within a range of storage temperatures $-0\cdot5°$ to 32·5°C the rate of change in the modulus when the bread contained no fat (hydrogenated vegetable shortening) was influenced by temperature but not E_∞, which was about 155 kilodyne/cm^2 at all storage temperatures. However, t_x increased from 2·08 at the lower temperature limit to 5·47 at the upper limit. When fat was added to the bread only small differences were observed in the change of elastic modulus with time. E_∞ was now 125 kilodyne/cm^2, with $t_x = 1\cdot39$ at the lower temperature limit and 5·51 at the upper limit. Thus, even if fat influences the diffusion of moisture from starch to gluten during staling, a process which is believed to be associated with staling, it does not seem to affect the crumb modulus.

The rate of increase in modulus increased at the lower storage temperatures, and this "...evidence favors a physical process involving a more ordered arrangement of atoms or molecules, such as crystallization, as the principal factor concerned in these changes in crumb firmness. It suggests that it is not primarily a diffusion process, such as moisture transfer between starch and gluten, since this would have a positive temperature coefficient: nor a chemical change, since the equilibrium condition does not appear to be affected by temperature and the rate of transformation has a negative temperature coefficient" (Cornford *et al.*, 1964).

Emulsifying agents such as poloxyethylene stearate and vegetable oil monoglycerides retard the rate of decrease in compressibility in bread during storage, the former being more efficient in this respect than the latter (Skovholt and Dowdle, 1950). Since the effect of the emulsifying agents does not become apparent for some few hours after the bread has been baked this observation refutes the belief that emulsifying agents merely soften the bread.

The development of toughness in cakes during staling can be followed by measuring their tensile strength in a very simple way (Platt and Kratz, 1933), as for rubber, fibres, etc. Samples 4 in × 6 in × 1 in are prepared, free from crust, and a neck of $1\frac{1}{2}$ in width is cut in the middle of the length using a special electric cutter. Clamps are fitted to either end of the sample, and the clamp at the upper end is then hung on a support frame. A container is attached to the lower clamp, and water is run in at a rate of about 200 g/min until the limit of tensile strength is reached and the sample ruptures. At this point in time the weight acting on the sample is the weight of the container plus the weight of water collected therein plus the weight of the clamp. The

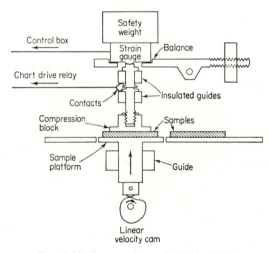

FIG. 4.18. Compression unit (Babb, 1965).

cross sectional area of the tear is measured, and the applied load is expressed in g/cm² of cross section. Compressibility and tensile strengths are interrelated in that the tensile strength increases as the compressibility decreases. These changes are also associated with a decrease in cake volume, which suggests some alteration in the basic foam structure.

For rapid routine measurements (24 slices per min) of firmness an automated compressibility device (Fig. 4.18) can be used (Babb, 1965). This instrument comprises two principal parts, viz. a powered compression unit consisting of a sample platform which is raised against a compression block by an interchangeable linear velocity cam whose speed can be varied between 3 and 27 rev/min, and a strain gauge measuring unit to which the forces developed are transmitted for conversion into electric signals. These signals are then fed via a control box into a high speed recorder. The output speed of the latter is controlled by low-voltage contacts which are actuated by the compression block, and they close as soon as compression begins.

Thick samples require a cam which raises the sample platform further than for a thin sample. With soft samples increased sensitivity is obtained by using a large diameter compression block. After the appropriate cam and compression block have been fitted, the height of the block is adjusted so that in operation it will give the required penetration. For most materials a 40 mm diameter compression block, and a cam giving a rise of 15 mm,

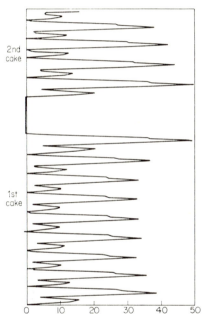

FIG. 4.19. Recorder chart of shear and compressibility tests on Madeira cake (Babb, 1965). (The force peaks are shown in pairs, the first, i.e. the small one, being the shear and the second the compression).

were suitable provided the block was adjusted to give a penetration of 5 mm when testing a 10 mm thick sample. Compression and breaking tangential (shear) forces can be measured simultaneously by using a special Perspex frame which takes up to 6 slices.

Figure 4.19 shows some typical compressibility traces for fresh Madeira cake, while Fig. 4.20 shows simultaneous compressibility and shear tests on cakes of different ages. The traces clearly show the variations in firmness across the fresh cake, and also the changes with time of storage. The forces developed increase with sample age, thus confirming the previously reported observations with the parallel plate viscoelastometer regarding the increase in firmness. Tests made on bread provided similar information.

Preliminary experiments, using a rounded plunger instead of the compression block, have indicated that this instrument is suitable also for measuring the brittleness of biscuits with different moisture contents (Fig. 4.21). The traces indicate that the biscuits became less brittle as their moisture content increased. A less sophisticated instrument for testing biscuits has been designed from a household kitchen scale (Bailey, 1934). The procedure is based on a method previously established (Davis, 1921) for cakes. A bis-

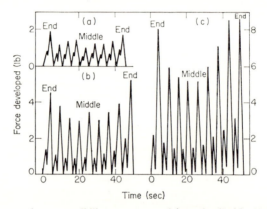

FIG. 4.20. Shear and compressibility tests on madeira cake (Babb, 1965). (a) fresh; (b) after 1 day; (c) after 3 days. The force peaks are shown in pairs, the first being the shear and the second the compression. (All slices of each cake, except crusts, were tested consecutively).

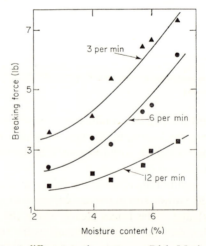

FIG. 4.21. Snap tests at different testing rates on Rich Marie biscuits (Babb, 1965).

cuit is laid over two supporting rails both of which are securely bolted to the scale platform. The rails, which are made from 6·5 mm diameter iron rod, are mounted in parallel with their centres 40 mm apart. A small motor fitted with reduction gears, so as to rotate at 45 rev/min, is used to apply a force to the centre of the biscuit via a flat metal rod, which is supported from a lever by a pin. To bring the rod in contact with the biscuit a simple windlass is employed. This is attached to the shaft of the motor, and is of such dimensions that a cord winds over it at the rate of about 120 mm/min when the shaft rotates. Following contact between the flat metal rod and the sample the applied force is registered by the scale dial. When the biscuit fractures the tension in the scale spring makes the scale platform jerk upwards very suddenly. This motion causes a pin which projects from the platform to strike a friction electric switch, and it is forced open. The switch is in series with one of the power lines to the motor so that the motor shaft stops rotating. At the moment when the biscuit fractures an optimum reading is

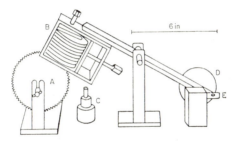

FIG. 4.22. Diagram of principal components of the texture meter (Wade, 1968). A Circular saw blade; B Biscuit sample holder; C Mechanical stop limiting depth of saw cut; D Counterweight on beam; E Mechanical latch retaining sample holder just clear of saw blade.

registered on the scale dial. When this has been noted the needle of the dial is returned to the zero in readiness for the next test.

The hardness of biscuits can be determined by measuring the time required to make a saw cut of standard dimensions into a stack of biscuits (Wade, 1968) using a circular saw. A 2 in stack of biscuits is clamped in a carriage at one end of a pivoted beam (Fig. 4.22). It is kept just clear of the saw teeth by means of a mechanical latch at the other end of the beam. To begin the test the saw blade is set in motion by switching on its motor control, and the latch is released. The blade then cuts through the stack of biscuits until it makes contact with a mechanical stop. A small brush close to the blade removes the crumbs produced during cutting.

The saw blade is 4 in diameter, it has 64 teeth, and its speed of revolution is 15 rev/min by means of a synchronous motor. The counterweight on the beam at the mechanical latch end is adjusted so that the weight on the biscuit carriage when empty is 100 g. In addition, the counterweight controls microswitches at the limits of travel, which in turn control an electric clock or a digital read-out seconds counter. Under these standardised conditions the time required to make the saw cut is 10–100 sec depending on the type of biscuit. Cream crackers required 13 sec and water biscuits 49 sec. Since biscuits are not homogenous products the mean of several readings is taken. In general it was found that the spread of individual results increased

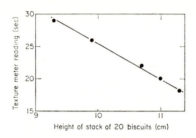

FIG. 4.23 Relationship between degree of aeration (measured by height of a stack of 20 biscuits of constant weight) and texture meter reading for a semi-sweet (Marie) type biscuit (Wade, 1968).

as the biscuits became harder. Factors influencing the meter readings were the degree of aeration of the biscuits (Fig. 4.23), and the protein content of the flour (Fig. 4.24) when flours containing 7–9% protein were used. Fats with different shortening effects had no significant effect on the meter readings, nor did variations in moisture content within the range 1·9–5·2% for semi-sweet biscuits and 2·4–5·8% for short sweet biscuits.

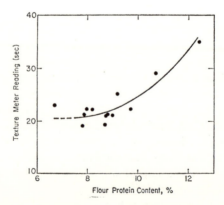

FIG. 4.24. Relationship of texture meter reading of semi-sweet biscuits to the protein content of the flour used in manufacture (Wade, 1968).

Correlations were attempted between meter readings and sensory assessment of hardness by a panel of laboratory personnel using ranking procedures. In all cases, irrespective of whether the biscuits were soft or hard or of intermediate hardness, the meter readings and the sensory assessments placed the samples in identical order of hardness. It was also found that panel members were able to distinguish between soft biscuits, but not hard biscuits, which had given differences in meter readings of 2–3 sec. Irrespective of whether the biscuits were soft or hard the panel members were always able to distinguish between samples showing meter reading differences of 5 sec over the range of hardness (meter readings of 11–49 sec) tested.

5. Cheese

Detailed studies of the consistency of cheese have been made using the principle of compression between parallel plates (Section 2.I.2, Chapter 2).

In one version of the procedure (Davis, 1937) a cylinder or block of cheese 3–6 cm long and 1·5–3·0 cm in diameter is placed on the base plate of the plastometer (Fig. 4.25). A circular disc is now gently brought into contact with the surface of the sample. This disc is connected by a vertical metal rod with a platform which is loaded with a weight of 100–200 g to ensure good contact between the disc and the sample surface. A moveable collar surrounding the rod bears an indicator arm, or long piece (44 cm) of untempered wire, which rests on a knife edge fulcrum affixed to the base plate. When the weight has been applied for 5 sec it is removed, and readings of the indicator arm on a graduated scale are checked until a constant value is recorded. A heavier weight (200–500 g) is now placed on the platform and scale readings are noted at 10 sec intervals over a minimum period of 60 sec. The weight is then removed and readings during the recovery period are

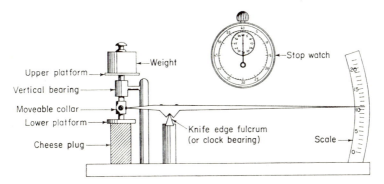

FIG. 4.25. Rheometer or apparatus for measuring deformations in cheese, butter and similar materials (Davis, 1937).

taken at similar time intervals as before. In order to make the recording procedure completely automatic a light quill pen was attached to the end of the indicator arm, and the scale readings were recorded on graph paper around a rotating drum.

Davis's (1937) summary of the interrelationships between sensory assessment of cheese consistency and the rheological parameters derived from compression-recovery tests at constant shearing stress is given in Table 4.17, as it is relevant to the assessment of other food materials also. This summary also refers to the crushing or compressive strength, which is determined by

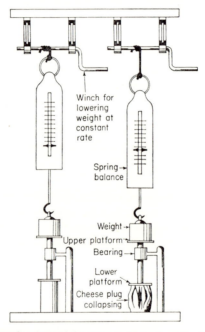

FIG. 4.26. Apparatus for determining the crushing strength of cheese (Davis, 1937).

compressing the sample using a much larger weight, for example 1000 g, than in the previous test. The total weight is not applied instantaneously in this case, but instead by means of a spring balance and a winch (Fig. 4.26) the load applied to the sample is gradually increased. The load required to crush the sample is derived from the reading on the spring balance. It is the weight employed, i.e. 1000 g, minus the balance reading. Since the weight required to crush the sample depends also on the rate at which the load is applied a standardised loading programme is required.

Viscosity values obtained from the first of any series of loadings are always much lower than the values obtained during subsequent loadings,

TABLE 4.17

Sensory assessment of cheese consistency and related rheological parameters (Davis, 1937)

Test	Sensation or result	Rheological properties concerned
Pressing with thumb into cheese, momentarily holding and then removing	(1) hardness	the sum of a function of viscosity and modulus (chiefly the latter for hard cheese)
	(2) spring (i.e. the amount of recovery)	time of relaxation
	(3) speed of recovery	elastic after-effect* (internal viscosity)
Applying a constant force against the cheese	increase in hardness as test proceeds	work hardening‡
Pressing with different forces	amount by which the cheese feels softer when it is pressed harder	structural viscosity†
Ordinary crushing	crumbliness ("shortness")	structural viscosity†(?)
Pressing on cheese (e.g. by a weight) for some time, removing and allowing full recovery	permanent change in shape	viscosity (external)
Applying increasingly heavy loads	(1) load required to crush	crushing or compressive strength
	(2) extent of deformation before crushing	plasticity(?)

$$* \text{ internal viscosity} = \frac{\text{shear stress} \times \text{time occupied by elastic recovery}}{\text{elastic strain}}$$

$$\dagger \text{ structural viscosity} = \frac{\text{viscosity at low stress}}{\text{viscosity at high stress}}$$

$$\ddagger \text{ work hardening} = \frac{\partial \text{ (viscosity)}}{\partial \text{ (non-recoverable deformation)}}$$

i.e. cheese shows work hardening. This greater flow in the initial loadings may be ". . . due to the air or gas spaces in the cheese mass and—the apparently greater flow is due to filling up of these spaces, which could obviously only be filled once" (Davis, 1937).

Data obtained for the apparent viscosity and shear modulus of several different types of cheese are listed in Table 4.18.

TABLE 4.18

Rheological data for various English cheeses (Davis, 1937)
using successive loadings of 50 g, 100 g, and 200 g.

Cheese	Viscosity (g cm^{-1} sec^{-1})	Shear modulus (g cm^{-1} sec^{-2})
Cheddar	141×10^6	$1 \cdot 15 \times 10^6$
Dunlop	167×10^6	$1 \cdot 13 \times 10^6$
Gloucester	155×10^6	$1 \cdot 15 \times 10^6$
Derby	110×10^6	$1 \cdot 07 \times 10^6$
Chester	56×10^6	$0 \cdot 79 \times 10^6$
Leicester	35×10^6	$0 \cdot 50 \times 10^6$

Expert graders classified the Cheddar, Dunlop, and Gloucester cheeses as being of maximum body, with Derby, Chester, and Leicester following in decreasing order. Thus, the rheological data and the sensory assessments give the same grading sequence.

Additional confirmation of the interrelationship between measured parameters and the sensory assessment of consistency was derived from double logarithmic plots of shear modulus against viscosity (Fig. 4.27). This form of plot was used so that the logarithms of the relaxation times, i.e. log (viscosity/shear modulus), fell on straight lines, this parameter being used to characterise consistency. Typical distributions of elasticity and viscosity for the characterisation of butter and hard cheese are also shown in Fig. 4.27, which indicates the influence of these two parameters on springiness, hardness etc. Interrelationships of this type will be discussed in further detail in the last chapter.

For routine assessment of cheese consistency, when less detailed information than that provided by Davis's (1937) technique will suffice, a spherical compression device can be used (Scott Blair and Coppen, 1941; Caffyn, 1945). A section of a metal sphere of diameter 1·5 in, which is attached by a metal rod to a handle, is brought into contact with the sample surface and a steady pressure is applied via the handle and a surrounding spring. Steady application of pressure is ensured by means of a plate, around the spherical body, which rests on the sample surface. A vernier in a frame around the rod is used to measure the depth of the indentation in the cheese surface. In a more sophisticated version of this instrument (Scott Blair and Baron, 1949) the pressure is applied by drawing a weight at a steady rate along a beam. As the sphere begins to penetrate the sample the stress falls rapidly, and the procedure is so arranged that the rate of which the load increases eventually compensates for the increasing area of contact between the indenting body and the sample. In this way the stress is kept almost constant for samples with similar consistencies. The deformation (d_i) at time t following application of the initial load is related to the stress p by

$$\log d_i = a_i + b_i \log t \qquad (4.20)$$

provided the stress remains constant, where a_i and b_i are constants.

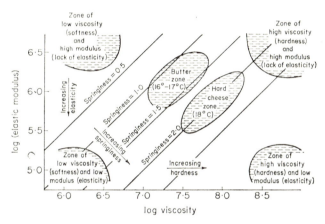

Fig. 4.27. Graph showing the relationship between modulus or elasticity, viscosity, and springiness. Plugs 3 cm long by 1·7 cm diameter, load 100 g, time of stress 30 sec (Davis, 1937).

Caffyn (1945) also assessed consistency using a sectilometer which consisted of a taut wire attached to a container by a vertical rod. Lead shot was introduced into the container at a steady rate, and the time required for the wire to cut through sample blocks of standardised dimensions was taken as an index of their "sectility".

Harper and Baron (1948) analysed six test procedures which are " . . . directed towards measuring essentially different properties of . . . Cheddar cheeses". These were (1) total penetration of a sphere under standard load into the surface after 30 sec, (2) percentage of this penetration remaining unrecovered 30 sec after removing the load, (3) penetration of a cylindrical borer under standard load after 10 sec, (4) ratio of the 10 sec reading to the 60 sec reading of the borer (taken negatively), (5) total penetration of a sphere following a series of increments of loads, and (6) the slope of the log deformation/log load curve derived from test (5). Analysis indicated that (1) and (3) showed high correlations with hardness, and (5) showed a less significant correlation. None of the tests showed a statistically significant correlation with either "springiness" or "hardening".

In recent years there has been a growing interest in the consistency of processed cheese. Detailed studies of the factors influencing consistency have been made by Szabo (1966) using the principle of compression between parallel plates. The raw cheese mixture consisted of 65% Emmenthal cheese, 20% Trappist cheese, 10% young hard cheese, and 5% additives. Five combinations of emulsifying salts were used, a processing temperature of 80°C, and a holding time of 2 min. The processing temperature was reached in 4 min. After processing the cheeses contained 48–56% water. When the fat content was increased from 20% through 32% to 45% (expressed as a % of the total solids), the elasticity decreased 25–35%, depending on the emulsifier used, and the viscosity decreased by an approximate factor of 3. Variations in the additive concentration range 0–10% increased the elasticity by approximately 9%, and the viscosity by 60–100%, but further increases in additive concentration up to 15% resulted in no further change. When the processing temperature was increased from 80°C to 95°C the elasticity was not affected, but the viscosity of the processed cheese at 18°C dropped by almost 30%. A general survey of the data in conjunction with sensory assessment indicated that the production of good consistency cheese requires the incorporation of pyro- and meta-phosphates with the emulsifying salts. A complete rheological description (Szabo, 1966b) of the samples involved 5 parameters, but provided the total deformation was not more than 10% of the original height of each sample two parameters associated with elasticity and the viscosity coefficient of a Kelvin–Voigt unit could be omitted from consideration. Consistency was then defined by two parameters, viz. a relative elasticity coefficient which was calculated from the ratio of elastic deformation to total deformation, and a quasi-viscosity which was calculated from the residual deformation. Data for these two parameters, as derived from creep compliance–time curves, were analysed in conjunction with sensory assessment responses. For the latter an arbitrary six point scale was used ranging from a score of 1 for soft consistency, 2 for spreadable and pasty, 3 for good spreadability, 4 for intermediate between spreadable and sliceable, 5 for sliceable, and 6 for hard and rubbery. Figure 4.28 shows the interrelationship between the two measured parameters and the mean scores, and indicates that the respective ranges of elasticity coefficient and quasi-viscosity for each point on the sensory scale are 270–420 and 10 000–28 000, 200–290 and 17 000–30 000, 180–280 and 25 000–55 000, 165–265 and 43 000–83 000, 160–240 and 65 000–120 000, and $\ngtr$ 175 and 85 000–

125 000. These data indicate reasonably well defined differences in the viscosity ranges associated with each division on the scale, but this is not the case for the elasticity coefficient ranges, and in particular for those elasticity ranges which are associated with scores of 2–5.

The consistency of processed cheese has been measured with a sectilometer also. The sample is introduced into a slotted container, and it is pressed down with a 5 lb steel weight which fits snugly into the container. A wire cutter is then driven through the sample at a constant speed. The cutting device can be in the form of 0·016 in stainless steel wire stretched tautly across a U-shaped frame (Emmons and Price, 1959), or in the form of three parallel wires, 0·4 mm diameter and 1·0 cm apart, which are held in a circular frame (Voisey and Emmons, 1966). By placing the sample

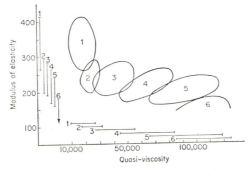

Fig. 4.28. Relationship between the rheological properties of processed cheese and its organoleptic properties (Szabo, 1966b).

container on an ordinary kitchen scale the resistance offered by the sample, during downward movement of the cutting surface, can be registered on the scale dial. Alternatively, the force can be recorded continuously via a transducer, strain gauges, and an amplifier on a strip chart recorder. When using a kitchen scale consistency is defined in terms of "firmness", which is the mean of the largest and lowest values recorded on the dial. With the electronic method of recording "firmness" can be expressed as resistance in gm/cm of wire length cutting the sample, or as before. Panel tests in which the firmness was scored using an arbitrary 5 point scale corresponding to 1 for too soft, 2 for soft, 3 for desirable firmness, 4 for firm, and 5 for too firm, showed a highly significant correlation (correlation coefficient 0·96–0·97) with "firmness" measurements by the sectilometer (Emmons and Price, 1959).

6. Fruit and Vegetables

The rheological properties of fruit and vegetables determine their ability to withstand mechanical damage during harvesting, mechanical handling, storage, transportation etc. (Somers, 1965; Mohsenin and Morrow, 1968). Many studies have been made using small slices or rectangular pieces, but the question remains as to how data obtained with such samples can be used to deduce the properties of the whole fruit or vegetable.

The Instron tester has been used to study the viscoelastic properties of potatoes, peas, and apples (Somers, 1965; Morrow and Mohsenin, 1966; Mohsenin and Morrow, 1968). In the case of Somer's (1965) studies with potatoes they were stored at 5°C until about one day before testing, when they were placed in water at the same temperature to ensure crispness. With the aid of a hand microtome slices 1 mm thick $\times$ 1 cm wide $\times$ 4 cm length were prepared, and their ends were clamped in the 1 cm apart jaws of the Instron. Two types of measurement were made, viz. tension and compression.

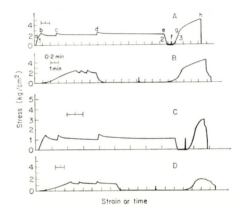

FIG. 4.29. Stresses resulting from cyclic elongation (strain) and stress relaxation of slices of potato tubers. See text for cycling details. In all cases the units for the abscissa are presented to the left above the tracing. In each case the line represents 0·2 min or 1 min, depending upon whether strain or stress relaxation is being observed. The treatments were A, in H_2O at 6°C; B, acetone −15°C and then rehydrated; C, acetone; and D, frozen in H_2O at −15°C and thawed; all overnight (about 18 h). Attention is called to the fact that the scale for C differs markedly from that of the other portions of the figure (Somers, 1965).

In the tension measurements the sample was elongated by downward movement of the crosshead, to which the lower jaw was attached, at a uniform speed of 0·5 cm/min. A sensitive strain gauge, attached to the upper jaw, recorded the sample response, and this was amplified and fed to a strip chart recorder. The test procedure followed a definite pattern. The sample was first elongated to produce the stress shown as (a) in Fig. 4.29. After a particular stress had been developed, which was somewhere below the breaking stress, the crosshead was stopped (b) and the stress was allowed to relax 20%. The sample was then again elongated so that the strain increased to (c), thus giving the same stress as at (b). Once again the crosshead was stopped, and the stress was allowed to relax 20%, after which the strain programme was repeated once again (d–e). Following this the crosshead was returned to its original position at a rate of 0·5 cm/min. The crosshead was then set in motion once again so that the sample elongated at the same uniform rate as before, and this was continued (f–g–h) until the sample ruptured. It was observed that with each strain cycle the relaxation time became progressively longer (A in Fig. 4.26), and that in the first cycle as the strain increased the stress-strain curve (1) was approximately linear initially, and then it became curvilinear, thus suggesting a plastic flow component.

Potato slices were treated in various ways to alter their turgidity. Curves B, C, and D in Fig. 4.29 show the corresponding modifications to the stress-strain behavior. Slices stored in acetone at −15°C and then rehydrated became flaccid so that stress developed slowly initially on elongation, whereas slices which were not rehydrated showed stress-strain behavior similar to that in curve A. Slices which were frozen in water and thawed (curve D) gave a stress–strain pattern similar to that in curve B. Apparent values of Young's modulus (E) were calculated from these curves in accordance with

$$E = \frac{F/A}{\Delta l/l} \qquad (4.21)$$

where F is the force (dyne/cm^2 initial cross sectional area), A is the initial area of the sample, l is the original length of the sample between the two jaws of the Instron, and Δl is the extension. Calculated apparent values of E, and other relevant data derived from the curves, are given in Tables 4.19 and 4.20. It is readily apparent that the various treatments have a marked effect on the response of the samples to cyclic elongation.

In Morrow and Mohsenin's (1966) and Mohsenin and Morrow's (1968) studies with the Instron tester the potatoes were subjected to uniaxial compression, instead of elongation, and whole potatoes were used instead of slices. To ensure good contact between the curved surface of each potato and the compression unit a cylindrical die was employed. Compression tests were also made with a spherical indenter and a cylindrical plug. Because there is no well established theory for the distribution of stress and strain within convex bodies two approaches were developed to this problem. The first approach, which was applied when using the cylindrical die, is based on Boussinesq's (1885) treatment for semi-infinite bodies which are subjected to concentrated compressive loads. The pressure distribution (P_d) at the surface of the body is given by

$$P_d = \frac{m}{2\pi r_d \, (r_d^{\,2} - a_d^{\,2})^{\frac{1}{2}}} \qquad (4.22)$$

TABLE 4.19

Summary of data derived from Fig (4.23) (Sommers, 1965)

Treatment	Curve	Magnitude of stress relaxation† (kg/cm²)	Time for 20% relaxation in stress (sec)			Elongation (mm)	
			First cycle	Second cycle	Third cycle	origin to b	f to h
water at 6°C	A	2·5–2·0	20 ±2·8	70 ±8·6	114 ±6·1	0·79±0·07	4·4±0·26
acetone at −15°C; and rehydrated	B	2·5–2·0	5·4±0·38	7·9±0·97	13 ±2·4	4·7 ±0·17	8·8±0·39
acetone at −15°C; no rehydration	C	1·5–1·2	2·0±0·13	42 ±0·38	9·2±0·93	0·55±0·05	1·7±0·20
water at −15°C; and thawed	D	1·5–1·2	8·2±0·48	33 ±10·3	71 ±24	4·4 ±0·19	6·8±0·40

* The first value quoted corresponds to position b in Fig. 4.23; the second value is the one to which the stress was allowed to relax.

TABLE 4.20

Apparent values of Young's modulus of elasticity obtained by cyclic application of tension to potato slices (Sommers, 1965)

Treatment	Curve	Apparent value of E (10^7 dyne/cm²)			Breaking load (kg/cm²)
		Initial strain (1)	Return to origin (2)	Final strain (3)	
water at 6°C	A	6·3 ±0·35	5·9±0·35	4·2±0·33	5·3±0·22
acetone at −15°C, and rehydrated	B	0·81±0·04	5·4±0·09	3·5±0·11	4·5±0·41
acetone at −15°C no rehydration	C	4·1 ±0·22	9·1±0·51	5·6±0·26	3·0±0·24
water at −15°C, and thawed	D	0·55±0·03	3·4±0·07	24±0·09	2·2±0·25

where a_d is the distance from the centre of the area over which the die is acting, r_d is the radius of the die, and m is the total load (lb) on the die. From this pressure distribution Young's modulus is derived

$$E = \frac{m(1-\mu^2)}{2r_d s'} \tag{4.23}$$

where s' is the deformation. Equations (4.22) and (4.23) are valid provided the die and the sample have very different diameters. The second approach to the treatment of convex bodies is based on the treatment of Hertz (1896), and it was applied when samples were compressed by a flat plate or a spherical indenter. In its original form this treatment deals with contact stresses between two convex bodies, but it has been generalised (Timoshenko and Goodier, 1951; Kozma and Cunningham, 1962) and extended to compression by flat plates. Young's modulus is given now by

$$E = \frac{0.5\,(1-\mu^2)\,F}{(s')^{3/2}} \left[\frac{1}{r_1} + \frac{1}{r_1'} + \frac{1}{r_2} + \frac{1}{r_2'} \right]^{\frac{1}{2}} \tag{4.24}$$

where s' is the displacement or approach of the centres of the two bodies, and r_1 and r_2 refer to the primary and secondary convex bodies. When compressing the sample by a steel plate $r_2 = r_2' = \infty$, and when using a spherical indenter $r_2 = r_2' = r_a$ where r_a is the radius of the indenter. Equation (4.24) is valid when the radius of the circle of contact between the indenter and the sample is much smaller than the radii of curvature of the two convex bodies, i.e. the indenter and the sample.

In order to calculate E from either Eq. (4.23) or Eq. (4.24) it is necessary to know the value of μ for potatoes. This is derived from the relationship between E and the last term in Eq. (1.20), using the linear region of the load-deformation curve to obtain K. For potato $\mu = 0.49$ (Finney et al., 1964).

According to Mohsenin and Morrow (1968) analysis of the load-deformation curves in terms of creep compliance with time indicated that the latter could be represented by a three element model consisting of a Hookean spring in series with a Kelvin–Voigt element, so that

$$J(t) = J_0 + J_m\,[1 - \exp\,(-t/\tau_m)] \tag{4.25}$$

where the J terms now refer to uniaxial creep compliance. However, comparison of the experimental and theoretical curves given for apples (Fig. 4.30) suggests that the right hand of Eq. (4.25) may well include a t/η_N term. Relaxation data fitted a model of two Maxwell elements in parallel more satisfactorily (Fig. 4.31), so that

$$G(t) = G_1 \exp\,(-t/\tau_{R_1}) + G_2 \exp\,(-t/\tau_{R_2}) \tag{4.26}$$

Table 4.21 shows data for potatoes derived from Eqs (4.25) and (4.26), with the moduli expressed in lb/in^2. For the tests on potato (tuber) the alternative types of loading used, viz. spherical indenter or cylindrical die, gave rather different values of the derived parameters. The creep moduli for potato (flesh) were larger than for potato (tuber), and the relaxation

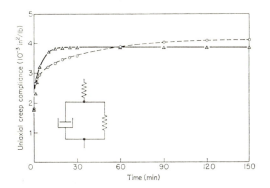

FIG. 4.30. Uniaxial creep compliance relations for McIntosh apple fruits (Mohsenin and Morrow, 1968). Loaded by a 0·312 in diam spherical indenter. o—o experimental values; △——△ values fitted, using inset model.

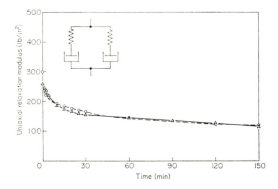

FIG. 4.31. Uniaxial relaxation modulus relations for McIntosh apple fruits (Mohsenin and Morrow, 1968). Loaded by a 0·312 in diam spherical indenter. o—o experimental values; △▬▬△ values fitted, using inset model.

moduli were lower, which suggests that the skin confers stiffness on the whole potato. However, the retardation times suggest the opposite, although they are not too dissimilar.

Some relevant data for apples, obtained in creep compliance and relaxation studies, are shown in Table 4.21. While there are no striking differences between the J_0 and J_m values for whole apple and whole potato, the rele-

TABLE 4.21

Creep and relaxation parameters for potato and apple as derived from uniaxial compression and relaxation tests
(Mohsenin and Morrow, 1968)

Sample	Types of loading	Creep parameters			Relaxation parameters			
		J_0 $(10^{-3}$ in^2/lb)	J_m $(10^{-3}$ in^2/lb)	τ_m (min)	G_1 (lb/in^2)	τ_{R1} (min)	G_2 (lb/in^2)	τ_{R2} (min)
Potato tuber	Spherical indenter (0·31 in diameter)	1·54	2·69	42·42	137·0	4·1	258·6	156·1
	cylindrical die (0·25 in diameter)	3·17	2·85	48·63	55·4	10·0	110·0	471·6
Potato (flesh)	cylindrical plugs (0·81 in × 0·81 in)	8·18	10·00	52·24	38·7	7·0	100·0	58·7
McIntosh apple (fruit)	Spherical indenter (0·31 in diameter)	1·79	2·05	3·45	91·0	8·2	167·1	454·7
	cylindrical die (0·25 in diameter)	2·64	3·61	10·00	116·8	10·0	281·0	170·9
	parallel plates	2·54	1·67	6·67	143·7	2·3	364·8	56·1
McIntosh apple (flesh)	cylindrical plugs (0·81 in × 0·81 in)	4·92	5·35	41·49	95·5	24·9	115·1	124·9

vant values for potato (flesh) are much higher than for apple (flesh), so that the latter has a firmer consistency. G_1, G_2, and τ_{R_1} and τ_{R_2}, are much lower for potato (flesh) than for apple (flesh).

Values of E_0 and $(G_1 + G_2)_{t=0}$ are compared in Table 4.22. In some cases the agreement is quite satisfactory, but in others the differences are quite large. Mohsenin and Morrow (1968) conclude that the application of Eqs (4.22)–(4.24) to convex bodies requires further investigation.

TABLE 4.22

Comparison of elastic moduli obtained by relaxation and uniaxial compression tests (Mohsenin and Morrow, 1968)

Sample	Type of loading	$G_1 + G_2$ (lb/in^2)	E_0 (lb/in^2)
McIntosh apple (fruit)	Spherical indenter (0·31 in diameter)	258	559
	Cylindrical die (0·25 in diameter)	398	379
	Parallel plates	509	394
McIntosh apple (flesh)	Cylindrical plugs	211	203
Potato (tuber)	Spherical indenter (0·31 in diameter)	396	649
	Cylindrical die (0·25 in diameter)	165	315
Potato (flesh)	Cylindrical plugs (0·81 in × 0·81 in)	139	122

Bulk compression studies have been made on fruit and vegetables (Morrow and Mohsenin, 1966; Mohsenin and Morrow, 1968) at both very low and very high hydrostatic pressures. For the former test condition the sample is confined within a sealed chamber (Fig. 4.32) which is filled with water. Compressed air is introduced into the chamber, and a graduated side-arm records the change in sample volume at different hydrostatic pressures. When working with very hard materials hydrostatic pressures up to 4000 lb/in^2 may be required (White, 1966). In such cases the sample was confined within fluid contained in a compression chamber which was designed to withstand very high hydrostatic pressures, and which was fitted with a metal plunger. Pressure was developed by the motion of the plunger in the fluid, and the change in volume was derived by multiplying the plunger cross-sectional area by its displacement, the latter being recorded continuously by a displacement transducer, transducer meter, and strip chart recorder. The change in pressure was followed by a pressure transducer and strip chart recorder. The bulk modulus is defined by Eq. (1.11), and the

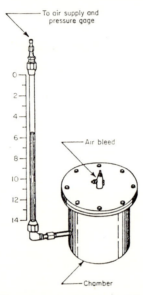

FIG. 4.32. Low-pressure bulk compression apparatus (Morrow and Mohsenin, 1966).

compressibility $= 1/K = (\partial V/V)/\partial P^a$. Three factors contribute to the change in volume—the sample (S), the fluid (F), and the expansion of the compression chamber (C); so that

$$\frac{\partial V}{\partial P} = \frac{1}{K_s} V_s + \frac{1}{K_F} V_F + \frac{1}{K_c} V_c \qquad (4.27)$$

where V is volume. The product $(1/K_c) V_c$ and $1/K_F$ were determined from compression tests at given pressures, with mercury in the chamber. Since V_F and V_S could be measured, the only remaining unknown quantity, $1/K_s$ was thus identified.

Information regarding the behavior of fruit when subjected to a rapid application of stress (Mohsenin and Göhlich, 1962), e.g. under impact conditions, has been obtained using the equipment shown in Fig. 4.33. A ball-bearing joint supports an arm at its pivot point, while the sample (A) is attached near the other end (B). The impact is created by the sample falling on to another sample, or on to a bearing surface (C) in the form of a steel plunger or plate. A linear variable differential transformer (D) records the maximum initial deformation and the final permanent set, and those data are amplified and recorded on an oscillograph chart. The energy of impact was calculated from the height of the fall and the total falling mass of the fruit and arm. Impact deformations from 0·001 in to 0·500 in can be followed with this apparatus.

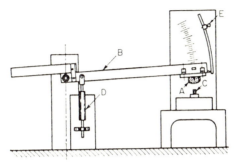

Fig. 4.33. Impact testing apparatus (Mohsenin and Göhlich, 1962). (A) Specimen; (B) Impact arm; (C) Plunger or bearing surface; (D) LVDT transducer; (E) Impact arm release.

The resistance of the skin of various fruits to shearing forces can be examined by placing a sample, after complete removal of fruit flesh, between the two sections of a shear press (Fig. 4.34). The spring-loaded bolts of the fixture (B) are tightened so as to keep the skin (A) taut, and the flat end of a plunger (C) then perforates the skin. A load cell in conjunction with an oscillograph chart record the force involved.

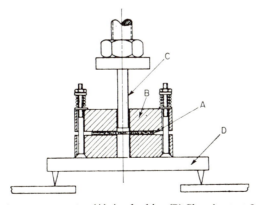

Fig. 4.34. Shearing test apparatus (A) Apple skin; (B) Shearing test fixture; (C) Plunger (D) Load cell platform. (Mohsenin and Göhlich, 1962).

According to Somers (1965) "In interpreting viscoelastic properties of plant tissues it is generally assumed that the properties reflect properties of the walls of the cells in the tissues used. It is also recognised that these walls are composed essentially of cellulose microfibrils embedded in an amorphous matrix. Admittedly, the cell walls represent a structural system which is complex and difficult to analyze in terms of the properties of the cell walls themselves. Yet, in quantitative terms they are presumably the principal viscoelastic components of plant tissue. They form a network of interacting units which, by interaction with the cell contents inside the walls, give the tissues their over-all viscoelastic properties. The cytoplasm is also visco-elastic, but the cytoplasmic connections among cells are so tenuous that they probably contribute little to an interacting viscoelastic network. The cell contents, including the protoplasm, are essentially liquid, and hence are essentially incompressible at the pressures used . . . . However the hydro-static pressures (turgor) of the cell contents, interacting with the viscoelastic cell walls, are of prime importance in determining the magnitude of the viscoelastic properties of plant tissues."

Dynamic moduli of elasticity have been determined for potatoes, apples, and peas, using vibrational techniques (Falk *et al.*, 1958; Drake, 1962). Figure 4.35 shows a block diagram of the equipment used by Drake (1962). A low frequency generator (1 Hz–5000 Hz) (Osc), with a motor (Mot) connected to the double potentiometer of a Wien-bridge arrangement, was linked with a transistor amplifier (Amp), which fed two electromagnets, and with a frequency meter (Fre) having an identical range. The latter was connected to an X–Y recorder (Rec). The electromagnets were situated on either side of a piece of iron which had been inserted into the rectangular sample. An optical-phototube-amplifier arrangement measured the amplitude of vibration of the sample-iron combination, and this was reproduced on the recorder (Rec) as a function of frequency.

Relevant information was derived from the amplitude-frequency curve (Fig. 4.36) by drawing a vertical line from the apex (amplitude H) to the frequency axis, and then drawing a horizontal line through it at an amplitude $h = \sqrt{2}H$. The width Δn of the peak cut off by this line was measured and also the distance n_2 between the vertical line and the line $n = o$, where n is the relative frequency ($= \omega/\omega_0$, where ω is the angular frequency, and ω_0 is the resonance frequency for the same oscillator without damping). The ratio $\Delta n/n_2$ gave a measure of the peak width, and for low damping it equals the damping factor β. For high damping $\Delta n/n_2$ is given by a more complex relationship

$$\frac{\Delta n}{n_2} = \left[1 + \frac{\beta(1 - \beta^2/4)^{\frac{1}{2}}}{(1 - \beta^2/2)}\right]^{\frac{1}{2}} - \left[1 - \frac{\beta(1 - \beta^2/4)^{\frac{1}{2}}}{(1 - \beta^2/2)}\right]^{\frac{1}{2}}. \qquad (4.28)$$

The dynamic modulus of elasticity was calculated from

$$E = \frac{48\pi^2 \rho l^4 v^2}{m_f^{\,4} t_s^{\,2}} \qquad (4.29)$$

where l and t_s are the length and thickness of the sample respectively, and m_f is a factor whose value depends on the mode of vibration. For the fundamental frequency $m_f^4 = 12 \cdot 36$. Food materials with a very high moisture content have $\rho \to 1$. Since $E \propto l^4$ it was extremely important that the samples had uniform dimensions throughout, and that l was measured very accurately. Samples of the same material which had different lengths gave different

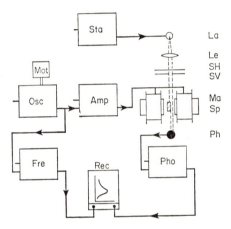

FIG. 4.35(a). Block diagram of vibrational test kit (Drake, 1962). Sta = stabiliser for the lamp; Mot = sweep motor; Osc = low frequency oscillator; Amp = amplifier for the electromagnets; Fre = frequency meter; Rec = X–Y recorder; Pho = phototube amplifier; La = lamp; Le = lens; SH = horizontal slit; SV = vertical slit; Ma = 2 electromagnets; Sp = test specimen; Ph = phototube.

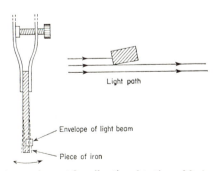

FIG. 4.35(b). Clamping equipment for vibrational testing of fruit and vegetables (Drake, 1962).

values of E, but when an increment of length equal to the sample thicknesses was added to their respective lengths all the calculated values of E were in reasonable agreement. Table 4.23 gives data for peas, apples, potatoes, and three foods for comparison, but conclusions cannot be drawn because detailed studies were not made of sample to sample variation on the same material. In addition, it was observed that E changed with time from the moment that the sample was cut. It appears, therefore, that tests should be made at various times after sample preparation, and that the data should then be extrapolated back to zero time. E values for apple, pear, and potato decreased with time from the moment that the samples were cut, whereas the E values for the cheeses and fish pudding increased with time.

Changes in the consistency of pears from the unripe condition to being overripe have been followed with a spring-loaded penetrometer (Isherwood,

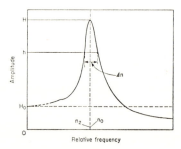

FIG. 4.36. Evaluation of a (smoothed) experimental curve for a soft foam rubber (Drake, 1962).

TABLE 4.23

Experimental data† at room temperature from vibrational testing of fruit, vegetables, and other food materials (Drake, 1962)

Material	l (cm)	t_s (cm)	ν (Hz)	E/ρ (relative values)	β
Apple (Jonathan)	4·3	0·7	36·6	156	0·075
Pear (Pecknam)	4·4	0·55	18·8	71	0·195
Potato	5·3	0·6	15·0	76	0·17
Fish pudding	4·9	0·6	5·8	8·5	0·13
Herrgärd cheese	5·8	0·6	14·0	90	0·33
Tilci cheese	5·3	0·6	7·1	17	0·35

† Specimen width perpendicular to vibration direction, 1·2 cm

1960). The instrument was pressed into the samples until penetration was achieved. In one series of tests at 15°C on Conference pears the penetrometer reading decreased curvilinearly with time from ~ 9 kg at the unripe stage to ~ 2 kg when overripe.

Drake (1963) introduced a completely new approach to rheological measurement of foods following his suggestion that the recording of food crushing sounds might provide useful data for correlation with panel assessment of consistency. A technique of this type should be particularly useful with crisp materials such as fruit and vegetables. The chewing sounds were picked up by a low impedance directional type microphone, or a hearing aid earphone of a high impedance extended-range type. In order to prevent the earphone from making vibrational contact with the head of the subject it was connected to the ear canal by a short plastic tube. The microphone was either pressed against the cheek on the side of the jaws where the chewing was done, or it was held at a distance of about 2 in from the open mouth, while the sample was chewed between the front teeth. When the hearing aid earphone was used its plug was inserted in one ear canal of the subject in such a way that the orifice was closed to the outside, and the sample was chewed on the same side of the jaw as the earphone. In a few instances chewing tests were not used, but instead the sample was pressed between a block of wood and a table top while a microphone was held about 2 in from the edge of the block.

The chewing sounds picked up by the microphone or earphone were recorded on a tape recorder set for 2400 ft of $\frac{1}{4}$-in half-track magnetic tape with speeds of 15 and $7\frac{1}{2}$ in/sec, and provided with a spring-loaded pulley so that endless tape loops could be used instead of conventional continuous (non-loop) tape. Alternatively, a less versatile sound recorder-reproducer was used which had speeds of $7\frac{1}{2}$ in or $3\frac{3}{4}$ in/sec. The low impedance output from the recorders was transmitted through a shielded cable to the amplifier input of a frequency analyser, which was the audio spectro-meter part of an automatic spectrum recorder (31 channels, of which 27 gave $\frac{1}{3}$-octave bands for the range 40–16 000 Hz, 3 gave weighted responses corresponding to standards for objective sound-level meters, and 1 gave a linear response corresponding to the total sound). Strip chart recordings of variations in amplitude with time were made from the tapes using three alternative instruments, viz. a Honeywell strip chart recorder for 1 mV d.c. with 1 sec full-scale pen travel, a vacuum tube voltmeter with a d.c. component, or an attenuator consisting of a decade resistance box in series with two resistors of 33 and $6\cdot 8$ kΩ.

Semi-logarithmic plots of sound amplitude against frequency were derived from direct measurement of the frequency bands on the strip-chart recordings. Figure 4.37 shows typical plots for apple and other foods obtained with the microphone pressed against the cheek etc. on the side where the sample was chewed. Sounds produced by a subject closing his jaws without anything between his teeth proved to be of much shorter duration than the sounds obtained when a food sample was chewed.

Correlation coefficients were calculated (Drake, 1965) between sound amplitudes and sensory assessment of tenderness by a panel of 12 subjects.

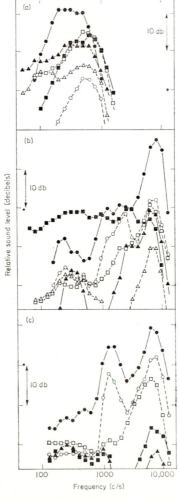

Fig. 4.37. Chewing-sound spectra obtained for six foodstuffs with the following techniques: (A) "through cheek"; (B) "open mouth"; and (C) "wood block". o = crisp white bread; □ = crisp head lettuce; △ = ham; • = crisp brown bread; ■ = apple; ▲ = sausage. A sound level of 20 db above a "general noise level" is indicated by an asterisk (Drake, 1963).

For 1250 Hz and 1600z H the values of the coefficient were $-0 \cdot 26$ and $-0 \cdot 28$ respectively, for which a $0 \cdot 01$ level of significance was claimed. Subsequent studies (1965) indicated that for vegetables there was a highly significant correlation between sensory evaluation of hardness and the average height of the main band, bite period time, i.e. the distance (mm) on the recorded trace between the start of two successive bites, bite time recalculated to 1/100 sec, i.e. the distance (mm) on the recorded trace between the beginning and end of a single bite recalculated to 1/100 sec, and the "chatteriness" pattern of the recorded sound. For the latter tests the technique for recording the chewing sounds was modified somewhat, because it had been found previously that extraneous sounds such as saliva, opening of the mouth etc. influenced the recorded pattern. A higher degree of resolution was achieved by using a "visible speech" technique. The chewing sounds were transmitted

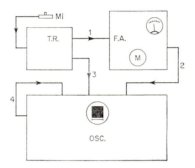

Fig. 4.38. A simple set-up for "visible speech" analysis (Drake, 1965). Mi = microphone; T. R. = tape recorder (using tape loops); F.A. = frequency analyser with sweep motor M; Osc. = oscilloscope with two sweep generators. Connections: 1 = sound track to F.A. input; 2 = F.A. output to upper light spot intensity control and to lower vertical deflection input; 3 = trigger signal track to start of sweep generators; 4 = output of lower sweep generator to upper vertical deflection input.

via a high-impedance extended range hearing aid earphone, located in a special holder, to a magnetic tape recorder equipped for running $\frac{1}{4}$-in two-track magnetic tape at a speed of $7\frac{1}{2}$ in/sec. Frequency bands of the sound signals were transmitted from the diode output (Fig. 4.38) of the tape recorder to a double beam oscilloscope after filtering and amplification by a sound and vibration analyser. When a continuous signal was received from the frequency analyser, or when the background light intensity was adjusted to relatively dark, the oscilloscope screen showed 47 horizontal lines each of 8 cm length. During chewing the signal fluctuated, and the recorded trace lit up in proportion to the recorded sound. The height of the picture on the oscillograph screen recorded frequency, the horizontal position recorded time, and the light intensity indicated amplitude of sound. These pictures were photographed by a camera placed behind the screen. Figure 4.39 compares typical traces for crisp brown bread, biscuit, and raw carrot.

7. DOUGH

In recent years the rheological properties of dough have been studied principally with a cone-plate viscometer, by creep compliance–time studies during extension of a cylinder of dough, and in more qualitative fashion using a Brabender Extensograph. Data relating to dough consistency are necessary because of its important role in determining the texture of bread crumb and also bread volume.

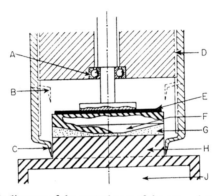

FIG. 4.40. Schematic diagram of the central part of the cone-plate viscometer (Bloksma, 1962). A = Ball bearing; B = Circular knife in high position; C = Circular knife in low position; D = Screw thread; E = Thermal insulation; F = Truncated cone; G = Test-Piece; H = Table; J = Thermostat water.

Bloksma (1962) has pointed out that when dough is tested either in a Brabender Extensograph or a Chopin Extensograph sample deformation is not homogeneous. Both stress and strain are complicated functions of time, and the deformation proceeds at a very high rate. Dough behaves non-linearly when a stress is applied, as do bread and cake (Section 4 of this chapter). Thus, information about the way in which dough behaves at high rates of deformation is not directly relevant to its behavior when slowly deformed, as in breadmaking. The latter condition can most readily be studied by creep compliance–time studies or viscometry at low shearing stresses and shear rates.

The cone-plate viscometer used for studies on dough (Bloksma, 1962, 1968), had a truncated cone. A circular knife (Fig. 4.40) which fitted closely around the cone and plate, and the table supporting the plates when the instrument was in operation, served to trim the edges of the sample after it had been deposited on to the plate. It also minimised water evaporation

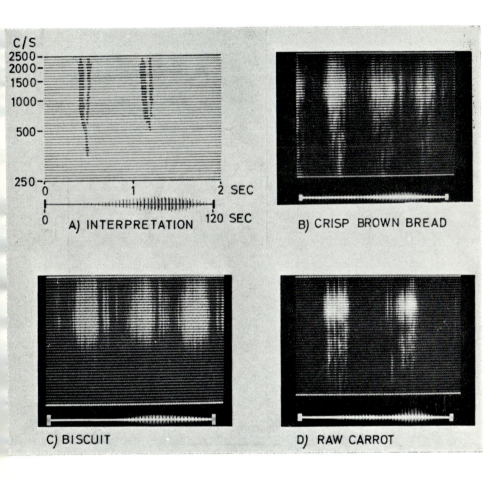

Fig. 4.39. Three time-frequency analyses of chewing sound amplitudes and their interpretation (Drake, 1965). Upper parts: time plotted horizontally and frequency vertically, with sound amplitude visualised as light intensity. Lower parts: time horizontally, sound amplitude vertically.

Facing p. 252

from the dough during the test. When the viscometer was not in use the knife was raised to its rest position just above the cone. In order to develop small stresses as low as 100 dyne/cm^2 an attempt was made to apply the torque by a pulley-weight system, while using an optical system to follow the shear. Unfortunately, the starting friction on this set-up was as high as 20 010 dyne/cm so that it was not suitable for low stress work. This difficulty was overcome by attaching the steel cone (B) to a bow (A) (Fig. 4.41) which was supported by two vertical torsion wires (6) between the bow and the top and base of the instrument. A torque was applied to the cone by rotating

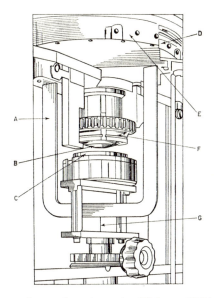

FIG. 4.41. Central part of cone-plate viscometer (Bloksma, 1962). A = Bow; B = Truncated cone; C = Table; D = Upper disc; E = Lower disc; F = Circular knife in high position; G = Torsion wire.

the upper and lower ends of the torsion wires, these ends being coupled by shafts and cog wheels so that they rotated over identical angles in the same direction. The periferies of two horizontal discs, one (E) mounted on top of the bow that carried the truncated cone and the other (D) which was situated a little higher and was coupled with the top of the upper torsion wire, carried the components of a differential transformer. When the truncated cone rotated slightly a similar motion was induced in the lower disc (E), and this produced an electrical signal from the transformer. This signal was amplified and caused a servomotor to rotate the upper and lower ends of the torsion wire. This rotation restored the original torsion of the wire, and since it was identical with the rotation of the cone it was a measure of shear. The shear was transmitted to a counter, and by using a potentiometer coupled magnetically to the shaft that made the ends of the torsion wires rotate a voltage recorder was able to draw a curve of rotation or shear versus time at a constant applied stress. By reversing the operation it was possible to study stress relaxation. Shear stress and shear strain were calculated in accordance with the formulae for an untruncated cone-plate given in Table 2.1, as it was considered that any correction which was necessary was insignificant.

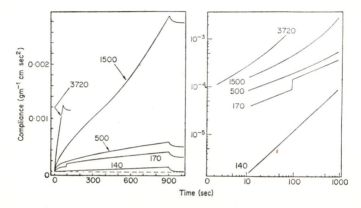

Fig. 4.42. Representative set of creep curves obtained with doughs from lower grade flour, using shear stresses of 140, 170, 500, 1500 and 3720 dyne/cm². The left hand part shows the curves on a linear scale, the right hand part the same curves on a double logarithmic scale. After 900 seconds (with 3720 dyne/cm² after 67 secs) the shear stress was removed and the elastic recovery was recorded. The latter is indicated only in the left hand part (Bloksma, 1962).

Figure 4.42 shows some typical creep compliance–time curves obtained on dough prepared from a low grade flour at shearing stresses ranging from 140 dyne/cm² to 3720 dyne/cm². The series of curves on the left hand side are normal plots, and the curves on the right are double logarithmic plots. The normal plots do not show a well marked linear region at longer times as do viscoelastic foodstuffs such as ice cream, margarine, cheese, etc. By arbitrarily selecting certain times on Fig. 4.42, the influence of shear stress on the compliance was derived for these times (Fig. 4.43). The data show clearly that dough is a non-linear material, the creep compliance increasing by at least one order of magnitude when the shear stress increased from 200 dyne/cm² to 500 dyne/cm². At a stress of 140 dyne/cm² recovery was almost complete when the shearing stress was removed, but when the latter was increased only slightly to 170 dyne/cm² the recovery was only 50% which indicated a yield value of ~ 150 dyne/cm². The elastic recovery was actually proportional to the shearing stress, and could be defined in the case of the bread flour by a constant modulus of $1 \cdot 2 \times 10^4$ ($\pm$ 25%) dyne/cm^2 and for the lower grade flour by a constant modulus of $0 \cdot 9 \times 10^4$ ($\pm$ 20%) dyne/cm^2. A retardation time calculated from the recovery curves was $\sim$ 20 secs. The values of the moduli and the retardation time agreed reasonably well with data derived from extension tests on dough cylinders (Schofield and Scott Blair, 1933).

At very low stresses the creep compliance curve was almost rectilinear, which indicated almost elastic behavior (Fig. 4.42). A similar relationship was observed also at very high stresses which typifies predominantly viscous behavior. At intermediate stresses the lower part of each curve was concave towards the time axis, and at the higher end of this stress region (e.g. 1500 dyne/cm^2) an inflexion point was occasionally observed.

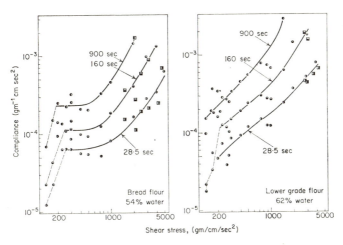

Fig. 4.43. Compliance after 28·5, 160 and 900 sec as a function of stress for doughs from two flours (Bloksma, 1962).

Creep compliance curves resembling those in Fig. 4.43 have been derived from extension tests on dough cylinders (Glücklich and Shelef, 1962; Funt et al,. 1968; Lerchenthal and Funt, 1968). The technique used was a modification of that developed previously by Schofield and Scott Blair (1933), and Halton and Scott Blair (1936). In the original method a dough cylinder was floated on a mercury bath, and a small graduated scale was attached by cotton to each end of the cylinder. A calibrated steel spring was then attached to one of the scales, while the other scale was attached by cotton to a small winch. When the winch was wound up the dough cylinder and both scales moved in the direction of the winch.

The stress was proportional to the spring extension, and inversely proportional to the cross-section of the sample. A direct measure of the stress applied to the dough was obtained from the position of the scale linked with the steel spring. In the amended procedure (Fig. 4.44) an extruded dough cylinder (A) which was approximately 4 cm in length had a paper cuff attached to each end. One end was then fixed to a threaded rod (B) in such a way that the cylinder shifted axially when a wheel (C) was rotated. The other end of the sample was affixed to one end of a thin rubber strand (D) which had its other end attached to a nut (L) with an indicator moving along another threaded rod (E). By rotating a graduated wheel (F) the rubber strand was stretched to any desired length, and the position of the indicator was read on a scale (G). Two travelling microscopes (H and J) focused on the paper cuffs at the extremities of the dough cylinder allowed displacements as small as 0·001 cm to be measured. An additional scale on

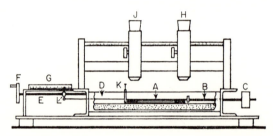

Fig. 4.44. Dough Extensimeter. A, dough cylinder; B, E, threaded rods; C, wheel; D, rubber band; F, wheel; G, scale; H, J, traveling microscopes; K, stop; L, nut (Glücklich and Shelef, 1962a).

microscope J objective, which was graduated in 0·004 cm, facilitated the reading of rapid displacements without having to shift the microscope. The stop K permitted the instantaneous transfer of load from the rubber strand to the dough sample. Extension of the rubber strand provided a direct measure of stress. That part of the dough cylinder which made contact with the mercury bath was coated with glycerine to minimise friction.

Four types of study were made with this extensimeter. These were measurement of instantaneous elasticity, deformation–time relationships at constant load, stress–deformation relationships at constant rate of stress, and stress relaxation. For the first mentioned test an instantaneous load was applied by stretching the rubber strand with the stop K in position, and subsequently it was released. The instantaneous elongation was measured by microscope J; the load was then removed and the instantaneous recovery was measured. Deformation–time relationships at constant load involved the application of an instantaneous load as in the first test, but in this case the load was kept constant by adjusting wheel C after the stop was released. Microscope H was used to determine elongation after selected time intervals. In stress relaxation studies the instantaneous elongation was kept constant by adjusting wheel F so as to relax the load. Readings of the latter were noted on scale G at various time intervals. For the fourth listed test the nut L was adjusted so that the rubber strand D was just free of stress. Wheel C was then rotated so that both ends of the sample were moved to the right, the left hand one at a constant rate. Readings with microscope J indicated the applied load and the difference between the readings of the two microscopes gave sample elongation. Stress and strain were calculated as described in Section 2.I.3 of Chapter 2. Figures 4.45–4.48 show typical data for the four types of test. With the loads used the curves shown in Fig. 4.46 resemble more closely those obtained for viscoelastic foods than do the curves in Fig. 4.42 in that they clearly exhibit instantaneous and delayed elastic regions, and a non-

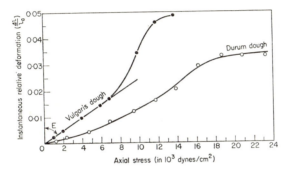

FIG. 4.45. Instantaneous elasticity of dough. Each point represents deformation both on loading and unloading, the two branches being coincident (Glücklich and Shelef, 1962a).

recoverable viscous region. Relevant parameters calculated for doughs made from two different types of flour are given in Table 4.24, which also includes bulk elasticity data for one of the doughs. These data were obtained by a technique similar in principle to that employed to determine the bulk modulus of fruit and vegetables (Section 6, this chapter). A dough cylinder was enclosed in a rubber bag and inserted in a transparent cell filled with glycerine. Variable pressures were applied to the glycerine, and the associated volumetric compression of the dough was derived from the drop in the glycerine level in a burette connected by a rigid tube to the compression cell.

According to Schofield and Scott Blair (1933), the behavior of dough under stress is defined by the Maxwellian relationship for rate of elongation plus an extra term da_e/dt, where a_e defines the influence of the elastic after-effect

$$\frac{d\varepsilon}{dt} = \frac{1}{E}\left(\frac{dp}{dt} - \frac{da_e}{dt}\right) + \frac{1}{\eta_N}\cdot p \qquad (4.30)$$

к

TABLE 4.24

Rheological parameters of dough at zero stress (Glücklich and Shelef, 1962a)

Parameter	Dough made from Triticum vulgare	Dough made from Triticum durum
Elastic bulk modulus (dyne/cm²)	$1\cdot4 \times 10^7$	—
Modulus of elasticity (dyne/cm²)	$4\cdot2 \times 10^5$	9×10^5
Poisson's ratio	$0\cdot5$	—
η_N (poise)	18×10^6	36×10^5
Relaxation time (sec)	43	40
Modulus of retarded elasticity (dyne/cm²)	$2\cdot7 \times 10^5$	9×10^5
Retardation time (sec)	9	20
Solid coefficient of viscosity† (poise)	$2\cdot4 \times 10^6$	18×10^6

† Refers to the viscosity associated with the retarded elastic compliance

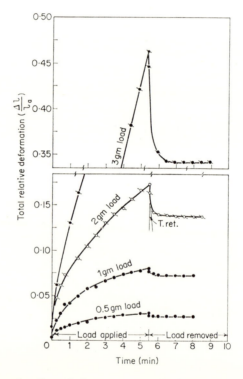

FIG. 4.46. Deformation-time relations at constant load, for vulgare dough. Each curve is a mean of 4 to 5 tests (Glücklich and Shelef, 1962a).

Poisson's ratio was assumed equal to 0·5, so that shear stresses were assumed to be one third of the measured tensile stresses. Glücklich and Shelef (1962b) believe that Eq. (4.30) does not explain the way in which da_e/dt varies with stress and strain. As an alternative they proposed a mechanical model containing a linear Hookean element, a St. Venant body (a weight resting on a table top, with solid friction between them, and a string attached to the weight) to represent the change in elastic modulus when a certain shearing stress is exceeded, a Newtonian dashpot to depict the observed non-linear viscous deformation, and a Kelvin–Voigt element to represent the observed retarded elasticity. Figure 4.49 shows how the various elements are believed to be linked up together in the simplest representation of one unit of the model. In practice, of course, each unit should be replaced by an infinite number of such units representing a non-linear spectrum of parameters, each of which has three quite independent yield values. According to Glücklich and Shelef (1962b) "The first St. Venant element (in part 2 of the unit) corresponds to the stress at which the elasticity ceases to be Hookean. The

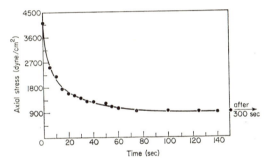

FIG. 4.47. Stress relaxation, vulgare dough. The curve is a mean of three tests. (Glücklich and Shelef, 1962a).

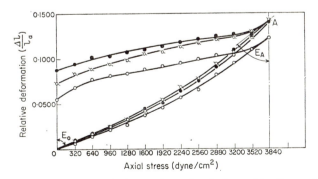

FIG. 4.48. Deformation-stress relations, vulgare dough. Rate of loading and unloading: 640 dyne/cm² per minute (Glücklich and Shelef, 1962a).

third (in part 4) represents the yield value in relaxation and, doubtless, in delayed elasticity". The second St. Venant body was inferred from deformation–time plots at different stresses.

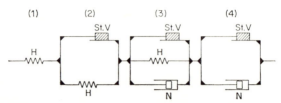

FIG. 4.49. One unit of model to illustrate rheological behaviour of flour dough (Glückiich and Shelef, 1962b).

By making suitable modifications the extensimeter was used to study the stress relaxation of dough (Shelef and Bousso, 1964). The paper cuffs were now attached to the sample cylinder of dough by silk threads, 20 cm in length. The right hand thread (Fig. 4.50) passed through an opening in the bath wall, and was held in position by a pin (4), while the other end of the thread which was outside the bath was attached to a balance pan (5) mounted on a knife edge by means of an arm (6). The left hand thread passed through an opening in the opposite wall of the bath and was also attached to an arm (7). A metal ball (8) placed on arm (7) sealed nozzle (9) when tension was applied to the silk thread. Air was supplied to the nozzle from a constant high pressure air main (10) through a throttling valve (14). Pressure before throttling was checked by means of an indicator (11). The pressure at the nozzle was measured with an alcohol manometer (12). Deformation of the sample was followed with a travelling microscope which was focused on a point at the right hand end of the sample.

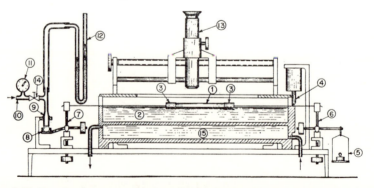

FIG. 4.50. Dough relaxometer (Shelef and Bousso, 1964). 1. dough cylinder; 2. mercury bath; 3. adhesive tape cuffs; 4. pin; 5. balance pan; 6. 7. arms; 8. metal ball; 9. nozzle; 10. air main; 11. pressure indicator; 12. alcohol manometer; 13. travelling microscope; 14. throttling valve; 15. water jacket.

In operation, following calibration of the instrument with various weights, a weight was placed on the balance pan (5), and the desired constant elongation was obtained by retracting pin (4) for a short time and then restoring it to its original position. The tension in the sample caused a load L_n to act on the nozzle, and this in turn produced a reading h_g in the manometer. As stress relaxed in the sample the load on the metal ball decreased, and so did the manometer reading, in such a way that the ratio L_n/h_g remained almost constant. The decay of stress was derived from the way in which h_g changed with time.

A typical relaxation curve for a flour–water dough is shown in Fig. 4.51. The stress decayed very rapidly initially, and then at a progressively slower rate, but even after long time periods complete relaxation was never ob-

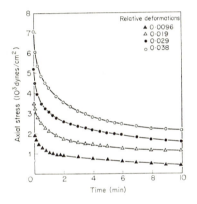

FIG. 4.51. Relaxation curve for flour 1 (Shelef and Bousso, 1964b).

served. This suggests a yield point related to the Newtonian element of part 4 of the mechanical model depicted in Fig. 4.49. The yield point increased with increasing constant deformation, i.e. with the initial stress. Double logarithmic plots of stress versus time were approximately linear for times greater than 0.5 min, so that the "Central Limit Theorem" (Grogg and Melms, 1958; Scott Blair and Burnett, 1959) could be applied. This theorem is based upon the concept that if the ratio of 100 × the stress ($p(t)$) at different time intervals to the initial stress (p_0) gives a straight line when plotted in probability coordinates against log t, then the relaxation times show an approximately log–normal distribution. The ratio $p(t)/p_0$ is given (Wiechert, 1893; Feltham, 1955) by

$$\frac{p(t)}{p_0} = \frac{1}{2}\left\{1 + \operatorname{erf}\left[b_r \ln\left(\frac{\lambda'_t}{t}\right)\right]\right\} \tag{4.31}$$

where b_r is a constant, λ_t' is the most frequently occurring relaxation time, and "erf" signifies the error function which can be obtained from special tables in standard works of reference. Provided $t \geqslant \lambda_t'$, then

$$\operatorname{erf}\left[b_r \ln\left(\frac{\lambda_t'}{t}\right)\right] = -\operatorname{erf}\left[b \ln t \left(\frac{t}{\lambda_t'}\right)\right]$$

so that the geometric mean relaxation time (Grogg and Melms, 1958) is approximately equal to the half relaxation period, and the geometric standard deviation is derived by extrapolation to $100 p(t)/p_0 = 16\%$. Plots of $p(t)/p_0$ in probability coordinates against log t (Shelef and Bousso, 1964) gave satisfactory linear relationships (Fig. 4.52) for times up to 10 min. Values of the geometric mean relaxation time, and geometric standard deviation, for doughs prepared from four different flours are shown in Table 4.25. The values did not change to any extent when the constant

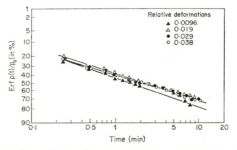

FIG. 4.52. Distribution of relaxation times for flour 1 (Shelef and Bousso, 1964).

TABLE 4.25

Relaxation time data for doughs prepared from 4 flours (Shelef and Bousso, 1964)

Graphical procedure	Parameter	Flour 1	Flour 2	Flour 3	Flour 4
Plot of $100\, p(t)/p_0$ versus log t	geometric mean relaxation time (sec)	114	102	168	198
	geometric standard deviation	16	25·5	20·3	27·6
Plot of $\dfrac{100\,[p(t)-p_r]}{p_0-p_r}$ versus log t	geometric mean relaxation time (sec)	37	24	28	31
	geometric standard deviation	6·4	8·8	5·8	6·7
	loaf volume (cm³)	2775	2200	2125	1650

deformation was altered. Also included in the table are values for the two parameters which were derived by plotting $\{100[p(t)-p_r]/(p_0-p_r)\}$ against log t, where p_r is the residual stress after 10 min. The latter set of data show little variation from one flour to another, and are in sharp disagreement with the first set of data. Shelef and Bousso (1964) suggested that the first graphical procedure was more appropriate, and that on this basis there appeared to be some relationship between geometric mean relaxation time and flour strength with stronger flours showing lower mean relaxation times. In addition, the ratio p_r/p_0 showed a linear correlation with the volumes of loaves prepared from the flour doughs.

Qualitative information about the consistency of dough can be obtained with the Farinograph or the Extensograph. The former instrument is essentially an automatic dough mixer which records the torque exerted by the dough on the mixing blades. The Extensograph stretches samples of dough,

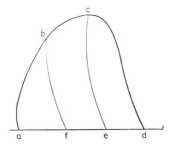

FIG. 4.53. Typical load extension curve for dough with Brabender Extensograph (Muller *et. al.*, 1961). ad = Total extensibility; ae = Extensibility to maximum resistance; ce = Maximum resistance; bf = Resistance at constant extensibility (or "constant sample deformation"); abcef = Area to maximum resistance; abcdef = Total area.

which have been moulded into cylinders under specified conditions in a metal cradle, by means of a hook which moves at a constant rate. A kymograph registers the load on the dough, thus giving an empirical load–extension curve (Fig. 4.53). This information can be translated into approximate values of stress, strain, modulus of elasticity, and viscosity, provided the Extensograph is calibrated by adding known weights to the dough sample in the cradle. A calibration of this form has been carried out on a Brabender Extensograph (Muller *et al.*, 1961). Assuming that the dough sample takes a V shape when extended (Fig. 4.54) the stress developed was derived from the relationship

$$F = 2T_e \cos \theta_e \qquad (4.32)$$

where F is the force applied by the hook, T_e is the tension when the half strand length of the dough sample is l_e and $\cos\theta_e = h_e/l_e$ (Fig. 4.51), where h_e is the hook descent, and

$$l_e = [h_e^2 + (3\cdot75/2)^2]^{\frac{1}{2}} \tag{4.33}$$

with 3·75 cm equal to the horizontal distance between the lower ridges of the dough cradle (A–B). The cross section area is the effective volume of dough/$2 \times l_e$, so that

$$\text{stress} = \frac{T_e}{(m/\rho) \times (1/2l_e)} \tag{4.34}$$

where m is the effective mass of dough, and ρ is its density (taken as 1·17 g/cm²). Variation in m during extension of the dough sample was determined by cutting the dough along the inner edges of the cradle, and removing and weighing the middle piece at different stages of the extension.

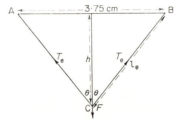

FIG. 4.54. Hypothetical V shape assumed by dough in Brabender Extensograph (Muller et al., 1961).

However, half the dough mass should be added to F to allow for gravity, so that Eq. (4.32) becomes

$$F + 0\cdot5m = 2T_e' \cos\theta_e \tag{4.35}$$

where T_e' is the average tension. Thus,

$$T_e' = \frac{F + 0\cdot5m}{2\cos\theta_e}.$$

The average stress (p') is then

$$p' = \frac{F + 0\cdot5m}{2\cos\theta_e} \left/ \frac{m}{2\rho l_e} \right. \tag{4.36}$$

or, since $\cos\theta = h_e/l_e$

$$p' = \frac{l_e^2 (2F + m)\rho}{2h_e m} \text{ (g/cm}^2\text{)} = \frac{(2F + m)\rho(h_e^2 + 3\cdot52)g}{2h_e m} \text{ (dyne/cm}^2\text{)}.$$

If it can be assumed that the extension of the dough sample occurs between the inner edges of the dough cradle, the strain (ε) is given by

$$\varepsilon = \frac{2l_{e(t)} - 2l_{e(0)}}{2l_{e(0)}} = \frac{l_{e(t)} - l_{e(0)}}{l_{e(0)}} \tag{4.37}$$

where $l_{e(t)}$ is the half length of the sample at time t, and $l_{e(0)}$ is the original half length, i.e. half the distance between the inner edges of the cradle. Values of $l_{e(t)}$ were derived by Eq. (4.33) using the kymograph readings of extensibility, and also

$$h_e = 2.20 \times \text{chart length (cm)} - 1.74 r_k \times 10^{-3} \tag{4.38}$$

where r_k is the kymograph reading in extensograph units, and $(1.74\, r_k \times 10^{-3})$ represents the cradle depression.

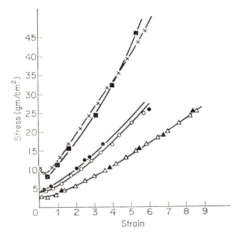

FIG. 4.55. Stress-strain curves for flour doughs (Muller *et al.*, 1961). Strong flour, x one mass ■ individual masses; Bread flour, o one mass • individual masses; Biscuit flour, △ one mass ▲ individual masses.

The principal source of error in the calculation of stress and strain from Extensograph data is the variation in the effective mass of the sample, thus leading to stress variation during extension. Weaker flours show greater flow, and this increases the effective dough mass, so that calculated differences between strong and weak flours may be misrepresentations. Figure 4.55 shows some calculated stress–strain curves for doughs prepared from strong flour, bread flour, and biscuit flour. The doughs were rested for 45 min after preparation and prior to testing. At any given stress within the experimental limits, biscuit flour dough showed a much larger strain than either of the other two doughs.

In order to calculate the elastic modulus of dough (Muller *et al.*, 1962) it was assumed that it is the ratio of elastic stress/elastic strain, although the elastic component of dough may not exhibit Hookean elasticity. The elastic strain was not calculated, however, from the extension of the dough sample. Instead, it was determined from the elastic recovery following removal of the stress after extension, which was assumed to involve the final length of the sample. Then

$$\text{elastic strain} = \frac{(h_{e(2)}^2 + 3 \cdot 52)^{\frac{1}{2}} - (h_{e(3)}^2 + 3 \cdot 52)^{\frac{1}{2}}}{(h_{e(3)}^2 + 3 \cdot 52)^{\frac{1}{2}}} \tag{4.39}$$

where $h_{e(3)} = 2 \cdot 2\, e_1$, $h_{e(2)} = 2 \cdot 2\,(e_1 + e_2) - 1 \cdot 74 \times 10^{-3}$ extensograph units, $(e_1 + e_2)$ is the length of the dough sample after elongation, and e_1 is the length of the sample when contracted following removal of the stress.

During time t the stress increases by an amount $\rho g F (h_{e(2)}^2 + 3 \cdot 52)/3m h_{e(2)}$ and the effective average stress acting during time t, which is used to calculate a coefficient of viscosity, was taken as half of this stress increase. From Eq. (4.33), assuming the viscous extension to be the non-recoverable part of the total extension,

$$\text{viscous strain} = \frac{(h_{e(3)} + 3 \cdot 52)^{\frac{1}{2}} - (3 \cdot 75/2)}{(3 \cdot 75/2)}. \tag{4.40}$$

The instantaneous rate of viscous strain was derived as the gradient of plots of viscous strain against time for different degrees of extension.

Calculated values of stress, strain, elastic modulus, and viscosity for bread and biscuit flour doughs containing 55% water are shown in Table 4.26. A comparison of data of this type with data obtained by extension of dough cylinders on a mercury bath, and by other techniques which have not been discussed, indicated that in spite of the relatively large errors to which Extensometer data are subject the calculated moduli of elasticity and the viscosities are of the right order of magnitude. All the tests were carried out within approximately the same range of shear stress.

The total work done on a sample of dough when it is extended consists of two parts, viz. recoverable work and non-recoverable work. There coverable work is conserved as potential energy and it is released when the dough contracts following removal of the stress, but the non-recoverable work is used up in overcoming internal friction. Tests with the Brabender Extensograph (Muller *et al.*, 1963) have been used to estimate the respective contributions of these two components to the total work done during extension of the dough sample. The work (dW_e) done by a mean force T_e' to produce a very small extension de_x follows from Eq. (4.35).

TABLE 4.26

Rheological data for bread and biscuit flour doughs derived from Brabender Extensograph tests (Muller *et al.*, 1962)

Extensograph units	Mass (g)	$e_1 + e_2$ (cm)	e_1 (cm)	Elastic strain	Elastic stress difference (dyne/cm² × 10⁴)	Elastic modulus (c.g.s. units × 10⁴)	Time (sec)	Viscous strain	Rate of strain (sec⁻¹)	Mean viscous stress	Coefficient of viscosity (c.g.s. units × 10⁴)
(a) Bread flour 55% water absorption											
200	67·5	2·05	0·0	1·43	0·72	0·50	3·13	0·0	0·52	0·36	0·69
255	73·5	3·30	0·35	2·48	1·23	0·49	5·04	0·08	0·52	0·62	1·2
300	79	4·50	2·10	0·90	1·79	2·0	6·87	1·67	0·52	0·90	1·7
360	81	5·60	3·05	0·70	2·58	3·7	8·55	2·72	0·52	1·29	2·5
400	88·5	8·00	4·50	0·69	3·74	5·4	12·21	4·37	0·52	1·87	3·6
(b) Biscuit flour 55% water absorption											
100	64	1·18	0·58	0·35	0·29	0·83	1·80	0·21	0·59	0·15	0·26
140	67	2·28	1·57	0·31	0·56	1·8	3·48	1·1	0·59	0·28	0·47
159	69·5	2·81	2·18	0·20	0·72	3·6	4·29	1·75	0·59	0·36	0·61
180	75·5	4·55	3·55	0·23	1·16	5·1	6·95	3·28	0·59	0·58	0·98
200	75	5·48	4·2	0·26	1·55	6·0	8·37	4·03	0·59	0·78	1·3

$$dW_e = \frac{(F+0\cdot5m)\,g\,.\,dx_e}{2\times0\cdot79\cos\theta_e} \text{ (erg)} \tag{4.41}$$

where $0\cdot79$ is the conversion factor for Extensograph units to grammes. If the sample is extended by $2x_e$, then the work (erg)/g is

$$W_e = \frac{1}{m}\int dW = \frac{1}{m}\int_0^{2x_e} \frac{(F+0\cdot5m)\,g\,.\,dx_e}{2\times0\cdot79\cos\theta} = \frac{1}{m}\int_0^{x_{e'}} \frac{(F+0\cdot5m)\,g\,.\,dx_e}{0\cdot79\cos\theta}. \tag{4.42}$$

Similarly, the work (erg)/g required to produce an elastic extension is

$$W_e' = \frac{1}{m}\int_0^{x_{e'}} \frac{(F+0\cdot5m)\,g\,.\,dx_{e'}}{0\cdot79\cos\theta}. \tag{4.43}$$

The irrecoverable work is then

$$W_e - W_{e'} = \frac{1}{m}\left[\int_0^{x_e} \frac{(F+0\cdot5m)\,g\,.\,dx_e}{0\cdot79\cos\theta} - \int_0^{x_{e'}} \frac{(F+0\cdot5m)\,g\,.\,dx_{e'}}{0\cdot79\cos\theta}\right] \tag{4.44}$$

From Eq. (4.39),

$$\text{total extension (cm)} = (h_{e(2)}^2 + 3\cdot52)^{\frac{1}{2}} - 1\cdot88 \tag{4.45}$$

where $1\cdot88$ cm is half the distance between the cradle edges, and

$$\text{elastic extension} = (h_{e(2)} + 3\cdot52)^{\frac{1}{2}} - (h_{e(3)} + 3\cdot52)^{\frac{1}{2}}. \tag{4.46}$$

The total and elastic extensions were plotted independently against $(F+0\cdot5m)/(\cos\theta_e)$, and the resultant graphs were used to derive the total work and the recoverable work from the relationships

$$\text{total work (erg/g)} = \frac{\text{(total area of graph)} \times 63\cdot3\,g}{m} \tag{4.47}$$

$$\text{recoverable work (erg/g)} = \frac{\text{(area of elastic region of graph)} \times 63\cdot3}{m} \tag{4.48}$$

The constant $63\cdot3$ depends on the scale of the area graph and the conversion of Extensometer units into grammes.

To obtain the non-recoverable work the recoverable work was subtracted from the total work. Figure 4.56 shows plots of the recoverable work and the non-recoverable work against total extension for doughs containing

52, 55, 56, and 58% added water, and Table 4.27 gives data for an arbitrary 4 cm extension which were derived from these curves. At this constant extension the total work reduced by about 50% as the water content increased from 52% to 58%, whereas the recoverable work decreased more than 60%. The non-recoverable work was affected least, and decreased by about 40%. When biscuit flour, low protein cake flour, high protein cake flour, and a strong bread flour were all tested at 54% added water, it was found that the various curves of recoverable work/g against total extension diverged more than the curves of non-recoverable work/g against extension. The latter set of curves were almost coincident for extensions up to 4 cm. At

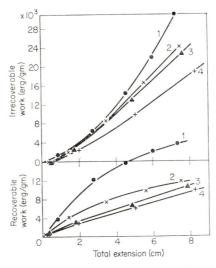

Fig. 4.56. Recoverable and irrecoverable work/g plotted against total extension for doughs of different water content (Muller *et al.*, 1963). (Water; 1 = 52%, 2 = 55%, 3 = 56%, 4 = 58%).

TABLE 4.27

Influence of water content on the recoverable and non-recoverable work at 4·10 cm extension (Muller *et al.*, 1963)

% Water added to dough	Total work (erg/g)	(%)	Recoverable work (erg/g)	(%)	Non-recoverable work (erg/g)	(%)
52	26·6	100·0	14·6	100·0	12·0	100·0
55	18·6	70·0	8·0	54·8	10·6	88·2
56	16·7	62·8	6·7	45·9	10·0	83·2
58	12·4	46·7	5·3	36·3	7·1	59·2

an arbitrary selected extension of 3 cm (Tabel 4.28) the chlorinated cake flours showed a much greater degree of recoverable work than the other flours, the highest value being shown by the cake flour with a low protein content.

The time for which the dough sample was allowed to rest, following preparation and before testing, had a greater influence at small extensions (4–5 cm) on the recoverable work than on the non-recoverable work. Rest times up to 120 min were employed with doughs prepared from unbleached, untreated, bread flour with 14% moisture, to which 55% water was added. After a rest time of 45 min most of the stress introduced during moulding of the dough sample had relaxed. Similar studies with doughs containing starches (Heaps and Coppock, 1968) indicated that starch exerted an important effect, probably through its associating with the protein and lipid constituents of the flour.

TABLE 4.28

Influence of flour type on recoverable and non-recoverable work at 3 cm extension (Muller *et al.*, 1963)

Flour	Protein (%)	Total work/g (erg/g × 10³)	Recoverable work/g (erg/g × 10³)	Non-recoverable work (erg/g × 10³)
Biscuit flour	8·5	6·9	2·3	4·6
Low-protein cake flour (chlorinated)	8·0	24·6	19·4	5·2
High-protein cake flour (chlorinated)	11·75	45·3	38·9	6·4
Bread flour	12·0	14·7	7·8	6·9
Strong bread flour	12·5	14·4	7·8	6·6

The problem of rest time has been studied in detail by Hlynka and his associates (see, for example, Hlynka, 1964) using a method analogous to that employed for rubber and other elastomers (Andrews *et al.*, 1948). From kymograph chart recordings obtained during extension tests with the Brabender Extensograph the load supported by each dough sample at an arbitrary extension of 11 cm was derived. The particular extension value selected did not significantly influence the interpretation of the data. Doughs with a water content equivalent to 63% water absorbed, and 1% sodium chloride, which were moulded into test pieces immediately after they had been prepared, i.e. zero reaction time, showed a marked change in the calculated load with rest time. These changes reflected the rate at which stresses developed in the dough during moulding were dissipated subsequently. With both untreated doughs and doughs treated with potassium bromate the load decreased at almost identical rates with increasing rest time. For short rest times the two series of curves were very similar, but at longer rest times the load in bromated doughs decreased to a steady value whereas in untreated doughs the load continued to decrease almost linearly with increasing rest time.

The general form of the load–rest time curves for bromated doughs suggested that they followed an exponential decay law, although the decay did not continue to zero load. If the loads derived from the kymograph curves were corrected by subtracting the appropriate final steady values of the loads, then the logarithmic plots of the corrected load for each potassium bromate concentration and each reaction time against rest time was linear over rest times as large as 60 min to 180 min. Thus, the behavior followed an equation of the form

$$L_{11(t)} = L_{11(0)} \exp\left(-k_e t\right) \tag{4.49}$$

where $L_{11(t)}$ and $L_{11(0)}$ are the loads at any time t and zero time respectively, and k_e is the rate at which the load decreased. At a bromate concentration of 3 mg %, doughs which had reaction times of 0, 1, 2, 3, and 4 hours showed values of k_e equivalent to 0·053, 0·044, 0·042, 0·036, and 0·032 respectively. When the bromate concentration was increased to 5 mg % the values of k_e were 0·057, 0·037, 0·028, and 0·010 for reaction times of 0, 1, 2, and 3 h.

By analogy with rheological theory for rubber and other elastomers, Andrews et al. (1948) assumed that because dough is partially elastic it contains flexible long chain molecules, presumably of protein. A three-dimensional network structure develops as result of cross-linking between neighboring molecules. "The cross-links may be primary covalent bonds, but are more probably points of strong intermolecular or secondary valence forces between polar groups of adjacent molecules. Sections of the long molecules between cross-links assume randomly kinked or crumpled configurations. But the structure is dynamic so that the shape and degree of kinking in individual molecular segments changes readily. In rested dough the lengths of molecular segments between cross-links of the network are considered to be normally distributed about a 'most probable length,' i.e. the segments are randomly kinked and randomly oriented with respect to each other. A certain minimum number of polar groups on adjacent molecules are considered to be involved in intermolecular cross-linkages. These cross links are labile; cross links between adjacent molecules are not always in the same position, but the average number of bonds remains the same. This system is characterised by a certain equilibrium energy the magnitude of which determines the number of cross-links in the postulated network."

When rested dough is shaped the mean length of the molecular segments is increased by mechanical unkinking, and previously non-bonded polar groups in adjacent molecules are brought into juxtaposition. Intermolecular forces between these groups establish additional cross-links in the network. Internal stresses are thus set up in dough by working and a considerable force is required to stretch the dough. Work done on dough by shaping, together with the energy available from bond formation results in an increase in the energy of the system. When the thermal kinetic energy of the molecular segments is greater than the polar bond energy, then rupture of some of the cross links formed during shaping occurs. Bond rupture continues until the average kinetic energy of the segments is of the order of the bond energy. An equilibrium will then exist between the number of bonds breaking and reforming. Thus, during a rest period following shaping, the thermal kinetic motions of the molecular segments causes the network to return to the equilibrium state. As the internal stresses relax thus, the load required to stretch a dough will decrease. The rate of relaxation is indicative of the rate at which the postulated network structure returns from a partly oriented, highly cross-linked, arrangement induced by working dough to the equilibrium arrangement. The rate of relaxation of internal stresses is determined by the extent of cross-linking of the network structure. Changes in k_e thus indicate changes in the degree of cross-linking. The bromate reaction progressively introduces relatively non-labile cross-links into the network. "Bromate . . . creates potentially reactive groups in certain molecules of the network. These groups interact to form cross-links between adjacent molecules only when brought together when the dough is worked" (Hlynka and Matsuo, 1960; Bloksma and Hlynka, 1960).

Stress relaxation studies on doughs stretched to a preset distance in a relaxometer (Cunningham et al., 1953) within the temperature range 13–35°C showed that the shape of the relaxation curve on a log time plot did not change with temperature. Instead, the curves shifted laterally along the log time axis (Cunningham and Hlynka, 1954). From these curves master curves were drawn showing the distribution of relaxation times. The rest time exerted its main influence on the long relaxation times, the effect decreasing with increasing rest time. The short relaxation times were influenced by water content when within the range 74·4–86·0% (dry basis). Arrhenius plots gave activation energies of 11–24 kcal, and indicated that the precise value depended on both temperature and rest time.

Bloksma (1962) calculated the rate of extension in fermenting dough resting in an open rectangular tin due to the excess pressure developed in the gas cells. The extension (e_x) is given by

$$e_x \equiv \ln\left(\frac{h_x}{h_{x(i)}}\right) = \ln\left(\frac{v_g}{v_{g(i)}}\right) \qquad (4.50)$$

where h_x and $v_{g(i)}$ are the initial height and volume of the dough sample respectively in the tin, v_g is the volume at any time t, and $h_{x(i)}$ is the height at the start of the last proof under bakery conditions.

Now

$$v_g = v_{g(i)} + k_d t$$

where k_d is the rate constant for the increase in dough volume, so that Eq. (4.50) becomes

$$e_x = \ln\left(1 + \frac{k_d t}{v_{g(i)}}\right).$$

The rate of extension is then

$$\frac{de_x}{dt} = \frac{1}{(v_{g(i)}/k_d) + t}. \tag{4.51}$$

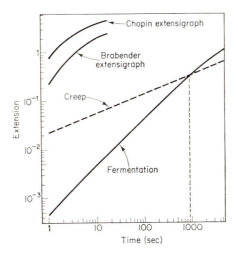

FIG. 4.57. Extension e_x of a fermenting dough during last proof under normal bakery conditions according to Eq. (4.50) (Bloksma, 1957). For comparison the straight line with a slope of 0·4 through the point $t = 900$ sec, $e_x = 0·35$, represents the creep experiment which has been chosen as a model for the extension during fermentation.

Figure 4.57 shows the double logarithmic plot of extension against fermentation time based upon Eq. (4.50). Using the cone-plate viscometer to determine creep compliance–time behavior, or using the Brabender or Chopin Extensograph, it is not possible to select a stress such that the extension curve coincides with the theoretical curve. Taking the extension in the creep compliance test after 15 min application of a constant stress of ~ 1700 dyne/cm^2 as 0·35, which is equal to the extension after 15 min fermentation, the slope of the extension–time curve was 0·4. Extension–time relationships for tests with the Brabender Extensograph were calculated from

$$e_x \equiv \ln\left(\frac{l_1}{l_2}\right) = \ln\left[\frac{l_2 - l_3 + 2\,(l_3^2/4 + v^2 t^2)^{\frac{1}{2}}}{l_2}\right] \qquad (4.52)$$

with $l_2/l_3 = 1$, where l_1 is the length of the cord to which the hook is attached, l_2 ($3\cdot7\,\text{cm} < l_2 < 8\cdot0\,\text{cm}$) is the distance between the two fixed points to which the cord is attached, l_3 ($= 3\cdot7\,\text{cm}$) is the distance between the two small pulleys which support the cord, and v ($= 1\cdot40\,\text{cm/sec}$) is the speed of descent of the hook. The extension at the top of the dough bubble in a Chopin Extensograph was calculated (Bloksma, 1957) by

$$e_x = \ln\,(\Delta_0/\Delta) = 2\ln\,(r_b^2 + h_f^2)/r_b^2 \qquad (4.53)$$

where Δ_0 and Δ are the thicknesses of the original sheet of dough and the wall of the dough bubble respectively, h_f is the height of the dough bubble, and r_b is the constant radius of the base of the bubble.

If Eqs (4.50), (4.52) and (4.53) are essentially correct then Fig. 4.57 indicates that the maximum rates of extension in the Chopin and Brabender Extensographs are about 1000 times higher than in fermenting dough. Furthermore, extension data derived from creep compliance–time tests are nearer to practical extension conditions than data derived by the Extensographs. This is particularly true for longer extension times.

8. Meat

Meat can be regarded as consisting of a large number of bundles of muscle fibres which lie approximately parallel to one another. The bundles are held together by a matrix of connective tissue. Tough meat contains a high proportion of connective tissue which behaves like a rubbery gel, while tender meat has far less, or weaker, connective tissue so that the meat is broken down more easily when chewed (Sale, 1960).

There have been few attempts to study the consistency of meat quantitatively. The main emphasis has been on simulating the principle forces to which meat is subjected when chewed, viz. shearing, penetration, biting, mincing, etc. (Szczesniak and Torgeson, 1965), the first type of test listed appearing to be the one which is used most often.

On a more quantitative level preliminary tests have been made on the dynamic rheological properties of meat (Lawrie, 1968) by exposing muscle cores of standard size to vibrational frequencies of 230 Hz. The muscle core was contained in a metal cylinder, and vibrations were imposed by means of a flat metal plate which was attached to the shaft of a vibrator powered by an oscillator. The amplitude of vibration in the absence or presence of a meat sample was determined by photographing the displacement of a thin rod which was attached to the shaft of the vibrator and normal to the axis of vibration. Muscle tissue was found to have the same order of resilience as a gelatin gel containing 30% water, which is about 25% larger than for the relatively tough musculature from the shin area.

Extension of psoas muscle (tenderloin) was used originally (Bate-Smith, 1939) to study the changes which occur during rigor mortis, following slaughter. The ends of muscle strips, which were several inches long, were wrapped in adhesive tape to prevent tearing and clips were then attached. One of the clips was permanently fixed, while the other was attached to one arm of a balance. By placing a weight on the pan attached to the other arm a stretching load was applied to the muscle strip. The change in length of the strip was recorded by a pen writing on a smoked drum, the arrangement giving a seven-fold magnification. At a later date (Bate-Smith and Bendall, 1949; Bendall, 1951; Bendall and Davey, 1957) the procedure was modified so that load application, and its removal, were carried out by an electrically operated arm. In tests on the shortening of rabbit muscle during rigor mortis loads of 50–100 g/cm^2 were applied intermittently by periodic removal for 15 sec, and the change in length of the muscle was recorded on a smoked drum of 6 in diameter which revolved once in 12 h. The work (W_f) done during shortening of the muscle strips was defined by

$$W_f \, (\text{g cm/g}) = \frac{m \times h_m}{w_m} \qquad (4.54)$$

where m is the load (g), h_m is the distance (cm) through which the weight was lifted by the sample, and w_m was the weight of the muscle strip.

During rigor mortis at 37°C the muscle shortened at a faster rate with time than at 17°C (Bendall, 1951). The chemical process most closely related to rigor mortis is the disappearance of adenosine triphosphate and creatine phosphate (Bate-Smith, 1939), and at 37°C the rate of disappearance was greater than at 17°C. Shortening of the muscle strip at 37°C began when the adenosine triphosphate (ATP) had fallen below 85% of the value before slaughter of the animal in one group of samples with a final pH of 6·5, and when it had fallen below 80% of the original value in another group of samples with a final pH of 6·0. At 17°C shortening began when the ATP fell below 85% of the original value in one group of samples with a final pH of 6·6, but when the final pH was 6·1 shortening did not commence until the ATP content fell to 60% of its original value.

The large influence of final pH, as well as temperature, is seen in the observations that at 37°C the average rate of shortening was 1·45 ± 0·2% of the total/min at final pH = 5·8, 1·3%/min at final pH = 6·2, and 1·02 ± 0·06%/min at final pH = 6·35–6·80. At 17°C the average rate was 0·72%/min at final pH = 6·0, and 0·50%/min at final pH = 6·50.

Similar observations have been made on strips of the large breast muscle (Pectoral major) of chicken (De Fremery and Pool, 1960). Strips of approximately 1 cm² cross-section and 4–6 cm length were subjected to 4 cycles of loading and unloading during a one hour period, the load being applied or removed for 7·5 min at a time by means of a time operated mechanism. When about 66% of the ATP content had disappeared the strips began to lose their extensibility, and shortening continued for as long as 8 h in some cases. In the pre-rigor condition the modulus of elasticity was about 1×10^3–2×10^3 g/cm², and this value increased 5–10 fold when rigor was fully established.

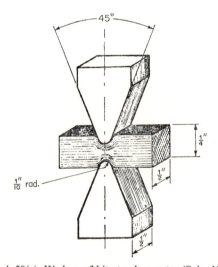

FIG. 4.58(a). Wedges of bite tenderometer (Sale, 1960).

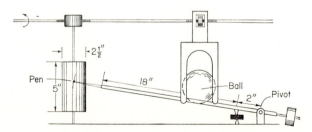

FIG. 4.58(b). Bite tenderometer (Sale, 1960).

The biting action involved during mastication has been simulated by means of a pair of wedges with blunt ends (Volodkevich, 1938; Winkler, 1939; Sale, 1960). In the original form of the instrument the lower wedge was fixed on a frame while the upper one was able to move in a vertical direction by means of two levers. A sample of meat was placed between the two wedges, and its resistance to the force applied by downward motion of the upper wedge was registered on a revolving drum as a function of the distance separating the two wedges. Forces of about 10–120 kg were used. Volod-kevich (1938) found that replacing the wedges by dentures resulted in less satisfactory data. Rounding the edges of the wedges to a fixed radius of curvature (Sale 1960) provided wedges which were easily reproduced and which preserved their shape. Rectangular-sectioned samples of meat were placed between the wedges (Fig. 4.58a), which were fitted with plates on either side to prevent the sample from being squeezed out from between the wedges. A force was applied to the meat by moving a steel ball (2 in or 3 in diameter) along a beam to which one of the wedges was attached (Fig. 4.58b). In this way the force on the wedges was increased until eventually they made contact. Penetration of the sample by the wedges was recorded on a drum chart driven in step with the steel ball, so that the trace represented a force-penetration record. Figure 4.59 shows some typical traces. Meat samples were usually cut so that the wedges bit across the muscle fibres, since this gave the best correlation between toughness and the force required to "bite" through the samples. Curves A and B are for tender meat, C and D are for "rubbery" samples which were difficult to break down, and curves E, F, and G represent tough meat which required much force to chew them.

The most popular instrument for cutting meat in simple shear is the Warner–Bratzler equipment (Warner, 1927; Bratzler, 1932, 1933, 1949). It comprises a blade of about 1 mm thickness with a hole large enough to hold a cylindrical ($\frac{1}{2}$–$1\frac{1}{8}$ in diameter) sample of meat (Fig. 4.60), which is sampled

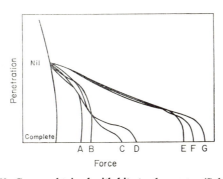

FIG. 4.59. Curves obtained with bite tenderometer (Sale, 1960).

from the test material with an instrument resembling a cork borer. Following insertion of the sample in the blade, the latter is introduced into a narrow gap between two shear bars. A force is now applied to the blade or the shear bars, and this force, which is recorded on a dynamometer, builds up to a maximum value when the sample fractures. The maximum value of the recorded force is usually taken as a measure of the toughness of the meat, but the slope of the force-time curve may be more informative (Hurwicz and Tischer, 1954). A large number of published papers have reported correlations between data obtained with the Warner–Bratzler apparatus and panel assessments of meat tenderness. The correlations vary over a wide range from poor to very good, but the majority of the studies appear to indicate a significant correlation.

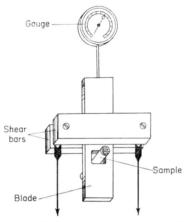

Fig. 4.60. Basis of Warner-Bratzler apparatus (Sale, 1960).

The L.E.E.–Kramer Shear press (Kramer *et al.*, 1951; Kramer and Backinger, 1959; Kramer and Twigg, 1959) appears to provide highly significant correlations with tenderness assessment by panels. In its most recent form the press consists of a test cell, a hydraulic drive system, and a dynamometer. The cell has 10 bars (0·124 in thick) which are 0·126 in apart. These bars pass through a sample holding box which has an equivalent number of slots in its base. A sample is placed across the slots, and the bars are driven at a preselected rate through the sample by a piston and simultaneously extrude parts of it. The force involved in shearing the sample is measured by compression of a proving ring dynamometer functioning within the range 100–5000 lb, and it is registered by a calibrated scale. Instead of recording the maximum force a recorder may be linked up to the shear press in order to obtain a force–time curve. Cutting or compression cells are provided for use in addition to the shear cells.

Both the Warner–Bratzler and L.E.E.–Kramer shear presses suffer from the disadvantage that the results depend on the sharpness of the shear blades or shear bars (Sale, 1960). Because of this, and other sources of error, Miyoda and Tappell (1956) proposed two other equipments for testing meat, the Christel texturemeter and the Hamilton–Beach food grinder. Using the texturemeter, which was developed originally for measuring the hardness of peas, 25 rods of $\frac{3}{16}$ in diameter are pushed into a cylindrical sample (4·8 cm diameter) contained in a box. In order to reduce the shear rate to 0·32 mm/sec, and to obtain a steady rate of shear, the texturemeter was attached to an electric motor and reduction gears to provide a 0·75 rev/min drive. As soon as the rods touch the sample a pressure gauge begins to record the force. Readings of the gauge are recorded at 2 sec intervals as a result of the drive mechanism activating a bell at the appropriate times. When the rods have passed through the sample, the drive mechanism is reversed and the rods are raised in readiness for the next test. Force (lb)–distance of penetration curves are then computed, and the area under the curve is measured. The penetration distances are derived from the time intervals at which readings are taken and the speed of the drive unit. A commercially available food grinder was used for the other test on the strength of the argument that both chewing and grinding involve squeezing, tearing, and cutting etc. The analogy can be extended by using "bite-size" pieces of meat and grinding them through a plate (36 holes, 5 mm diameter) which produces "chewed-size" pieces. The electric current consumption by the grinder is then proportional to the toughness of the meat, and the total energy used in grinding the meat is represented by the area under the curve of power consumption (watts) as a function of time. Coefficients of variation of about 2% were obtained with the Christel texturemeter and the Hamilton–Beach food grinder, whereas the Warner–Bratzler shear press gave coefficients of variation equal to 4·79, 7·41, and 9·0%. On the strength of these observations it was concluded that the texturemeter and food grinder are more accurate than the Warner–Bratzler shear press. The first two mentioned instruments have the additional advantage that the meat does not have to be aligned with its grain perpendicular to the shearing head.

The consistency of meat has been investigated by various types of penetrometer. Using a penetrometer (Tressler et al., 1932; Tressler and Murray, 1932) fitted with a needle of 1·375 in length and 0·15 in diameter, and with a rounded tip of radius 0·07 in, firmness was defined as the distance (mm) of penetration during 15 sec when the penetrating device was loaded with a 225 g weight. The meat sample was contained in a 1 in rectangular box having eight 0·375 in holes in its upper surface through which the needle could be inserted. Although this penetrometer gave more uniform results than the Warner–Bratzler shear press, the data did not correlate well with panel assessments of tenderness, which is not surprising. In another study (Pilkington *et al.*, 1961) a ball penetrometer was employed to measure firmness, and a Warner–Bratzler shear press and a panel to evaluate tenderness. A low, but significant, positive correlation was obtained between firmness and tenderness. Firmness was highly dependent on the fat content of the meat.

A compact, inexpensive, shear apparatus (Fig. 4.61) based upon the Warner–Bratzler principle has been proposed by Voisey and Hansen (1967). A 0·02 h.p. 1725 rev/min synchronous motor (A) drives two threaded shafts (C) by gears (M). A bar contains two nuts (B) which engage the threads on

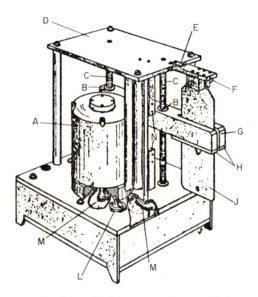

FIG. 4.61. Shear apparatus for evaluating meat tenderness (Voisey and Hansen, 1967). A, synchronous motor; B, nuts; C, threaded shaft; D, top plate of main frame; E, strain gages; F, pin supporting blade at end of beam; G, slot in bar; H, stainless-steel inserts; J, shearing blade; L, motor control switch; M, gears driving shafts; N, micro-switches.

the shaft. When the motor operates the bar moves at a constant speed of 9·0 in/min. A slot (G) at one end of the bar passes over the shearing blade (J) which has a triangular hole to hold the sample. Two stainless steel inserts (H) of thickness 0·25 in, which are attached to the base of the bar, reduce the width of the slot to clear the blade by exactly 0·005 in. A switch (L) controls and reverses the motor, so that the slot can be raised or lowered over the blade. The blade is supported by a cantilever beam integral with, and projecting from, the upper plate of the main frame (D). It can be removed from the apparatus by withdrawing a pin (F). Microswitches (N), at either end of the travel of the bar, stop the motor at these points until it is reversed. The shearing force on the blade is detected by four strain gages (E) which are mounted in pairs on either side of the cantilever beam, and which are connected electrically to form a Wheatstone bridge. The bridge circuit is connected to an amplifier which converts the output from the strain gages to a millivolt signal for recording which is proportional to the force on the blade. Calibration of the equipment is done by hanging known weights on the lower end of the blade.

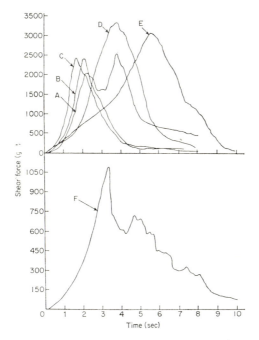

FIG. 4.62. Typical records of shear-force for various meats using a potentiometer-type recorder at a sensitivity of 5000 g. (Voisey and Hansen, 1967). A, goose breast, 1·0-cm-diameter core, showing effect of inhomogeneous sample; B, goose breast, 1·0-cm-diameter core; C, goose breast, 1·0-cm square of slices; D, beef, 1·5-cm-diameter core; E, pork, 2·5-cm-diameter core; F, wiener (sensitivity 1500 g).

Typical shear force–time curves obtained with various types of meat are shown in Fig. 4.52. They included goose breast, beef, pork, and wieners. The traces show the influence of sample inhomogeneity, particularly for the goose breast sample A and for the wieners. Both of these meats, and beef and pork in addition, show large areas under the curves, i.e. a large amount of energy is adsorbed in shearing the samples. The area under the curves can be determined accurately using a planimeter, or by electronic integration of the millivolt signal. By analogy with studies on Cottage cheese using this instrument the sensitivity can be improved by using three blades instead of one.

The behavior of meat in compression can be studied by raising a platform, on which the sample rests, against a stationary body (Voisey *et al.*, 1967). The mobile platform (F) is suspended between the top and base plates of the instrument (Fig. 4.63) by four 0·375 in diameter screws. It travels along the shafts (J) on linear-recirculating ball-sleeves (K). The four screws (I) drive the mobile plate by nuts (E) made of phosphor bronze. A 0·125 h.p. 110-volt motor (B) below the base drives the four screws by sprockets (C) and a chain. The motor can be rotated in either direction to raise or lower

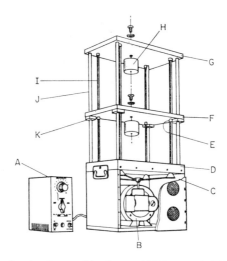

Fig. 4.63. Compression–tension machine for meat (Voisey *et al.*, 1967). A, Motor control; B, motor; C, sprockets; D, base plate; E, nut; F, crosshead; G, top plate; H, compression boss (shown in two of four possible locations); I, threaded shaft; J, shaft; K, recirculating ball sleeve.

the mobile plate, and can be braked to a stop by the electronic control (A). Its speed can be varied over the range 1·8–63 rev/min, and this produces a speed range of 0·077–3·37 in/min for the mobile plate. The force involved in compression of the sample against H is registered by a strain gage transducer which has been calibrated against known weights. Deformation of the sample is calculated from the time of compression and the speed of the mobile plate. An amplifier converts the transducer output into a form suitable for registering on a high frequency recorder. Cyclic deformation of the sample at a selected rate can be achieved by using a timer to operate the motor control. When the mobile plate moves at a speed below 0·30 in/min there may be a difference of 5% between the up and down speeds, and this is overcome by using an increased ratio between the motor and four screw sprockets, so that the motor speed is higher.

The apparatus can be adapted to shear meat samples as in the L.E.E.-Kramer and Warner–Bratzler shear tests by introducing suitable sample holders and transducers.

Proctor *et al.* (1955, 1956) attempted to simulate chewing of food in the mouth by designing a tenderometer which employed human dentures, and produced a frequency of chewing and simultaneous grinding and crushing as in normal masticatory processes. The forces involved were registered by strain gages. Figure 4.64 shows a block diagram of the equipment. The sample is placed in the masticator which has three basic forms of motion, viz. opening and closing or vertical motion, lateral or sideways motion, and forward motion. A Hanan model H articulator (Fig. 4.65) was used for this purpose. It consists of plastic dentures (a) which are fastened securely to the articulator, which is fitted with simulated cheeks, lips, and tongue built up from resilient plastic material to keep the food between the teeth during the test. The articulator is set in motion by a special transmission device connected with a geared-down electric motor. It is free to swivel at (c), lateral

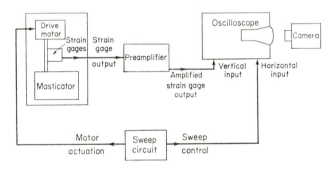

FIG. 4.64. Block diagram illustrating electrical and mechanical connections of components of strain-gage denture tenderometer (Procter *et al.*, 1955).

movements (~ 2 mm) are actuated by beam (d), and vertical movements are actuated by beam (e). Condylar adjustments (h), both left and right, left and right lateral adjustments (j), and left and right centric adjustments (k) permit variation of the motion. Special springs (f) and (g) prevent excessive damaging pressure building up on the teeth by taking up any excess forces on the beams. The teeth do not actually meet during chewing of the sample because there is an open vertical space of 1 mm or 22 mm between the upper and lower first molars when the jaws are in the "fully closed" and "fully open" positions respectively. This lack of contact between the teeth was justified on the grounds that it purposely avoids complete shearing of the sample on the first bite, and also because it has been pointed out (Yurkstas and Curby, 1953) that the occlusal surfaces of the teeth rarely make contact during mastification. When the masticator is in motion the upper jaw moves relative to the lower jaw instead of the reverse, but this was not believed to introduce any significant error.

A switch actuates the motor, which drives the masticator so as to produce a chewing frequency of 45 cycle/min. Changes in the resistance to chewing which are encountered by the masticator are registered as changes in strain on a mechanical drive mechanism connecting with the masticator. These strains are translated by electrical resistance type strain gages into changes in resistance, which, by means of the preamplifier, are amplified as changes of potential and fed into the vertical deflection circuit of a cathode ray oscilloscope with a sensitivity of 100 mV d.c. full scale deflection, which is

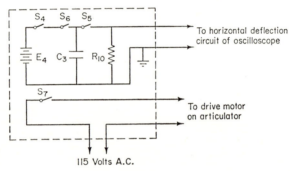

FIG. 4.66. Horizontal sweep circuit and motor power switch of masticator (Proctor *et al.*, 1955).

equivalent to a deflection factor of 25 peak-to-peak mV/in. The electron beam is driven vertically across the cathode ray tube as a result, and simultaneously an electronic sweep mechanism is actuated, and feeds a time signal into the horizontal input circuit of the oscilloscope to drive the electron beam horizontally across the cathode ray screen. Thus, the trace on the face of the cathode ray tube has strain and time as the vertical and horizontal coordinates respectively. A camera attached to the oscilloscope records the trace continuously during the test.

In order to correct for any changes in the sensitivity of the equipment during different series of tests several cycles of "blank" runs are made. This is done by checking the amplifier and adjusting for electrical balance, and adjusting the oscilloscope so that the spot on the tube face is on the left side of the screen. The shutter of the camera is then opened, and the mechanical movement is actuated by throwing a switch S_5–S_6–S_7 (Fig. 4.66),

Fig. 4.65. Details of articulator of tenderometer showing its mechanical adaptation for continuous chewing motions (Proctor *et al.*, 1955). *a*, plastic dentures; *b*, plastic material; *c*, swivel point; *d* and *e*, beams activating lateral and vertical movements, respectively, of articulator; *f* and *g*, springs to take up any excess force on beams; *h*, condylar adjustments; *j*, lateral adjustments, and *k*, centric adjustments to vary lateral motion of articulator; *l*, strain gages; *m*, incisal pin.

Facing p. 284

which at the same time actuates the horizontal sweep circuit so that the spot moves from left to right at a velocity depending on the choice of R_{10} and C_3 in the sweep circuit. All measurements are normalised to a standard blank run in accordance with

$$\text{(normalised deflection)} = \text{(observed deflection)} \times \left(\frac{\text{standard blank deflection}}{\text{observed blank deflection}}\right)$$

(4.55)

where "observed deflection" refers to the actual deflection during a test with a food sample. The normalised sample deflection minus the standard blank deflection is then the deflection produced by the sample. All measurements with food samples were made on the left side of the model jaw.

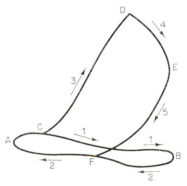

FIG. 4.67. Force–penetration diagram obtained with masticator by using strain gage output as vertical input and position indicator output as horizontal input to oscilloscope (Proctor *et al.*, 1956a). AB (1 and 2) = free run; C = point of initial contact with sample; 3 = "bite-down" to D = point of maximum deflection; 4 = "bite-down" to E = point of maximum penetration; 5 = "bite-up" to F = loss of contact with food.

By using a position transmitter, i.e. a potentiometer which has a 360° tapped linear resistance with double sliding contacts fixed 180° apart attached to the continuously rotating shaft, and replacing the plate and filament batteries of the preamplifier by the stabilised power supply of the oscilloscope, it was possible to obtain simultaneous records of the forces and penetration associated with mastication (Proctor *et al.*, 1956a). Individual traces of force and penetration outputs with respect to time were obtained using a dual channel amplifier, and using the two signals to actuate both horizontal and vertical deflections of the electron beam on the face of the oscilloscope cathode ray tube. In this way a continuous plot of force and penetration was derived during the chewing cycle. Figure 4.67 shows a typical photographic record for a complete cycle during a single bite. Tests were made with a wide range of food materials including apples, peas, potatoes, carrots, cheese, celery, hard boiled eggs, bread, and meat. The meat samples studied were raw steak and braised steak.

Proctor et al. (1956a) state that the force–penetration diagrams should be analysed in the following manner

"(1) Measurement should be made on the 'bite-down curve'. The angle of ascent to maximum force (or any unique shape or characteristic if this curve is non-linear), the maximum force, and the distance of penetration to which the maximum force appears should be measured. The angle of 'bite through' following maximum force to the point of maximum penetration (or any unique shape or characteristic of this curve, if non-linear) should be measured.

(2) The significance of the 'bite-up' curve (which represents the force exerted against the teeth when the mouth is opening) is possibly not as great as the 'bite-down' curve, and its variability is considerable because of the fact that the food has already been broken up. However, in certain applications, these characteristics should be observed. The area bounded by the two curves is an indication of the change which has occurred in the food.

(3) Measurements should be normalized to correspond to differences resulting from size of sample, and/or overall sensitivity".

For a crisp material such as celery an appreciable force is required initially to break through the food, but after this the force involved in completing the bite is much lower. The curve for celery, in the form of a 0·7 cm cube, (Fig. 4.68) therefore shows a steep rise in the force to a maximum, followed by a sharp decrease to the point of maximum penetration. With 1·0 cm cubes of raw steak the force rises much less steeply to a maximum, and the pattern of decrease follows a similar pattern in reverse. The area between the up and down parts of the curve is very much smaller than for celery. The latter material also shows a greater difference in the shape of the up and down parts of the curve. Broiled steak, in 1·5 cm cubes, showed a similar pattern to raw steak with a gradual rise in the force with penetration followed by a gradual decrease to maximum penetration, but the maximum force developed was lower than for raw steak.

No detailed studies appear to have been made by Proctor et al. on the relationship between tenderometer data and panel assessment of texture. In the one such study reported (Proctor et al., 1956b) a high correlation coefficient of 0·96 was found for peas, the panel using an 11-point scale ranging from a score of 10 for "soft like an olive" to 0 for "hard like a peanut."

Friedman et al. (1963) experienced difficulty in keeping samples between the teeth of the denture tenderometer during the chewing process, and they proposed that the dentures should be replaced by a platform, on which a cup was mounted to contain the sample, and a plunger. Cups and plungers of various sizes were constructed so as to suit a wide range of food materials. In addition, location of the sensing element on the activating arm of the upper jaw of the denture tenderometer disturbed the force–penetration curve due to inertia of the arm movement. The strain gage unit was therefore re-

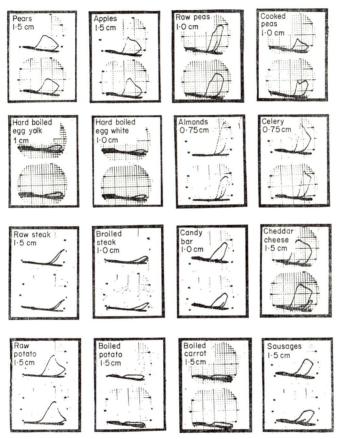

FIG. 4.68. Force–penetration oscillograms of representative foods. The measurements in cm indicate the size of the sample used in each test (Proctor *et al.*, 1955).

located on the stationary arm of the sample plate in the modified version. This resulted in clearer recorded signals and improved reproducibility of results. Two sets of tooth gears were also introduced so that four rates of chewing were possible instead of only one. It was observed that a speed of 42 bites/min gave reproducible results. The force–distance records correlated very well with sensory evaluation by panels of a wide range of attributes associated with texture. These correlations, and their significance, will be discussed in some detail in the final chapter. Estimates of the hardness, cohesiveness, elasticity, and chewiness of meat derived for modified texturometer curves showed highly significant correlations with panel evaluation of these attributes (Szczesniak *et al.*, 1963), the precise value of each correlation depending on the type of meat.

The technique of recording food crushing sounds, which was mentioned in Section 6 of this chapter, has been used (Drake, 1965) to evaluate meat texture. Sensory evaluations of the tenderness of four samples of tough meat and four samples of tender meat correlated significantly with the average sound amplitudes recorded at 1250 Hz and at 1600 Hz. The statistical level of significance for each type of meat was stated to be 0·01.

9. HYDROCOLLOIDS

A complete characterisation of the texture of hydrocolloid gels involves the analysis of several characteristics (Kramer and Hawbecker, 1966). These include deformation, since a gel is by definition a material which has structure and elasticity so that it can withstand a certain amount of change in its shape before it breaks, gel strength, uniformity of structure, since gel rupture is essentially a surface phenomenon and provides little information about bulk structure, and adhesiveness. Gel strength is the characteristic which most of the available instruments measure.

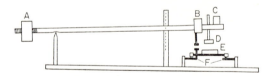

Fig. 4.69. Static and impact loading beam for pectin gels (Watson, 1966). A, counterweight; B, LVDT; C, loading platform; D, loading plunger; E, specimen; F, cantilever beams.

The rheological properties of viscoelastic pectin gels have been studied (Watson, 1966) by measuring the stress-strain behavior with three different types of loading—impact, compression, and static. Static and impact loads were applied by means of a 2 ft balance arm mounted on knife edges (Fig. 4.69). A weight was placed on a platform at the free end of the beam, which was initially held at a fixed distance above the sample by means of a pin. When the pin was removed the beam moved downwards at an average velocity of 155 in/min, and the plunger head (1 in or 1½ in diameter) struck the sample. At impact the terminal velocity was 300 in/min, as estimated from oscilloscope records. The stress exerted on the sample was measured by an assembly of 4 cantilever beams, and the deflection of each beam was sensed by two strain gages, amplified, and then recorded using a 2-channel dual beam oscilloscope with a camera or an X–Y recorder. Compression loading was done by a Bellows hydrocheck compression unit, the force being applied to the gel through a circular plunger (1 in or $1\frac{1}{2}$ in diameter).

Typical stress–strain curves at various rates of strain for standard pectin gels, at pH 2·6, containing 65% sucrose are shown in Fig. 4.70, while Figs 4.71 and 4.72 show the influence of pH and sucrose content, respectively. Over the range of experimental conditions utilised the stress increased with

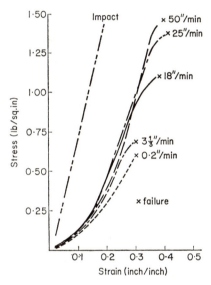

FIG. 4.70. Stress–strain relationships at various rates of strain. Each curve represents the average values for five tests on 65% sucrose gel pH = 2·6. Standard pectin (Watson, 1966).

increased rate of loading, decreasing pH, and increasing sucrose concentration at a given strain. The stress/strain ratios provided only apparent values of E, since no allowance was made for stress concentration under the plunger. Figure 4.73 shows how the apparent E at failure was influenced by pectin content for several rates of loading. Plots of creep compliance against time gave the characteristic form for viscoelastic behavior (Fig. 4.74) until failure occurred, when the compliance suddenly increased at a more rapid rate.

In general it was found that, although the pectin gels were always produced under the same standardised conditions, the results showed considerable variation. This was attributed to the sag of the sample and its time

L

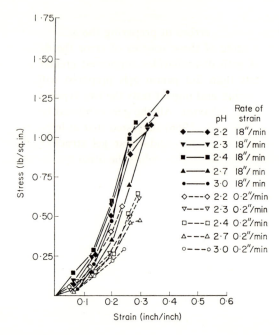

FIG. 4.71. Stress–strain relationships for 65%–sucrose gels. Standard quantities of slow set pectin. (Watson, 1966).

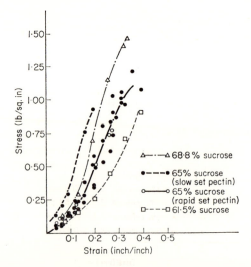

FIG. 4.72. Stress–strain relationships for sucrose gels at a rate of strain of 18 in/min, pH = 2·6. Standard quantities of slow-set pectin except where otherwise noted (Watson, 1966).

dependence, work hardening of the sample before testing, the friction and inertia in the equipment, errors in preparing the gels, and errors in zeroing the equipment. In spite of these sources of error the data clearly indicated that gels prepared with disaccharides supported greater stresses for longer times in creep tests than did pectin gels prepared with monosaccharides, whereas in compression and impact tests the two types of gels showed much smaller differences in behavior. All the gels exhibited brittle failure. On the basis of these data Watson (1966) suggested that at least two types of bonding are involved in the pectin–acid–sugar gel structure. One type is very weak and reforms after disruption, while the other type is very much stronger

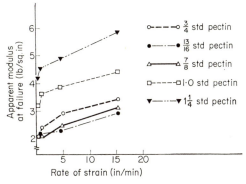

FIG. 4.73. Apparent modulus at failure for 65%-sucrose gels, pH = 2·6 (Watson, 1966).

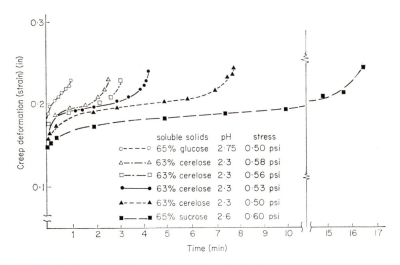

FIG. 4.74. Typical creep deformation (to failure) of gels containing standard quantities of slow set pectin (Watson, 1966).

and confers the elastic properties observed at shearing stresses above 0.25 lb/in^2. This arbitrary division of bonds into weak and strong categories resembles the interpretation proposed for the rheological behavior of margarine (Section 1 of this chapter) in terms of primary and secondary bonds.

Harvey (1960) also studied the viscoelastic properties of pectin gels, but he employed the F.I.R.A. jelly tester. This instrument is described later in the chapter. The main point of interest in these studies was the effect of neutral salts at constant pH, and at variable acid pH, in addition to the effect of sugar. In the presence of buffer solutions from hydrochloric acid, or the chlorides of sodium, potassium, magnesium, or calcium, the stress–strain behavior of the pectin gels did not alter. At variable low pH using different calcium and magnesium ion concentrations the strain for a given stress decreased with increasing pH. Dissociation of the carboxylic acid groups of the pectin molecules is suppressed at pH $\sim$ 2, and lower. Metal

TABLE 4.29

Rigidity of 5·5% gelatin gels as a function of maximum strain
(Saunders and Ward, 1954)

Modulus of rigidity (dyne/cm$^2 \times 10^{-4}$)	6·56	6·58	6·61	6·64	6·66	6·69	6·75	6·80	6·84
Maximum strain	0·066	0·084	0·103	0·122	0·162	0·182	0·202	0·220	0·242

ions can only attach themselves to ionised carboxylic acid groups, and when such sites are absent no metal ion attachment can take place and the rheological characteristics of the pectin gels are not affected. Sugar was regarded as having a two-fold action, but no experiments were made, e.g. such as those of Watson (1966), to determine whether different sugars had different effects on gel formation and gel properties. The sugar had a plasticising effect, which resulted in greater separation of the cross-linked network of elongated, chain-like, pectin molecules, and also it reduced the solubility of the pectin. An observation similar to the second effect has been reported for the influence of sucrose on the solubility of the amylose fraction of starch in water (Campbell and Briant, 1957).

Gelatin–water gels have been studied by several methods including the Saunders and Ward (1954, 1958) tube for determining modulus of rigidity, their load-extension behavior (Ward and Cobbett, 1968), and their behavior in a coaxial cylinder viscometer (Roscoe, 1965). The modulus of rigidity of a 5·5% gelatin gel was found (Table 4·29) to increase slightly with increasing maximum strain (Saunders and Ward, 1964), so that it is not truly Hookean. Thus, complete characterisation of the gel requires that the rigidity be measured over a wide range of shear strains, but in practice when only comparing different gels it is sufficient to determine their rigidities at a suitable maximum strain. A maximum strain of 0·15 is quite suitable, and according to Table 4·29 the error should not be greater than about 1% when comparing different samples.

When gelatin samples were demineralised prior to preparing gels it was found that the modulus of rigidity of gels containing not less than about 8% of the treated gelatin was always higher than the rigidity of gels containing an equivalent concentration of untreated gelatin (Table 4.30). At gelatin

TABLE 4.30

Concentration dependence of the rigidity of demineralised and untreated gelatins (Saunders and Ward, 1954)

Gelatin as supplied (pH 6·3)		Demineralised gelatin (pH 5·10)	
concentration (g/100 ml)	rigidity (dyne/cm$^2 \times 10^{-4}$)	concentration (g/100 ml)	rigidity (dyne/cm$^2 \times 10^{-4}$)
20·2	49·5	22·4	61·5
16·3	35·2	18·0	42·8
12·15	22·4	12·7	24·5
8·12	12·05	9·2	14·7
5·09	5·32	5·14	5·25
1·98	0·64	1·95	0·45
1·59	0·350	1·63	0·258
1·42	0·250	1·46	0·165
1·17	0·125	—	—

concentrations below about 8% the untreated gelatin gave higher rigidity values, but now the differences between the two series of gels were small. Demineralisation alters the ionic strength, so that a change in rigidity would be anticipated. To check this point different concentrations of sodium chloride were added to a gel prepared with demineralised gelatin. A gelatin concentration of 18·4% was used. The salt reduced the rigidity of the gel to the expected level (Table 4.31). With gels containing low concentrations of gelatin adding salt increases the rigidity, but only to the extent of one third of the difference between demineralised and untreated gelatins.

The breaking load for dilute gelatin gels subjected to high strains at a constant rate of extension depends mainly on the elastic modulus (Ward and Cobbett, 1968). Dumb-bell shaped test pieces of 0·324 cm thickness, with a parallel sided central section which was 6·35 cm long and 1·28 cm wide

(Fig. 4.75), were supported on a mercury surface and extended at a constant rate of 0.536 cm/sec in a way which bears some resemblance to the technique employed for extension tests on dough cylinders (Section 7, of this chapter). To facilitate clamping of the ends of the test pieces 2.5 cm square slips of wood were cast in the ends. Extension of the test strip was effected by a constant speed motor which was connected by a cord to one end of the strip. The loads were derived from the optically magnified deflexion of a stiff cantilever which was connected to the other end of the strip.

TABLE 4.31

Influence of salt content on the rigidity of an 18.4% demineralised gelatin gel (Saunders and Ward, 1954)

Rigidity (dyne/cm$^2 \times 10^{-4}$)	52.2	49.2	47.2	43.1
% salt	0	2.7	8.0	21

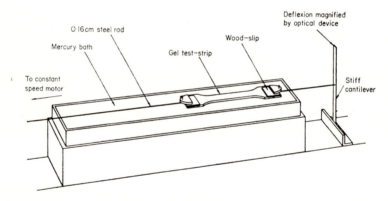

FIG. 4.75. Apparatus for linear extension of gelatin gels (Ward and Cobbett, 1968).

Most of the load during extension–overall extension plot was linear, and the Young's modulus E was calculated from the slope. Deviation of the plot from linearity near the point of rupture slightly alters the distribution of strain between the parallel sided region of the test piece and its ends, but it does not influence the results significantly. Almost all the deformation was elastic. Typical load–extension curves for gels prepared from three different gelatins are shown in Fig. 4.76. The load at rupture is related to the Bloom jelly strength (Fig. 4.77). Bloom jelly strength is determined arbitrarily by allowing gelatin to set in a wide necked bottle for 17 h at 10°C. A plunger (0.5 in diameter) is then forced to a depth of 4 mm into the gel by loading at the rate of 40 g/sec. The load required to give the necessary depth of penetration is quoted as the Bloom jelly strength. As the latter is related linearly to the elastic modulus, the load at rupture must also be related to the elastic modulus. On the basis of much published evidence (Herrmann *et al.*, 1930; Herrmann and Gerngross, 1932; Ferry, 1948; Boedtker and Doty, 1954; Hastewell and Roscoe, 1956; Saunders and Ward, 1958; Jopling,

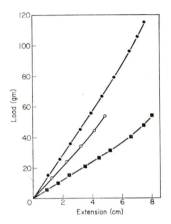

FIG. 4.76 Load/extension plots for three gelatin gels (Ward and Cobbett, 1968). o—o 85, acid-pigskin gelatin; •—• 130, limed-hide gelatin; ■—■ 148, limed-hide gelatin.

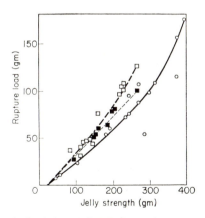

FIG. 4.77. Rupture load of gelatin gels in relation to jelly strength (Ward and Cobbett, 1968). o—o A, 10% gelatin; □—□ B, 6⅔% gelatin; ■—■ C, 6⅔% gelatin (η constant).

1958) gelatin gels are regarded as three dimensional networks of polypeptide chains which are held together by secondary attraction forces at certain widely separated points along each molecule, the free regions exhibiting rubber-like elasticity. It follows from the relationship between load at rupture and elastic modulus that the former is related to the concentration of inter-chain linkages. Increasing the gelatin concentration should alter the concentration of these linkages, and hence change the rupture load. This was confirmed experimentally, the rupture load for gels prepared from two types of gelatin increasing with increasing gelatin concentration up to 13·5% gelatin in one series of gels, and up to 12·5% gelatin in the other series (Table 4.32). However, the extension at rupture remained approximately constant. Gels prepared from gelatin fractions of different molecular weights showed an increase in both the rupture load and extension at break with increas-

TABLE 4.32

Influence of gelatin concentration on the rheological properties of gelatin gels (Ward and Cobbett, 1968)

Gelatin	concentration (%)	Solution viscosity (cp)	Bloom jelly strength (g)	Slope of load-extension plot (g/cm)	Load at rupture (g)	Extension at rupture (cm)
	5·0	2·27	71	1·78	18·1	7·27
	6·67	3·31	113	3·87	31·7	6·92
Acid-	8·5	4·87	177	6·12	50·9	7·13
pigskin	10·0	6·71	235	8·95	71·8	7·38
No. 215	11·5	8·80	308	11·10	86·3	7·15
	12·5	10·60	352	12·67	101·1	7·24
	13·5	12·84	405	14·7	111·6	7·10
	6·67	5·27	154	5·35	54·1	8·42
Limed-hide	10·0	12·12	311	11·40	108·1	8·74
No. 147	12·5	20·8	458	16·0	147·9	8·48

ing molecular weight. All the gels contained 6·67% gelatin. Ward and Cobbett (1968) point out that, nevertheless, the mean molecular weight of these gelatin fractions does not adequately define their molecular characteristics since a small proportion of very high molecular weight material would exert a disproportionately large influence on the extension at rupture. A "rigidity factor" is also involved (Stainsby et al., 1953; Saunders and Ward, 1955), which relates to the change in the maximum rigidity when the gelatin is degraded at elevated temperature, or by high pH. Degradation of gelatin in neutral or alkaline conditions not only reduces the average molecular weight but also the "rigidity factor", but in acid pH the latter is not significantly affected (Saunders and Ward, 1957).

The temperature dependence of the rigidity of gelatin gels has been studied by subjecting the gels to a constant strain and measuring the change in stress as the temperature changes (Jopling, 1958; Roscoe, 1965); also, by following the change in strain with temperature at constant stress (Roscoe, 1965). As Roscoe (1965) points out "There is some theoretical interest in the variation of rigidity with temperature in the absence of the effects normally produced by bond breakage and formation. Thus, on elementary ideas, if the deformation of the gel only involves a change in configurational entropy (as in the case of an ideal crosslinked polymer with free rotation of the chain elements) the temperature coefficient of rigidity should be positive and equal to the reciprocal of the absolute temperature. On the other hand, if the deformation only involves a change in internal energy, there should be very little variation of rigidity with temperature. Intermediate cases in which the temperature coefficient is positive but less than the reciprocal of the absolute temperature are also to be expected." When creep is negligible the temperature coefficient of rigidity (T_r) can be calculated from

$$T_r = -[1/(\varepsilon+\delta\varepsilon)]\,[\delta\varepsilon/\delta T] \qquad (4.56)$$

where ε is the initial strain, and $\delta\varepsilon$ is the change in strain when the temperature changes by δT. The principle of the constant stress method is that the gel is set between the inner and outer cylinders of a coaxial cylinder viscometer. The inner cylinder is suspended by a weak torsion wire so that a slight rotational movement of the cylinder, due to a change in temperature, does not change the stress on the gel to any significant degree.

Figure 4.78 shows an outline diagram of the equipment used, which is based on that previously employed by Hastewell and Roscoe (1956). It consisted of an outer cylindrical vessel (radius 1·59 cm) and an inner hollow cylinder (radius 1·30 cm) which was suspended by a 100 cm long steel or tungsten torsion wire. The height of the inner cylinder was 3·89 cm, and its base was 0·75 cm from the surface of the outer cylinder's base. Both cylinders were made from thin metal plated with rhodium to facilitate transmission of any alteration in temperature to the gel when the outer cylinder was immersed in a vessel containing hot or cold water. A plane mirror was fixed to the chuck which connected the torsion wire to a short steel needle projecting from the top of the inner cylinder. Light from a slit source was projected by a lens on to the mirror, and the image activated a photoelectric spot follower which gave a continuous record of the image displacement with an accuracy better than $\pm$ 0·02 cm. A displacement of 1 cm corresponded to a shear in the gel of 0·006. When the temperature of the gel was altered changes in the displacement of up to 1 cm were observed, which corresponded to a cylinder rotation of 0·06°. Thus, a constant stress was maintained to within 0·3%. A minor correction had to be made to the displacement readings because the steel needle suffered a certain degree of torsion. The correction was determined by clamping the inner cylinder and recording the deflections for given rotations of the torsion head. This technique can be applied to all gels which exhibit linear viscoelasticity up to the maximum stress employed.

Tests were made on dilute gelatin gels (1·5 g/100 ml), prepared from gelatin with a Bloom jelly strength of 220 g at 6·0% concentration, which were exposed to room temperature for 15 h before testing. The gels were subjected to a temperature drop of about 11°C followed by a rise of 9°C. Both temperature changes took place in 30 sec. The strain in these tests ranged from 0·02 to 0·14, and some typical results are shown in Fig. 4.79. No

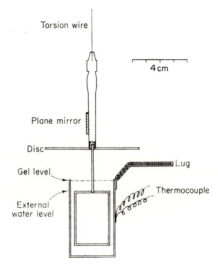

FIG. 4.78. Diagram of concentric-cyliner apparatus (Roscoe, 1965).

systematic interrelationship was observed between T_r and ε, but there appeared to be some dependence of T_r on the concentration of gelatin in the gel when the latter was increased up to 5·74 g/100 ml (Table 4.33). When the gels were subjected to a temperature drop of 3·5°C followed by a rise of 2·5°C, both temperature changes taking place in 30 sec, the results were significantly higher than those shown in Fig. 4.79, so that T_r depends to some extent on the temperature changes involved. Gels prepared from low concentrations of rice starch gave values of T_r similar to those for gelatin gels.

Although the values of T_r for gelatin gels are positive they are 2–3 times higher than were anticipated, and they are also higher than the values reported by Saunders and Ward (1958). The discrepancies were attributed to differences in the number of natural bonds in the respective gel structures.

Gelatin gels soften when exposed to temperatures from about 24°C upwards, which may be reached on a warm summer's day, and this presents problems regarding their use in meat pies and other jellied products. This temperature sensitivity can be reduced by incorporating a small amount of agar agar (Selby, 1955), which has a lower temperature sensitivity than

TABLE 4.33

Influence of gelatin concentration on T_r of gelatin gels (Roscoe, 1965)

gelatin concentration (g/100 ml)	T_r (°C^{-1})
0·75	0·0073 ± 0·0004
1·50	0·0076 ± 0·0004
2·97	0·0089 ± 0·0004
5·74	0·0107 ± 0·0008

gelatin. The agar agar also confers a less rubbery texture on the gelatin gels. Agar forms gels at much lower concentrations than gelatin does, but these gels are firm and rather brittle, and they do not melt readily on the palate.

The texture of agar gels has been studied in detail using the F.I.R.A. jelly tester (Jones, 1968). This instrument consists essentially of a small metal vane which is mounted on a shaft carrying a graduated scale and a counter-poised bucket which, on filling with water, produces rotation of the vane. The vane is inserted in a gel to a given depth, and water is run into the bucket at a selected speed until the vane has rotated through the desired angle. The time taken is noted with a stop watch. From the rate at which water flows

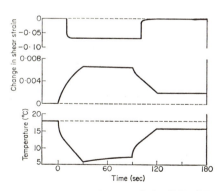

FIG. 4.79. Constant stress measurements on gelatin (1·5 g/100 ml). Upper curve is stress removal test (Roscoe, 1965).

into the bucket it is possible to calculate the load at any time, so that a load–deflexion graph can be composed. A typical curve for an agar–sucrose–water gel is shown in Fig. 4.80. Beyond point A the deflexion, which up to this point has increased almost linearly with increasing load, increases more rapidly as the load continues to rise, and at C it breaks down. The deflexion at C is proportional to the breaking strain, while the load at this point is

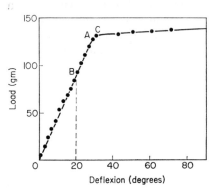

Fig. 4.80. Load/deflexion curve for an agar jelly (Jones, 1968). 0·5% New Zealand agar, 60% sucrose, 20·8°.

proportional to the ultimate strength. Although rigidity, breaking strain, and ultimate strength are interrelated, the slope of the linear portion of the curve can vary considerably before the gel breaks down and these three parameters can then vary independently of each other.

In order to remove the brittleness of agar gels sugar is added. Using a mixture of glucose and sucrose syrups in equal proportions gels of Japanese agar (Fig. 4.81) and Danish agar (Fig. 4.82) both lost the sharp break in

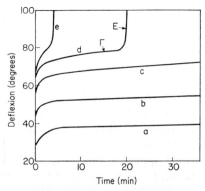

Fig. 4.81. Time deflexion curves under various loads for Japanese agar gels (Jones, 1968). 3·0% Japanese agar, 75% sucrose, 18°. *a*, 5 g; *b*, 10 g; *c*, 30 g; *d*, 32·5 g; *e*, 37·5 g.

their load–deflexion curves (Jones, 1968). Danish agar differs in its chemical composition from other agars. With increasing sugar content up to 65% the Japanese agar gels showed an increase in rigidity and ultimate strength, and also a slight increase in the breaking strain. At 71·4% sugar content the ultimate strength was still higher, and this was accompanied by a large increase in the breaking strain, but the slope of the initial part of the curve decreased. At 78·4% sugar the ultimate strength decreased, and at 80·5% sugar the slope was less than that for the gel which contained no sugar. Danish agar gels showed no fall in ultimate strength for sugar concentrations up to 85%, and also an increasing breaking strain with increasing sugar content. Creep compliance–time studies at constant stress indicated that although a large viscous element had developed there was still a large elastic element present also.

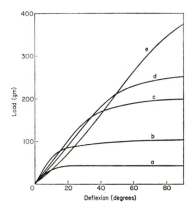

FIG. 4.82. Load/deflexion curves of Danish agar–sugar jellies (Jones, 1968). 0·6% agar, 7·5°. Curve a, 0% sugars; b, 71·5%; c, 75·9%; d, 80·5%; e, 85·5%.

Penetrometer studies on wheat starch pastes containing citric acid and sucrose indicated that the size of the starch granules influenced gel rigidity (Campbell and Briant, 1957). Gels prepared from small granule fractions of starch were weaker than those prepared from either medium or large granule fractions. In addition, both citric acid and sucrose interfered with gel formation, giving weaker gels as their concentration increased, but whereas the former additive increased the amount of amylose that went into solution the latter ingredient decreased the amylose in solution. Gel formation by starch is often associated with solution of amylose, so that the action of citric acid would appear to be contrary to this belief. It was also observed that the citric acid caused fragmentation of the starch granules, whereas sucrose did not. On the basis of the effect of the two additives on the starch granules and on gel formation it was suggested that there are at least two basic requirements for starch gel formation. First, sufficient relatively intact starch granules should be present to provide a minimum of rigidity. Second, enough amylose should be in solution to act as a "binding" between the granules. With sucrose the second requirement is probably not met, whereas with citric acid the first requirement is not fulfilled.

Phosphates can improve the water binding capacity of meat, and this appears to be associated with solubilisation of (acto) myosin, which then forms a gel when the meat is cooked. The rigidity of such gels has been

TABLE 4.34

Modulus of rigidity of heat-coagulated meat extracts (Sherman, 1961a)

Phosphate used	pH of 1·0% aqueous solution	modulus of rigidity (dyne/cm²) at the following pressures (dyne/cm²)	
		150 00	60 000
sodium metaphosphate	6·3	19 250	18 450
commercial phosphate	7·2	19 500	18 500
sodium tetrapolyphosphate	8·5	23 000	18 150
tetrasodium pyrophosphate	10·3	24 100	18 850

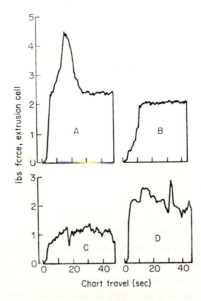

FIG. 4.83. Typical extrusion curves for different gels (Kramer and Hawbecker, 1966). A, jellied cranberry sauce; B, apple jelly; C, grape jelly; D, cellulose gum, type 7LP, 20 g/250 ml.

investigated (Sherman, 1961) by extracting the (acto) myosin solution from chopped meat–phosphate–water mixtures, which were aged at 0°C for 18 h and then centrifuged, and heating them in a Saunders and Ward (1954) rigidity tube until the extracts coagulated. Table 4.34 shows how phosphate type influenced the rigidity of the protein gels. The more alkaline phosphates gave gels of highest rigidity. These phosphates also caused larger concentrations of (acto) myosin to pass into solution in the water, so that rigidity increases as the dissolved (acto) myosin concentration increases.

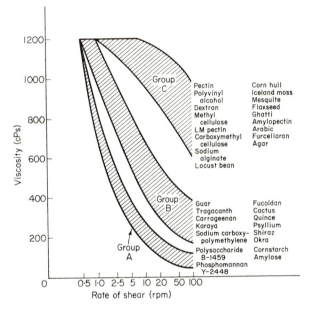

FIG. 4.84. Effect of shear rate on the viscosity of gum solutions (Szczesniak and Farkas, 1962).

Information relating not only to gel rigidity, but also to the uniformity of its structure and to adhesiveness, has been obtained using a L.E.E.–Kramer shear press (Section 8, of this chapter) with a modified extrusion cell (Kramer and Hawbecker, 1966). The container in which the gel was set formed the body of the extrusion cell, and the slots from the usual extrusion cell were fixed to the lower end of the container following removal of its base.

A cylindrical plunger replaced the usual set of metal bars used to extrude the sample. When dealing with weak gels a hollow cylinder with no opening in its base was used, along with a plunger which fitted snugly into the cylinder leaving a gap all round of $\frac{1}{8}$ in. Typical extrusion curves for some fruit gels are shown in Fig. 4.83. Grape gel is far less firm than cranberry, apple, and cellulose gum gels. In addition, the irregularities in the trace indicate that its structure is less uniform. Cranberry gel exhibited a much larger maximum extrusion force than the other samples. The adhesiveness of a gel is represented by the height and shape of that part of the curve which corresponds to the return stroke of the plunger. Ease of spreadability is probably related to the height of the maximum peak and also to the area under the curve.

TABLE 4.35

Interrelationship of mouthfeel of hydrocolloid solutions and their flow properties
(Szczesniak and Farkas, 1962)

| Hydrocolloid | % concentration | Sensory evaluation | | Viscosity versus rate of shear grouping |
		score	how slimy	
Starch	2·0	1	nil	A
Phosphomannan Y-2448	0·75	1	nil	A
Polysaccharide B-1459	0·15	1	nil	A
Sodium carboxy-polymethylene	0·3	2	very slight	B
Carrageenan	1·0	3	somewhat	B
Gum karava	1·0	3	somewhat	B
Gum tragacanth	1·0	2	very slight	B
Gum guar	0·6	2 − 3	very slight	B
Locust bean gum	0·7	3	somewhat	C
Carboxymethyl cellulose	1·0	4	moderate	C
Sodium alginate	1·3	5 − 6	slimy to very	C
Low-methoxy pectin	5·0	7	extremely	C
Methyl cellulose	2·6	6	very	C
Polyvinyl alcohol	7·0	6	very	C
Dextran B-512	18·0	7	extremely	C
Pectin	2·5	7	extremely	C

Viscometric studies on solutions of a large number of hydrocolloids over a range of shear rates (Szczesniak and Farkas, 1962) have shown that they fall into three distinct groupings, based upon the rate at which viscosity falls with increasing rate of shear in a semi-logarithmic plot (Fig. 4.84). Group A showed a very sharp drop in viscosity when shear rate was increased from 0·5 to 100 sec^{-1}, group B showed a less sharp drop, while group C exhibited the slowest change in viscosity with increasing shear rate. When the solutions were assessed on the palate by a panel some very interesting observations emerged; these are summarised in Table 4.35. Solutions falling within group C generally gave a very slimy sensation in the mouth, group A solutions exhibited no sliminess, and group B solutions gave a slight sensation of sliminess. Thus, it appears that sliminess is related to the degree of non-Newtonian behavior. The sliminess was not influenced to any degree by hydrocolloid molecular weight, so that it must be due to some other basic property of the solutions. Hydrocolloid concentration had little effect on the viscosity of those solutions which showed either slight or pronounced deviation from Newtonian behavior. Solutions falling at the limits of the three groups in Fig. 4.84 were able to move from one group to another when the hydrocolloid concentration was altered.

10. MILK AND CREAM

Although it has long been recognised that both milk and cream are fat-in-water emulsions, cream containing a much higher concentration of fat particles, surprisingly little attention appears to have been devoted to studying their basic structure. The pioneer work of Leviton and Leighton (1936), and the flow equation which they derived, have already been referred to in Section 3.A.2, Chapter 3. Since the particle size distribution may range from about $0 \cdot 1 \ \mu m$ to about 10 μm, with the highest proportion of particles around 3 μm, both the particle size distribution and the mean particle size should influence the flow properties. Also, the fat content (F_m) will exert a significant effect. Several investigators have shown that the viscosity of whole milk is higher than that of skim milk (Mohr and Oldenburg, 1929; Spöttel and Gneist, 1941–43; Tapernoux and Vuillaume, 1934). Others (Eilers et al., 1947; Torssel et al., 1949; Seidler and Elke, 1959; Spöttel and Gneist, 1941–43; Kooper, 1914) have shown that, at a given temperature, the viscosity of concentrated skim milk, skim milk, milk containing 3% fat, and whole raw milk increases with increasing total solids–not–fat concentration (S_n). Statistical analyses of these latter data have shown (Cox et al., 1959) that viscosity can be represented satisfactorily by an empirical relationship

$$\eta_m = A_1 + A_2 F_m + A_3 S_n + A_4 F_m^2 \qquad (4.57)$$

where A_1, A_2, A_3, and A_4 are constants, but for the individual sets of data there are large differences between the calculated values for any one constant.

Removal, or addition, of colloidal phosphate and citrate from milk (viscosity $1 \cdot 7$ cP at 20°C) produced a sharp change in its viscosity (Fig. 4.85). In both cases the viscosity increased significantly (McGann and Pyke, 1960). When the colloidal phosphate and citrate had been removed completely the viscosity was about 30% higher than for the original milk. Furthermore, milk is very stable to calcium ions, and it does not coagulate normally when concentrations of calcium chloride up to 1 mole/litre are added, within the temperature range 0·40°C, provided that the pH is not readjusted to its original value. Following removal of colloidal phosphate and citrate the milk was very sensitive to calcium ions, and it coagulated when a concentration of calcium chloride as low as 25 m–mole/litre was added. On the basis of these observations it appears that elimination of colloidal phosphate and citrate produces some structural alteration in the casein micelles, which involves not only changes in their size and shape, but also changes in the degree of interaction between the various casein components within the micelles. The micelles are normally approximately spherical, and about 100 Å diameter (Shimmin and Hill, 1964). The col-

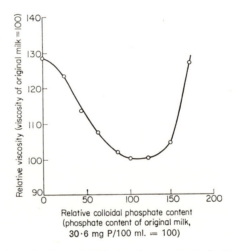

FIG. 4.85. Effect of altering the colloidal phosphate content of milk on the viscosity of the milk (McCann and Pyke, 1960).

loidal phosphate–citrate may preserve micellar stability by acting as a cross-link between some of the phosphate groups of the casein components, thereby reducing the number of groups which are able to form inter-micellar linkages, or it may act merely as a filler which influences the size and form of the micelles and consequently affects their stability to calcium ions. Experimental evidence suggests that the first explanation is more likely. The final high viscosity which was observed (Fig. 4.85) when the colloidal phosphate–citrate concentration was increased to double its normal level (it normally constitutes about two-thirds of the casein micelle calcium content) was probably due to coprecipitation of caseinate with colloidal phosphate and citrate, and not to the mechanism operating when all of the latter was removed.

When concentrated skim milk (25% total solids) was treated with calcium hydroxide at temperatures below 10°C after adjusting the pH to 11·7, its viscosity at 49·5–54 sec⁻¹ showed a marked change with time (Fig. 4.86). It reached a maximum value after about 3 min, and then decreased, initially at a rapid rate and then more slowly, until it achieved a steady value which was maintained for 10 days. The milk showed the greatest deviation from Newtonian behavior when the viscosity was at its maximum. This was followed by a sharp rise in viscosity, and the milk set to a gel (Kumetat and Beeby, 1954, 1958). The addition of urea or calcium ions at high pH produced similar increases in viscosity to a maximum value followed by a marked decrease (Beeby and Kumetat, 1959a). At low pH the viscosity showed a slow rate of increase, without a maximum developing, and eventually the milk gelled. Urea had little effect at low pH. In the absence of additives a

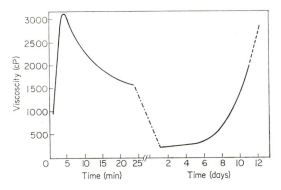

Fig. 4.86. The viscosity-time curve of concentrated skim-milk (22·7% t.s.) treated at 4°C. with calcium hydroxide (pH 11·7). (Beeby and Kumetat, 1959a).

milk solids concentration range of 19·1%–25·2% did not influence the time taken for the maximum viscosity to be reached when the rate of shear was ~ 50 sec⁻¹ (Beeby and Kumetat, 1959b). Rate of shear also had no influence on this time, within the range 11·84–67·99 sec⁻¹; neither did temperature within the range 3°–11°C, when the milk was treated with calcium hydroxide or sodium hydroxide. The viscosity rise to a maximum was attributed to swelling of the casein micelles, and also to interaction of the distended micelles in those milk samples containing the higher concentrations of milk fat. With further swelling the micelles disintegrate and the viscosity falls, since the total volume of the micelles has now decreased. At high pH there is a greater net charge on the micellar protein, and this increases the repulsion forces and promotes dispersion. Both calcium ions, by forming ionic bridges between casein fractions, and hydrogen bonds, are responsible collectively for the stability of casein micelles. Neither can maintain this stability in the absence of the other. They unite to counteract the dispersion tendency of the electrostatic repulsion forces.

Milk sets into a weak gel, or curd, under the influence of rennet, a commercial enzyme preparation. The modulus of rigidity of these gels is too low for measurements to be made with the Saunders and Ward (1953) rigidity tube. Instead, the modified version described in Chapter 2, Section 2.I, has been used (Scott Blair and Burnett, 1958a). Figure 4.87 shows the change in rigidity with time observed for four typical samples of liquid milk. Each

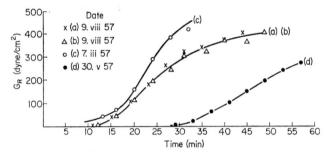

FIG. 4.87. Rigidity data for samples of liquid milk (Scott Blair and Burnett, 1958a)

curve was sigmoidal, and consisted of three distinctive regions. For the first few minutes there was no measurable rigidity, i.e. $G_R < 10$ dyne/cm^2, and then G_R increased very rapidly, to be followed by a much slower rate of increase. In the middle region G_R was independent of quite large changes in pressure, but in the final region G_R became pressure dependent. The major part of the setting process, with the exception of the first few minutes, proceeded at a rate which was inversely proportional to t, so that plots of G_R against log t gave very good straight lines (Fig. 4.88). At very low stresses

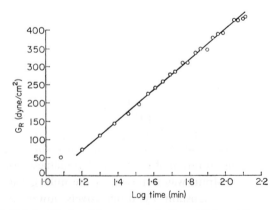

FIG. 4.88. Dried milk showing no syneresis (Scott Blair and Burnett, 1958a).

the curd showed viscoelasticity (Scott Blair and Burnett, 1959), and the creep compliance with time could be represented, in terms of the simplest mechanical model, by a Burgers body, i.e. a Hookean spring and a dashpot in series linked with a Hookean spring and a dashpot in parallel. Thus, four rheological parameters were involved, E_0, E_1, η_1, and η_N. Each of these parameters also gave a straight line when plotted against log t. Furthermore, plots of log $[G_{R(\infty)} - G_R/G_{R(\infty)}]$, where $G_{R(\infty)}$ was the value of G_R obtained by extrapolating (Fig. 4.88) to a time which was twice the duration of the experiment, against time t were linear, which suggested that in the later stages of setting the rate of increase in the value of each parameter was proportional to the concentration of residual unset casein.

An additional point of interest arising from the creep compliance–time curves was that the instantaneous elastic displacement was very much larger than the instantaneous elastic recovery. This phenomenon is known as "static fatigue", and is believed to arise from the fact that the instantaneous elastic compliance is not truly instantaneous but is really lightly damped. Thus, during this phase of the test some bonds are actually broken.

Homogenisation, separation and pasteurisation delayed the onset of rigidity in renneted milk (Scott Blair and Burnett, 1958b). Tables 4.36 and 4.37 show the intercepts on the log t axes obtained by extrapolating the linear plot of each rheological parameter against log t, and the corresponding logarithmic rates of increase respectively. The single values quoted after each set of three experimental values give the % increase of the parameter based on raw milk. Since the plots were not linear at very small values of t, the intercepts on the log t axes do not coincide exactly with the times at which the curds first show signs of setting, but the two times were closely related. All three treatments applied to the milk increased the values of the intercepts on the log t axis, i.e. the setting times. Homogenisation and separation increased the rate of setting once it had started, although the effect was rather small for separation, but pasteurisation decreased the rate of setting. The two first mentioned treatments also altered the balance between immediate and slow elastic recovery.

Addition of calcium chloride to reconstituted dried fat-free milk progressively reduced the time for rigidity to appear, but it did not alter the final setting rates. When lactic acid was also added, so as to prevent the calcium chloride from altering the pH, the differences between the setting curves were small. Once again plots of log $[G_{R(\infty)} - G_R/G_{R(\infty)}]$ against t were approximately linear, so that the rate at which the curd set was still proportional to the concentration of casein which remained unset at that time.

When a fairly large strain was imposed on the milk gel in five seconds, and it was maintained constant by progressively lowering the stress, a double logarithmic plot of stress and time (Fig. 4.89) gave a straight line

(Scott Blair and Burnett, 1959a.) Zero time was taken as the start of loading. The relaxation followed a power law relationship between stress and strain with an exponent of 0·40. On repeated loading and unloading hysteresis curves were obtained (Fig. 4.90) showing, respectively, a marked stiffening followed by a softening when the direction of loading was reversed. The distribution of relaxation times was studied (Scott Blair and Burnett, 1959b) by applying the Central Limit Theorem, as discussed in Section 7 of this chapter, and in particular by using Eq. (4.31). The data employed for this analysis were those for a one-hour-old gel which was strained eight times in

TABLE 4.36

Intercepts (min) on log t axis for extrapolated rheological parameters
(Scott Blair and Burnett, 1958b)

Parameter	Raw milk	Separated milk		Homogenised		Pasteurised	
	16·5	18		21·5		17·5	
G_R	18	21	+ 10	21	+ 17	19·5	+ 27
	19·5	23		21		31·5	
E_0	16	18		20		18	
	18	22	+ 18	20·5	+ 14	19·5	+ 30
	19·5	23		20·5		32	
	15·5	18		20		17	
E_1	17	21	+ 20	20·5	+ 17	19·5	+ 32
	19·5	23		20·5		3·2	
	17	17·5		21		19	
η_1	19·5	20·5	+ 17	25	+ 19	21	+ 27
	19·5	22·5		21		31·5	
	19·5	17·5		20		18	
η_N	17	21·5	+ 11	20	+ 9	21	+ 25
	19·5	23·5		21		31	

alternate directions, so that there was very little stiffening or softening. Since the data from the eight relaxations proved very similar mean values of p were used, and only one curve was plotted (Fig. 4.91). The graph was linear, so that a log–normal distribution of relaxation times could quite safely be assumed.

Returning to the earliest phase of the setting of milk with rennet, double logarithmic plots of stress and shear rate have proved linear (Fig. 4.92), but with different gradients for different times so that it appears that the

TABLE 4.37

Rates of increase in the rheological parameters of curd produced by various treatments of milk prior to renneting (Scott Blair and Burnett, 1958b)

Parameter	Raw milk	Separated milk	Homogenised	Pasteurised
	$G_R \times 10^{-2}$ dyne/cm²			
	2·52	2·82	4·47	2·27
$t\dot{G}_R$	2·36	2·82 } +6	4·06 } +23	1·75 } −14
	2·44	2·79	3·30	1·65
	5·25	5·85	8·20	5·52
$t\dot{E}_0$	5·25	6·22 } +3	5·87 } +6	5·08 } −7
	5·25	5·52	6·21	3·04
	4·48	5·22	8·20	3·58
$t\dot{E}_1$	4·48	5·22 } +3	5·88 } +12	3·23 } −11
	4·48	4·27	7·10	3·05
	$\eta \times 10^{-4}$ poise			
	3·14	3·30	5·23	2·82
$t\dot{\eta}_1$	3·29	3·15 } +2	5·53 } +23	2·27 } −11
	2·72	3·05	6·22	2·18
	1·83	1·78	2·64	1·19
$t\dot{\eta}_N$	1·72	1·90 } 0	2·27 } +24	1·33 } −27
	1·59	1·65	2·93	0·98

graphs eventually meet (Tuszynski and Scott Blair, 1967). Following the studies of Nedonchelle and Schutz (1967) with starch pastes each graph obeys a modified power law

$$\log p - \log \alpha_m = n_s (\log D - \log \beta_m) \qquad (4.58)$$

where α_m has the dimensions of a stress, and β_m has the dimensions of a shear rate. If p' is substituted for α_m, and D' for β_m, then

$$\frac{p}{p'} = \left(\frac{D}{D'}\right)^{n_s}$$

This version of Eq. (4.58) confirms that all the $p - D$ plots should meet in a point with coordinates (p', D') when extrapolated. It also, incidentally, overcomes the dimensional problem of the power law, as discussed in Chapter 3, Section 2.A.

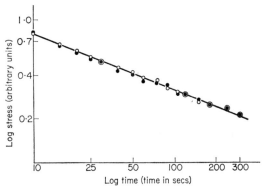

FIG. 4.89. Relaxation curves (milk curd) o; •, duplicate runs. (Scott Blair and Burnett 1959a).

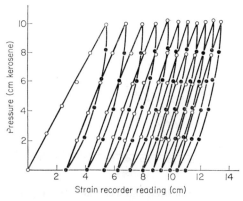

FIG. 4.90. Hysteresis curves for milk gels (Scott Blair and Burnett, 1959a).

Sweetened condensed milk, which is essentially a dispersion of milk protein and lactose crystals in a saturated solution of lactose and sucrose, thickens when aged (Samel and Muers, 1962). Fresh samples show only slight deviation from Newtonian behavior at rates of shear up to 49 sec^{-1},

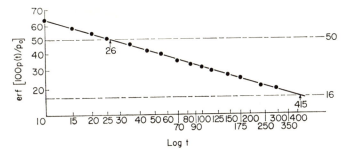

FIG. 4.91. Log-normal distribution of relaxation times for symmetrically stressed gel (Scott Blair and Burnett, 1959b).

but during aging the deviation becomes progressively larger. Furthermore, the area of the hysteresis loop obtained by increasing the shear rate gradually to a maximum value of 49 sec^{-1}, and then reversing the procedure, was quite small for fresh samples and much larger for aged samples (Fig. 4.93). As would be expected, the area of the hysteresis loop for aged samples

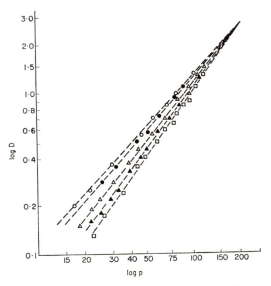

FIG. 4.92. Double log curves of stress and shear rates during early stages of coagulation. The five curves represent results 10 (o), 30 (•), 60 (△), 90 (▲) and 120 (□) min after reaching temperature equilibrium at 15°C. (Tuszynski and Scott Blair, 1967).

decreased as the rest period before reducing the shear rate from its maximum was increased. The stress corresponding to the lowest rate of shear employed ($4 \cdot 61 \ \text{sec}^{-1}$, on a portable Ferranti viscometer) was regarded as a measure of gel strength. When samples were stored at 39°C for periods up to 12 days, the area of the hysteresis loop derived from shear rate–stress determinations became particularly marked.

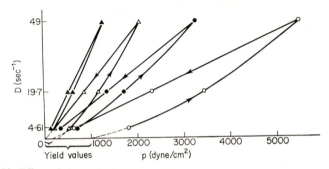

FIG. 4.93. Effect of time of stirring on flow curves of age-thickened condensed milk (Samel and Muers, 1962). Interval between viscometer readings, 2 sec o—o; 60 sec ●—● ; 30 min △—△. Fresh condensed milk, 30 min interval, ▲—▲.

Changes in the rheological properties of sweetened condensed milk when aged at 39°C after violent agitation in a Hobart mixer are due both to structure recovery and thickening. On the other hand when samples are stored at 5° and 18°C after such agitation any viscosity changes with time must be due to structure recovery only, since there is no significant age thickening at these temperatures. To test this hypothesis samples of condensed milk which had been stored for three weeks at 39°C to accelerate the age thickening were agitated in a Hobart mixer for 2 h and then subdivided into three equal portions. (Samel and Muers, 1962a). The portions were stored at 5°C, 18°C, and 39°C respectively. The viscosity of the samples stored at 5°C and 18°C increased slowly for about four days (Fig. 4.94), and then there was little further change. Only a small fraction of the initial viscosity prior to mixing in the Hobart was thus recovered. Storage at 39°C, however, resulted in a rapid increase in viscosity with time, and eventually the viscosity exceeded that of the original condensed milk before mixing in the Hobart. The magnitude of the viscosity rise at 39°C was greater than age thickening alone would produce, so that structure recovery is also greater at this temperature than at 5°C and 18°C.

The viscosity increase associated with age thickening of condensed milk could arise from either, or both, of the principal, components, viz. the milk protein and/or the continuous phase. This problem was investigated by studying the properties of model gelatin gels, some of which contained added sucrose and lactose. Both types of gel were thixotropic, and both exhibited a yield value which was not destroyed by stirring, but after agitation the gels containing added sugars recovered their viscosity much more slowly than the gels which had no sugar added. If these observations can be applied to dispersions of other proteins then it appears that while the continuous phase does not contribute significantly to structure changes on agitation, the viscosity of the continuous phase may influence the rate of structure recovery. Similarly, the influence of storage temperature on the rate of structure recovery could be due to the temperature dependence of the viscosity of the continuous phase.

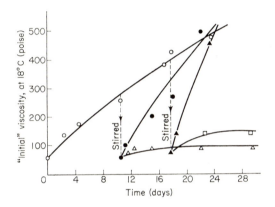

FIG. 4.94. Recovery of consistency by age-thickened condensed milk after stirring (Samel and Muers, 1962). Stored at 39°C, not stirred o—o. Stirred after 10½ days at 39°C, further storage at 39°C, ●—● ; 5°C, △—△. Stirred after 17½ days at 39°C, further storage at 39°C, ▲—▲ ; 18°C, □—□.

As with dispersions of colloidal particles (Section 5, Chapter 3) age thickening may be due to flocculation of casein micelles and the development of a network structure which is held together by weak attraction forces. This network holds a part of the continuous phase within its structure. In addition, there is a slow increase in that part of the total viscosity which is not destroyed by agitation. The progressive linking together of casein micelles would alter the size and shape of the flocculates, and consequently change the viscosity, and this could explain the latter observation.

Low shear studies in a coaxial cylinder viscometer on natural creams containing not less than 48 % fat indicated pronounced non-Newtonian behavior (Scott Blair and Prentice, 1966; Prentice, 1968) when the maximum shear rate did not exceed 1·77 sec^{-1}, and the minimum shear rate was $4·8 \times 10^{-3}$ sec^{-1}. Within this shear range all the creams conformed with Eq. (3.1), as indicated by the typical plot in Fig. 4.95, but at higher shear rates the creams became almost Newtonian and the power law relationship was no longer followed. At a steady state rate of shear, equivalent to the minimum shear rate obtainable, the creams exhibited the same increase of stress with time to a maximum followed by a fall to an equilibrium steady value that has been observed (Trapeznikov and Shalopalkina, 1957; Lucassen–Reynders, 1962; Sherman, 1967) with dispersions of solid particles and also with

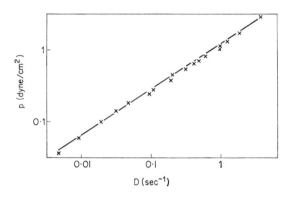

FIG. 4.95. Flow curve for a typical cream (Prentice, 1968).

emulsions. When oscillatory tests were made on the creams with the same viscometer some degree of elasticity was indicated. In terms of the basic mechanical elements, the equivalent Kelvin–Voigt body would have a viscosity of 0·80 poise and an elastic modulus of 0·52 c.g.s. units. An equivalent Maxwell body would have a viscosity of 0·82 poise with an elastic modulus of 1·57 c.g.s. units. It is rather surprising, however, that the interpretation was not based on a simple Burgers body with a Maxwell body and Kelvin–Voigt body in series.

BIBLIOGRAPHY

Andrews, R. D., Hofmann-Bang, N. and Tobolsky, A. V. (1948). *J. Polymer Sci.* **3**, 669.
Avrami, M. (1939). *J. Chem. Phys.* **7**, 1103.
Avrami, M. (1940). *J. Chem. Phys.* **8**, 212.
Avrami, M. (1941). *J. Chem. Phys.* **9**, 177.
Aylward, F. (1960). in "Texture in Foods", S.C.I. Monograph No. 7, p. 75.
Babb, A. T. S. (1965). *J. Sci. Fd. Agric.* **16**, 670.
Bailey, C. H. (1934). *Cereal Chem.* **11**, 160.
Bate-Smith, E. C. (1939). *J. Physiol.* **96**, 176.
Bate-Smith, E. C. and Bendall, J. R. (1949). *J. Physiol.* **110**, 47.
Bechtel, W. G. (1955). *Trans. Am. Assoc. Cereal Chemists* **13**, 108.
Bechtel, W. G. (1961). *Baker's Dig.* **35**, 48, 172, 174.

Beeby, R. and Kumetat, K. (1959a). *J. Dairy Res.* **26**, 248.
Beeby, R. and Kumetat, K. (1959b). *J. Dairy Res.* **26**, 258.
Bendall, J. R. (1951). *J. Physiol.* **114**, 71.
Bendall, J. R. and Davey, C. L. (1957). *Biochim. Biophys. Acta* **26**, 93.
Bice, C. W. and Geddes, W. F. (1949). *Cereal Chem.* **26**, 440.
Bingham, E. (1922). "Fluidity and Plasticity", McGraw-Hill, New York.
Bloksma, A. H. (1957). *Cereal Chem.* **34**, 126.
Bloksma, A. H. (1962). *Rheol. Acta*, **2**, 217.
Bloksma, A. H. (1968). in "Rheology and Texture of Foodstuffs", S.C.I. Monograph No. 27, p. 153.
Bloksma, A. H. and Hlynka, I. (1960). *Cereal Chem.* **37**, 352.
Boedtker, H. and Doty, P. (1954). *J. phys. Chem.* **58**, 968.
Boussinesq, J. (1885). "Applications des potentiels a l'étude de l'equilibre et du mouvement des solides élastiques" (Paris).
Bratzler, L. J. (1932). M.S. thesis, Kansas State College, U.S.A.
Bratzler, L. J. (1949). Proc. 2nd Ann. Reciprocal Meat Conf. p. 117.
Caffyn, J. E. (1945). *Dairy Industries* **10**, 257.
Campbell, A. M. and Briant, A. M. (1957). *Food Res.* **22**, 358.
Campbell, L. E. (1940). *J. Soc. Chem. Ind.* **59**, 71.
Cornford, S. J. (1963). B.B.I.R.A. Report No. 68.
Cornford, S. J., Axford, D. W. E. and Elton, G. A. H. (1964). *Cereal Chem.* **41**, 216.
Cox, C. P., Hosking, Z. D. and Posener, L. W. (1959). *J. Dairy Res.* **26**, 182.
Cunningham, J. R., Hlynka, I. and Anderson, G. W. (1953). *Can. J. Technol.* **31**, 98.
Cunningham, J. R. and Hlynka, I. (1954). *J. appl. Phys.* **25**, 1075.
Davis, C. E. (1921). *Ind. Eng. Chem.* **13**, 797.
Davis, J. G. (1937). *J. Dairy Res.* **8**, 245.
De Fremery, D. and Pool, M. F. (1960). *J. Food Sci.* **25**, 73.
De Man, J. M. (1963). *Food in Canada* **23**, 27.
De Man, J. M. and Wood, F. W. (1959). *J. Dairy Sci.* **42**, 56.
Dempster, C. J., Hlynka, I. and Winkler, C. A. (1952). *Cereal Chem.* **29**, 39.
Drake, B. K. (1962). *J. Food Sci.* **27**, 182.
Drake, B. K. (1963). *J. Food Sci.* **28**, 233.
Drake, B. K. (1965). *J. Food Sci.* **30**, 556.
Drake, B. K. (1965). *Biorheol.* **3**, 21.
Eilers, H., Saal, R. N. J. and van der Waarden, M. (1947). "Chemical and Physical Investigations in Dairy Products" Elsevier, Amsterdam.
Einstein, A. (1906). *Ann. Physik*, **19**, 289.
Emmons, D. B. and Price, W. V. (1959). *J. Dairy Sci.* **42**, 553.
Falk, S., Hertz, C. H. and Virgin, H. I. (1958). *Physiol. Plantarum* **11**, 802.
Feltham, P. (1955). *Brit. J. Appl. Phys.* **6**, 26.
Ferry, J. D. (1948). *Advs. Protein Chem.* **4**, 20.
Fincke, A. and Heinz, W. (1956). *Fette u. Seifen* **58**, 902.
Finney, E. E., Hall, C. W. and Mase, G. E. (1964). *J. agric. Engng. Res.* **9**, 307.
Frandsen, J. H. and Arbuckle, W. S. (1961). "Ice Cream and Related Products" Avi Publishing Co., Westport, Connecticut, p. 251.
Friedman, H. H., Whitney, J. E. and Szczesniak, A. S. (1963). *J. Food Sci.* **28**, 390.
Funt, C. B., Lerchenthal, C. H. and Muller, H. G. (1968) in "Rheology and Texture of Foodstuffs" S.C.I. Monograph No. 27, p. 197.

Glücklich, J. and Shelef, L. (1962a). *Cereal Chem.* **39**, 242.
Glücklich, J. and Shelef, L. (1962b). *Kolloid Zeit. u. Zeit. für Polym.* **181**, 29.
Grogg, B. and Melms, D. (1958). *Cereal Chem.* **35**, 189.
Guice, W. A., Lovegren, N. V. and Feuge, R. O. (1959). *J. Amer. Oil Chem.* **36**, 4
Guth, E. and Simha, R. (1936). *Kolloid Z.* **74**, 266.
Haighton, A. J. (1965). *J. Amer. Oil Chem. Soc.* **42**, 27.
Halton, P. and Scott Blair, G. W. (1936a). *J. Phys. Chem.* **40**, 561.
Halton, P. and Scott Blair, G. W. (1936b). *J. Phys. Chem.* **40**, 811.
Harbard, E. H. (1956). *Chem. Ind.* 491.
Harper, R. and Baron, M. (1948). *Nature* **162**, 821.
Harvey, H. G. (1960). in "Texture in Foods" S.C.I. Monograph No. 7, p. 29.
Hastewell, L. J. and Roscoe, R. (1956). *Brit. J. Appl. Phys.* **7**, 441.
Heaps, P. W. and Coppock, J. B. M. (1968). in "Rheology and Texture of Food-
    stuffs" S.C.I. Monograph No. 27, p. 168.
Heimann, W. and Fincke, A. (1962). *Zeit. Lebensm. u. Forschung* **117**, 93, 225, 297.
Heinz, W. (1959). *Materialprüf.* **1**, 311.
Hermann, K. and Gerngross, O. (1932). *Kautschuk* **8**, 181.
Hermann, K., Gerngross, O. and Abitz, W. (1930). *Z. physic. Chem.* **B10**, 371.
Hertz, H. (1896). *"Miscellaneous Papers"*. Macmillan, New York.
Hlynka, I. and Matsuo, R. R. (1960). *Cereal Chem.* **37**, 721.
Hlynka, I. (1964). "Wheat, Chemistry, and Technology", *Amer. Ass. Cereal Chem.*
Huebner, V. R. and Thomsen, L. C. (1957). *J. Dairy Sci.* **40**, 834.
Hurwicz, H. and Tischer, R. G. (1954). *Food Tech.* **8**, 391.
Isherwood, F. A. (1960). in "Texture in Foods", S.C.I. Monograph No. 7, p. 135.
Jellinek, H. G. and Brill, R. (1956). *J. appl. Phys.* **27**, 1198.
Jones, N. R. (1968). in "Rheology and Texture of Foodstuffs", S.C.I. Monograph
    No. 27. p. 91.
Jopling, D. W. (1958). *Rheol. Acta* **2/3**, 133.
Kooper, W. D. (1914). *Milchw. Zbl.* **43**, 169.
Kozma, A. and Cunningham, H. (1962). *J. Ind. Mathematics* **12**, 31.
Kramer, A. and Backinger, G. (1959). *Food* **28**, 85.
Kramer, A. Burckhardt, G. J. and Rogers, H. P. (1951). *Canner* **112**, 34.
Kramer, A. and Hawbecker, J. V. (1966). *Food Technol.* **20**, 111.
Kramer, A. and Twigg, B. A. (1959). *Advances in Food Research* **9**, 153.
Kumetat, K. and Beeby, R. (1954). *Dairy Ind.* **19**, 730.
Kumetat, K. and Beeby, R. (1958). *Dairy Ind.* **23**, 481.
Lawrie, R. A. (1968). in "Rheology and Texture of Foodstuffs", S.C.I. Mono-
    graph No. 27, p. 134.
Lerchenthal, C. H. and Funt, C. B. in "Rheology and Texture of Foodstuffs"
    S.C.I. Monograph No. 27, p. 203.
Leviton, A. and Leighton, A. (1936). *J. phys. Chem. Ithaca*, **40**, 71.
Lovegren, N. V., Guice, W. A. and Feuge, R. O. (1958). *J. Amer. Oil Chem. Soc.*
    35, 327.
Lucassen-Reynders, E. H. (1962). Doctoral dissertation, University of Utrecht,
    Holland.
McGann, T. C. A. and Pyke, G. T. (1960). *J. Dairy Res.* **27**, 403.
Miyoda, D. S. and Tappell, A. L. (1956). *Food Technol.* **10**, 142.
Mohr, W. and Haesing, J. (1949). *Milchwissenschaft* **4**, 255.
Mohr, W. and Oldenburg, F. (1929). *Milchw. Forsch.* **8**, 429, 576.
Mohsenin, N. N. and Göhlich, H. (1962). *J. agric. Eng. Res.* **7**, 300.

Mohsenin, N. N. and Morrow, T. (1968). in "Rheology and Texture of Food-stuffs" S.C.I. Monograph No. 27, p. 50.

Morrow, T. and Mohsenin, N. N. (1966). *J. Food Sci.* **31**, 686.

Muller, H. G., Williams, M. V., Russell Eggitt, P. W. and Coppock, J. B. M. (1961). *J. Sci. Food Agric.* **12**, 513.

Muller, H. G., Williams, M. V., Russell Eggitt, P. W. and Coppock, J. B. M. (1962). *J. Sci. Food Agric.* **13**, 572.

Muller, H. G., Williams, M. V., Russell Eggitt, P. W. and Coppock, J. B. M. (1963). *J. Sci. Food Agric.* **14**, 663.

Naudet, M. and Sambuc, E. (1959). *Rev. Franc. Corps Gras* **6**, 537.

Nederveen, C. J. (1963). *J. Colloid Sci.* **18**, 276.

Nedonchelle, Y. and Schutz, R. A. (1967). *C. R. Acad. Sci.* **265** (Séc. C), 16.

Pilkington, D. H., Walters, L. E. and Whiteman, J. V. (1961). *J. Animal Sci.* **20**, 1241 (Abstract).

Platt, W. (1930). *Cereal Chem.* **7**, 1.

Platt, W. and Kratz, P. D. (1933). *Cereal Chem.* **10**, 73.

Platt, W. and Powers, R. (1940). *Cereal Chem.* **17**, 601.

Prentice, J. H. (1952). B.F.M.I.R.A. Research Report No. 37.

Prentice, J. H. (1954). Laboratory Practice **3**, 186.

Prentice, J. H. (1956). B.F.M.I.R.A. Research Report No. 69.

Prentice, J. H. (1968). in "Rheology and Texture of Foodstuffs" S.C.I. Monograph No. 27, p. 265.

Proctor, B. E., Davison, S. Malecki, G. J. and Welch, M. (1955). *Food Technol.* **9**, 471.

Proctor, B. E., Davison, S. and Brody, A. L. (1956a). *Food Technol.* **10**, 327.

Proctor, B. E., Davison, S. and Brody, A. L. (1956b). *Food Technol.* **10**, 344.

Pyke, G. T. and McGann, T. C. A. (1960). *J. Dairy Res.* **27**, 403.

Robson, A. H. (1961). *J. Fd. Technol.* **1**, 291.

Roscoe, R. (1965). Proc. 4th Intern. Congr. Rheol. **3**, 593.

Rosenberg, G. F. von (1954). *Fette Seif. Anstrichm.* **56**, 214.

Sale, A. J. H. (1960). in "Texture in Foods", S.C.I. Monograph No. 7, p. 103.

Samel, R. and Muers, M. (1962a). *J. Dairy Res.* **29**, 249.

Samel, R. and Muers, M. (1962b). *J. Dairy Res.* **29**, 259.

Saunders, P. R. and Ward, A. G. (1953). Proc, 2nd Intern. Congr. Rheol. 284.

Saunders, P. R. and Ward, A. G. (1955). *Nature* **176**, 26.

Saunders, P. R. and Ward, A. G. (1957) in "Recent Advances in Gelatin and Glue Research" Pergamon Press, London.

Saunders, P. R. and Ward A. G. (1958). in "Rheology of Elastomers" p. 45. Pergamon Press, London.

Scherr, H. J. and Witnauer, L. P. (1967). *J. Amer. Oil Chem. Soc.* **44**, 275.

Schofield, R. K. and Scott Blair, G. W. (1933a). *Proc. Roy. Soc.* **A139**, 557.

Schofield, R. K. and Scott Blair, G. W. (1933b). *Proc. Roy. Soc.* **A141**, 72.

Scott Blair, G. W. and Baron, M. (1949). *Nature* **164**, 148.

Scott Blair, G. W. and Burnett, J. (1958a). *J. Dairy Res.* **25**, 297.

Scott Blair, G. W. and Burnett, J. (1958b). *J. Dairy Res.* **25**, 457.

Scott Blair, G. W. and Burnett, J. (1959a). *Brit. J. Appl. Phys.* **10**, 15.

Scott Blair, G. W. and Burnett, J. (1959b). *Brit. J. Appl. Phys.* **10**, 97.

Scott Blair, G. W. and Coppen, F. M. V. (1941). *J. Dairy Res.* **12**, 44.

Seidler, L. and Elke, M. (1949). *Milchwissenschaft* **4**, 105.

Selby, J. W. (1955). B.F.M.I.R.A. Research Report No. 65.

Shama, F. and Sherman, P. (1966). *J. Food Sci.* **31**, 699.
Shama, F. and Sherman, P. (1968). in "Rheology and Texture of Foodstuffs" S.C.I. Monograph No. 27, p. 77.
Shaw, D. J. (1963). in "Rheology of Emulsions" (ed. P. Sherman) Pergamon Press, London, p. 125.
Shelef, L. and Bousso, D. (1964). *Rheol. Acta* **3**, 168.
Sherman, P. (1961a). *Food Technol.* **15**, 79.
Sherman, P. (1961b). *Food Technol.* **15**, 394.
Sherman, P. (1965). *J. Food Sci.* **30**, 201.
Sherman, P. (1966). *J. Food Sci.* **31**, 707.
Sherman, P. (1967a). *J. Colloid Interf. Sci.* **24**, 67.
Sherman, P. (1967b). *J. Colloid Interf. Sci.* **24**, 107.
Sherman, P. (1967c). Proc. 4th Intern. Congr. Surface Activity II, 1199.
Shimmin, P. D. and Hill, R. D. (1964). *J. Dairy Res.* **31**, 121.
Skovholt, O. and Dowdle, R. L. (1950). *Cereal Chem.* **27**, 26.
Soltøft, P. (1947). Doctoral dissertation, Copenhagen. Bjorne-Kristensen, Bogtrykkeri.
Somers, F. (1965). *J. Food Sci.* **30**, 922.
Sone, T. (1961). *J. Phys. Soc. Japan* **16**, 961.
Sone, T., Fukushima, M. and Fukada, E. (1962). Proc XV Intern. Dairy Congr. III, 165.
Spöttel, W. and Gneist, K. (1941-43), *Milchw. Forsch.* **21**, 214.
Stainsby, G., Saunders, P. R. and Ward, A. G. (1953). Proc. 13th Intern. Congr. Pure App. Chem. (Upsala), Ward A. G. *Nature* **171**, 1099.
Steller, W. R. and Bailey, C. H. (1938). *Cereal Chem.* **15**, 391.
Steiner, E. H. (1958a). B.F.M.I.R.A. Research Report No. 88.
Steiner, E. H. (1958b). *Rev. Int. Choc.* **13**, 290.
Steiner, E. H. (1959). in "Rheology of Disperse Systems" (ed. C. C. Mills) Pergamon Press, London. p. 167.
Sterling, C., Shimazu, F. and Wuhrmann, J. J. (1960). *Food Res.* **25**, 630.
Sterling, C. and Wuhrmann, J. J. (1960). *Food Res.* **25**, 460.
Szabo, G. (1966a). 17th Intern. Dairy Congr. **4**, 251.
Szabo, G. (1966b). 17th Intern. Dairy Congr. **4**, 257.
Szczesniak, A. S. and Farkas, E. (1962). *J. Food Sci.* **27**, 381.
Szczesniak, A. S., Sloman, K., Brandt, M. and Skinner, E. Z. (1963). Proc. 15th Research Conf. Am. Meat Inst. Foundation, p. 121.
Szczesniak, A. S. and Torgeson, K. W. (1965). *Advances in Food Research* **14**, 33.
Tapernoux, A. and Vuillaume, R. (1934). *Lait* **14**, 449.
Torssel, H., Sanberg, V. and Thureson, L. E. (1949). Proc. 12th Intern. Dairy Congr. **2**, 246.
Trapeznikov, A. A. and Shalopalkina, T. G. (1957). *Colloid J.* (*USSR*) **19**, 243.
Tressler, D. K., Birdseye, C., and Murray, W. T. (1932). *Ind. Eng. Chem.* **24**, 242.
Tressler, D. K. and Murray, W. T. (1932). *Ind. Eng. Chem.* **24**, 890.
Timoshenko, S. and Goodier, J. N. (1951). "Theory of Elasticity" McGraw-Hill, New York.
Tobolsky, A. and Eyring, H. (1943). *J. Chem. Phys.* **11**, 125.
Tuszyński, W. and Scott Blair, G. W. (1967). *Nature* **216**, 367.
Vaeck, S. V. (1951). *Int. Choc. Rev.* **100**, 350.
Vaeck, S. V. (1952). *Int. Choc. Rev.* 325.
Vaeck, S. V. (1955). *Zucker u. Süssw, Wirtsch.* **8**, 718.

Van den Tempel, M. (1961). *J. Colloid Sci.* **16**, 284.

Vasić, I. and De Man, J. M. (1968). in "Rheology and Texture of Foodstuffs" S.C.I. Monograph No. 27, p. 251.

Voisey, P. W. and Emmons, D. B. (1966). *J. Dairy Sci.* **49**, 93.

Voisey, P. W. and Hansen, H. (1967a). *Food Technol.* **21**, 355.

Voisey, P. W. MacDonald, D. C. and Foster, W. (1967b). *Food Technol.* **21**, 361.

Volodkevich, N. N. (1938). *Food Res.* **3**, 221.

Wade, P. (1968). in "Rheology and Texture of Foodstuffs" S.C.I. Monograph No. 27, p. 225.

Ward, A. G. and Cobbett, W. G. (1968). in "Rheology and Texture of Foodstuffs" S.C.I. Monograph No. 27, p. 101.

Warner, K. F. (1927). U.S. Dept. Agr. Nat. Coop. Proj., Coop. Bur. Animal Ind. Rev. ed.

Watson, E. L. (1966). *J. Food Sci.* **31**, 373.

Wiechert, E. (1893). *Ann. Phys.* (Leipzig) **50**, 335.

White, R. K. (1966). M.S. dissertation, Pennsylvania State University, U.S.A.

Whitehead, J. and Sherman, P. (1967). *Food Technol.* **21**, 107.

Winkler, C. A. (1939). *Can. J. Res.* **17D**, 8.

Yurkstas, A. A. and Curby, W. A. (1953). *J. Prosthetic Dentistry* **3**, 82.

M

CHAPTER 5

Rheological Properties of Pharmaceutical and Cosmetic Products

1. Introduction 323
2. Rheological Studies on Ingredients of Pharmaceutical and Cosmetic
   Products 327
   A. Suspending Agents 327
   B. Ointment Bases 330
   C. Model Emulsions 335
3. Pharmaceutical and Cosmetic Products 353
   A. Lotions 353
   B. Penicillin Suspensions 358
   C. Vaccines 361
   D. Creams and Ointments 362
   E. Pressurised Foams 365
Bibliography 367

1. INTRODUCTION

Fewer rheological studies have been made with pharmaceutical and cosmetic products than with food products and, in addition, these studies have been often on a lower level of sophistication than those described in Chapter 4. The emphasis has not been concentrated on the commercial products themselves, but instead it has been distributed between actual products, the basic ingredients used therein, and model systems, with a greater emphasis on the last two mentioned categories.

Many pharmaceutical and cosmetic products are essentially emulsions, so that the discussions in Chapter 3 on the rheology of flow (Section 2), the factors which influence the rheological properties of emulsions (Sections 3 and 4), and the changes in the rheological properties of emulsions when aged (Section 5), have direct relevance to their formulation, manufacture, and application. Emulsification is a process for diluting oils so as to facilitate their use in several ways. In the pharmaceutical field a medicinal oil may not only be more acceptable orally when it is administered as an O/W

323

emulsion, but it may also be adsorbed more readily. Similarly, the application of an oil to the skin in emulsion form promotes easier spreading and adsorption. Carter (1962) enumerates the advantages of using emulsion based cosmetic products as follows ". . . oily materials can be incorporated into non-oily systems, thereby improving application properties. Another reason is that the cost of the final product can be reduced, perhaps by the elimination of solvents. Still another is that normally incompatible materials can often be incorporated in the same product, for example, oil-soluble and water-soluble ingredients. Often, emulsification increases the rate and extent of skin penetration of the cosmetic ingredients. An emulsified system sometimes offers a means of obtaining added cleansing action. Emulsion systems also aid in carrying water, an excellent softener, to the skin."

Until recently all rheological studies on pharmaceutical and cosmetic products took the form of viscosity measurements over relatively wide ranges of shear rate, but the minimum rate of shear employed was, nevertheless, such that the product suffered significant structural damage during the test. Since the aim of such studies was primarily to predict the performance of the material under normal usage conditions, such test conditions

TABLE 5.1

Summary of rheological instruments used to study the rheological properties of pharmaceutical and cosmetic product ingredients, and of model systems.

rheological instrument	material studied	authors reference
Ferranti–Shirley cone-plate viscometer	pharmaceutically interesting semi-solids in white petrolatum, petrolatum N.F., mineral oil, white wax, lanolin etc.	Boylan (1966), (1967)
Ferranti–Shirley cone-plate viscometer	sodium dodecyl sulphate— cetyl alcohol-water systems	Barry and Shotton (1967a)
Ferranti–Shirley cone-plate viscometer	O/W emulsions containing lauryl, oleyl, or cetostearyl alcohol	Talman et al. (1967)
Ferranti–Shirley cone-plate viscometer	O/W emulsions stabilised with gum acacia	Shotton and White (1960), (1963)
Ferranti–Shirley cone-plate viscometer	O/W emulsions stabilised with gelatin, sodium alginate, propylene glycol alginate, etc.	Shotton et al. (1967)
Ferranti–Shirley cone-plate viscometer	O/W emulsions stabilised with potassium arabate	Shotton and Davies (1968)
Haake "Rotovisko" viscometer	W/O emulsions	Fox and Shangraw (1966)
Epprecht "Rheomat" viscometer	vaselines	Van Ooteghem (1963), (1968)

TABLE 5.1 (continued)

rheological instrument	material studied	authors reference
Epprecht "Rheomat" viscometer	gum arabic mucilages, polyvinyl pyrrolidone, Plasdone, Luviskol K.	Münzel (1964)
cone-plate of Weissenberg rheogonimeter	sodium dodecyl sulphate-cetyl alcohol-water systems; (forced sinusoidal oscillations; small strain experiments)	Barry and Shotton (1967b)
Brookfield Synchro-Lectric viscometer	W/O emulsions	Wurdack (1959)
Brookfield LVT viscometer	W/O emulsions	Hammill (1965)
modified Brookfield viscometer	Veegum solutions	Wood et al. (1963)
automatic recording viscometer with rotating cup and stationary bob	corn oil emulsions (O/W) with methyl cellulose	Sheth et al. (1962)
Stormer viscometer	aqueous solutions of natural and synthetic gums	Kabre et al. (1964a)
Stormer viscometer	pseudoplastic suspending agents	Kabre et al. (1964b)
modified Stormer viscometer	white petrolatum—white wax mixtures; petrolatum—mineral oil mixtures ± zinc oxide; water-in-petrolatum emulsions	Kostenbauder and Martin (1954)
Cochius (falling sphere) viscometer	slime solutions of tragacanth, Manucol SS–LH or EA/KN, carboxymethylcellulose derivatives, etc.	Münzel and Schaub (1961)
cone penetrometer	vaselines	Van Ooteghem (1963), (1968)

were completely justified even though, as we shall see later, the magnitude of the shear conditions operating in practice are not known with any degree of certainty. The Brookfield range of viscometers appear to have been used very frequently in such studies, but because of the limitations of these instruments, as outlined in Table 2.2., the findings should be treated with some degree of caution. Other, more reliable, viscometers which have been employed include the Haake "Rotovisko", the Ferranti–Shirley cone-plate, the Epprecht "Rheomat", and the Stormer. Very recently some attention has been focused on the near stationary state structure, and studies on various materials have been made with a coaxial cylinder viscometer operating at a constant low shearing stress (Barry and Shotton, 1967; Warburton and Barry, 1968) as described in Section 2.E. of Chapter 2, and with a Weissenberg Rheogoniometer in oscillatory mode (Barry and Shotton, 1967).

Tables 5.1 and 5.2 list the main investigations during the past 15 years or so on the ingredients used in pharmaceutical and cosmetic products, and on the products themselves, respectively, along with the viscometers used in these studies. All discussion in the remainder of this chapter is based primarily on these studies and their findings. Rheological studies on ingredients, model emulsion systems, and products will be discussed in this order, and the observations will be correlated with principles elucidated in preceding chapters wherever this is possible.

TABLE 5.2

Summary of rheological instruments used to study the rheological properties of pharmaceutical and cosmetic products.

rheological instrument	material studied	authors reference
Ferranti–Shirley cone-plate viscometer	procaine Penicillin G suspension	Boylan and Robison (1968)
Ferranti–Shirley cone-plate viscometer	ointments	Boylan (1966)
Ferranti–Shirley cone-plante viscometer	ointments, creams, pastes	Lange and Langen-bucher (1966)
Haake "Rotovisko" viscometer	thixotropic lotions	Woodman and Marsden (1966)
Haake "Rotovisko" viscometer	ointments, creams, pastes	Lange and Langen-bucher (1966)
Haake "Viskowage" viscometer (capillary extrusion)	eye ointments and other ointments	Kragh et al. (1966) Münzel (1968)
Haake "Viskowage" (capillary extrusion)	hydrocarbons for eye ointments	Berneis and Münzel (1964)
Haake "Rotovisko" with device for viscoelastic measurements	pharmaceutical pastes	Berneis and Münzel (1964)
Epprecht "Rheomat" viscometer	thixotropic lotions	Woodman and Marsden (1966)
Epprecht "Rheomat" viscometer	ointments	Van Ooteghem (1963), (1968)
Epprecht "Rheomat" viscometer	ointments, creams, pastes	Lange and Langen-bucher (1966)
Brookfield viscometer	ointments	Fryklöf (1959)
Brookfield viscometer	fat-in-water ointments	Amman (1956)
Brookfield viscometer	ointments and creams	Neuwald (1967)
Brookfield viscometer	water-in-fat ointments	Zwicky (1956); Münzel and Zwicky (1956)

TABLE 5.2 (continued)

rheological instrument	material studied	authors reference
Brookfield viscometer	dermatologic lotions	Setnikar et al. (1968)
Hercules Hi-Shear viscometer	procaine penicillin G preparations	Ober et al. (1958)
modified Stormer viscometer	hydrophilic ointment	Kostenbauder and Martin (1954)
coaxial cylinder viscometer	pharmaceutical creams and pastes (small strain experiments)	Warburton and Barry (1968); Davis et al. (1968)
coaxial cylinder viscometer	W/O ointments (qualitative assessment of elasticity)	Zwicky (1956); Münzel and Zwicky (1956)
cone penetrometer	W/O ointments	Zwicky (1956); Münzel and Zwicky (1956); Fuller and Münzel (1962); Berneis et al. (1964).
cone penetrometer	ointments	Van Ooteghem (1963), (1968)
cone penetrometer	thixotropic ointments	Delonca et al. (1967)

2. RHEOLOGICAL STUDIES ON INGREDIENTS OF PHARMACEUTICAL AND COSMETIC PRODUCTS

A. Suspending Agents

Polymers are commonly used as suspending agents, e.g. viscosity raising fractions of natural gums, hydrocolloids such as cellulose and its derivatives, alginates, etc., and finely divided solids, e.g. calcium carbonate, magnesium aluminium silicate, etc. A primary requirement of a suspending agent when used in pharmacy is that it ". . . should be tolerant to and effectively stabilize both positively and negatively surface charged drugs" (Storz et al., 1965).

Kabre et al. (1964) studied the shear stress–rate of shear relationships for a wide range of natural and synthetic gums in aqueous solution. The materials examined included high viscosity sodium alginate (Kelcosol), medium viscosity sodium alginate (Kelgin), the potassium derivative of alginic acid (Kelmar), a microbiological polysaccharide from glucose (Kelzan), gum tragacanth, sodium carboxymethylcellulose, and methyl cellulose. All the solutions showed pseudoplastic flow behavior and conformed with the power law relationship given by Eq. (3.1) with $n_s > 1$ in all cases, its value increasing as the gum concentration increased. Equation (3.1) can be rewritten

$$\log D = n_s \log p + \log (1/\eta_p) \tag{5.1}$$

so that the value of l/η_p for each solution, and hence of η_p, could be derived by plotting log D v. log p. In addition a first order relationship was found between η_p and the concentrations (C_g) of suspending agent employed.

$$\eta_p = \exp\left(K_g C_g + B_g\right) \tag{5.2}$$

where K_g and B_g are constants with characteristic values for a particular gum (Table 5.3).

TABLE 5.3

Material constants for solutions of natural and synthetic gums (Kabre *et al.*, 1964)

	K_g	B_g
high viscosity sodium alginate	2·7042	−1·4533
medium viscosity sodium alginate	2·2744	−1·6153
sodium alginate	2·7046	−1·4424
potassium derivative of alginic acid	2·7271	−1·5443
microbiological polysaccharide from glucose	7·5840	−2·4834
gum tragacanth	2·2858	−3·6653
sodium carboxymethylcellulose	1·4222	−2·1351
methylcellulose	1·8363	−2·0426

The ability of aqueous solutions of these eight gums to maintain two insoluble drugs—zinc oxide USP, and sulfamethazine USP—in suspension was examined at four alternative values of η_p, viz. 1, 5, 10, and 20. As defined by Eq. (5.2) η_p does not have the dimensions of a viscosity. Sedimentation rates of the drugs were determined, and also the rates of visible separation from the aqueous media. When the value of η_p did not fall below 10 the drugs remained suspended for reasonable periods of time.

Another commonly used suspending agent is Veegum, a complex magnesium aluminium silicate which contains a small proportion of metal oxides. Its composition is intermediate between that of aluminium silicate and that of magnesium silicate. In suspension Veegum forms a strong gel when aged, the gel strength increasing with increasing temperature of preparation. Hysteresis D–p plots using a maximum shear rate of 45 sec^{-1} showed the initially prepared suspension to be fluid. It exhibited typical pseudoplastic behavior with both increasing and decreasing shear rate. During aging a static yield value developed, i.e. a minimum shear stress had to be applied to initiate flow, and its magnitude increased with increased aging time. Two per cent Veegum suspensions showed pseudoplastic behavior. A double logarithmic plot of shear stress against shear rate gave reasonably good straight lines in accordance with Eq. (3.1), the value of n_s ranging between 0·33 and 0·70 for aging times up to 3 h.

The temperature dependence of Veegum suspensions indicates that their apparent viscosities increase with aging time in accordance with the Arrhenius equation

$$\eta_{app} = A_b \exp(-E_A/RT) \qquad (5.3)$$

where A_b is a constant, and E_A the activation energy has a value of about 8000 cal/mole. This value represents the theoretical activation energy required to open up the crystal lattice of the slowly hydratable Veegum to water. Following Wood (1961) the build up of viscosity with aging time can be represented by

$$\log \eta_{app} = a_v \log t + b_s \qquad (5.4)$$

where a_v and b_s are constants whose values depend on temperature, aging time, and other factors such as shear rate. Because a_v is influenced by shear rate it was concluded that the static yield value increases at a faster rate with aging time than does the viscosity.

The simultaneous influence of aging time and temperature on the thickening of Veegum suspensions can be derived by combining Eqs (5.3) and (5.4)

$$2 \cdot 303 \log \eta_{app} = 2 \cdot 303 \, a_v \log t - \frac{E_A}{RT} + j_v \qquad (5.5)$$

or

$$\eta_{app} = h_v \, t^{a_v} \exp(-E_A/RT)$$

where h_v and j_v are constants.

Yakatan and Araujo (1968) found that a modified Williamson equation [Eq. (3.29)] satisfactorily defined the flow properties of CMC solutions containing $1 \cdot 6$–$2 \cdot 8 \%$ (wt/wt) of CMC. However, as the values of the constants increased with increasing CMC concentration the validity of Eq. (3.29) was by no means proven.

The viscosity–concentration relationships of solutions of many suspending agents, e.g. tragacanth, sodium alginate, propylene glycol ester of alginic acid, sodium carboxymethylcellulose, methyl cellulose, and polymeric polyvinyl pyrrolidone, can be represented in a rather different way to Eq. (5.2).

$$\log \left(\frac{\eta_{rel}}{c_h} \right) = K_h \qquad (5.6)$$

where c_h is the concentration of suspending agent and K_h is a constant (Münzel and Schaub, 1961). Using a capillary viscometer and a falling sphere viscometer to measure the viscosities of the suspending agent solutions, K_h was found to be between $0 \cdot 145$ and $4 \cdot 05$, the precise value depending on the molecular weight of the suspending agent employed.

B. Ointment Bases

Detailed rheological studies with a Ferranti–Shirley cone-plate viscometer have been made on white petrolatum and anhydrous lanolin between 20° and 35°C (Boylan, 1966; 1967). The effect of temperature on viscosity was studied, and also its effect on the change in shear stress as the rate of shear was increased to a maximum value of 1074 sec⁻¹ and then decreased to a minimum value. Upsweep and downsweep times of 10, 20, 40, 60, 120, 240, 480, and 600 sec were used. In each of these tests the upcurve and downcurve did not coincide, and a characteristic hysteresis loop (Fig. 5.1) was

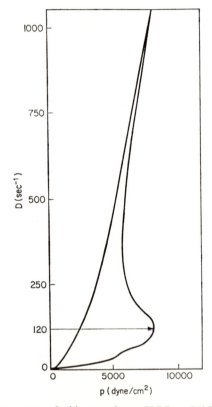

Fig. 5.1. Rheogram of white petrolatum U.S.P. at 25°C (Boylan, 1966).

obtained. It is often assumed that the area enclosed by the hysteresis loop is related to the degree to which sample structure is destroyed during the test. For example, a thixotropic index has been proposed (Kostenbauder and Martin, 1954) which is twice the difference between the apparent viscosities derived for two different maximum rates of shear divided by the natural logarithm of the square of the ratio of the higher maximum rate of shear to the lower maximum rate of shear. The significance of hysteresis loops has been the subject of much debate since the technique was proposed by Green and his co-workers (see Green, 1949). Obviously the area enclosed by the loop will depend on the rate at which bonds are ruptured in the sample, and also on the rate at which they are re-established. Both of these rates depend, in turn, on the precise nature of the times involved in increasing the shear rate to a maximum and then decreasing it, the time for which any rate of shear in the shear programme is held constant, the time for which the maximum rate of shear is maintained etc. Furthermore, it is possible for the hysteresis loop to be an artefact. For example, a Newtonian fluid may give a hysteresis loop if viscous heating raises its temperature, or alternatively, if the experiment is done too quickly and the inertial forces are large (Fredrickson, 1964). When the hysteresis curve indicates that viscosity decreases with time the sample is said to exhibit thixotropy. If the viscosity increases with time the sample exhibits anti-thixotropy.

Rheograms derived for white petrolatum by Boylan (1966, 1967) and Kostenbauder and Martin (1954) show that it exhibits thixotropic behavior. The former worker, however, observed a characteristic "spur" in the hysteresis loops at low rates of shear which does not appear to have been observed by the latter mentioned workers. In accordance with the comments

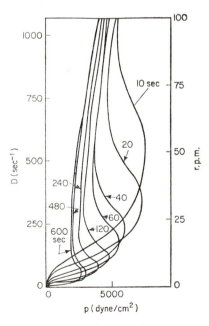

F FIG. 5.2. The effect of sweep time on the ascending portion of the white petrolatum U.S.P. rheogram (Boylan, 1967).

in the previous paragraph the areas enclosed within the hysteresis loops were greatly influenced by the time taken to increase the shear rate to its maximum value (Fig. 5.2) and then to reduce it to its minimal value (Fig. 5.3), with the downcurves being less influenced by shear time than the up-curves. Apart from the tests with a maximum shear rate of $1074 \, sec^{-1}$ (Boylan, 1966, 1967), which destroyed most of the petrolatum structure, other tests were run in which the maximum rate of shear achieved in the shear cycle was either $120 \, sec^{-1}$ or $270 \, sec^{-1}$, and the maximum shear rate was maintained for 40 sec. The time taken to reach the maximum rate of shear was very short (1–2·5 sec) so as to minimise structural breakdown before achieving maximum shear, but it introduced the danger of sample

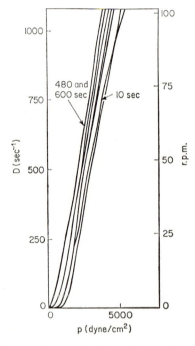

Fig. 5.3. The effect of sweep time on the descending portion of the white petrolatum U.S.P. rheogram (Boylan, 1967).

slippage. At both $120 \, sec^{-1}$ and $270 \, sec^{-1}$ maximum rate of shear most struc-tural breakdown occurred within the first 10–20 sec (Fig. 5.4), the time in-volved being longer at the lower maximum rate of shear. Anhydrous lanolin showed similar behavior at $120 \, sec^{-1}$ maximum rate of shear, but through-out the 40 sec shear it remained much stiffer than white petrolatum which was subjected to the same shear programme. The two curves for white petrolatum in Fig. 5.4 appear to be in the reverse order to what might be expected, i.e. at the higher maximum rate of shear the sample structure should be broken down to a greater extent. This apparent anomaly presumably arises because a longer time (2·5 sec) was taken to reach the maximum shear rate of 270 sec^{-1} than was taken (1 sec) to reach the maximum shear rate of 120 sec^{-1}. In the former test the sample therefore suffered less structure breakdown before the maximum shear rate was reached.

White petrolatum contains n-, iso-, and cyclic-paraffins, and its rheological behavior depends on the ratio of these three components (Schulte and Kassem, 1963). The n-paraffins are believed to be responsible for the network structure of petrolatum, while the iso-paraffins' function is to ensure that the network contains small crystallites. As the size of the crystallites decreases the viscosity of the petrolatum increases. Its thixotropic break-

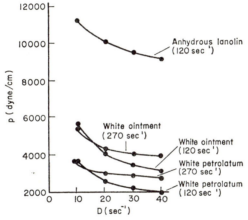

FIG. 5.4. Structural breakdown at constant shear. Data obtained using 120 sec.$^{-1}$ max and 270 sec.$^{-1}$ max (Boylan, 1967).

down does not resemble the rapid sol ⇌ gel transformation exhibited by fluid dispersed systems. Instead, it is a more permanent structural change, with the original structure and consistency not being re-established even after several months. The "spur" in the upcurve of the hysteresis loops for white petrolatum, and the pattern of structure breakdown at maximal rate of shear, is explained by Boylan (1967) in the following way. "At rest the petrolatum structure is probably a three-dimensional gel comprised of n-, iso-, and cyclic-paraffins attached through random entanglement and chemical bonding. As the shear rate increases the n-paraffins tend to align in the direction of shear, whereas the iso- and cyclic-paraffins cannot align as readily and serve to retain some of the three-dimensional character of the system. As the shear rate further increases, the paraffins become further aligned concurrent with rupture of some of the entangled paraffin chains. At this point, the difference between the upcurve and downcurve becomes minimal." Because of its content of high molecular weight components the shear behavior of anhydrous lanolin must also involve the alignment of long chain molecules in the direction of shear plus some rupture of inter-linking bonds.

Waxes are often added to ointments to make them harder. Rheological studies on the flow properties of white petrolatum containing up to 20% (wt/wt) white wax (Kostenbauder and Martin, 1954) using maximum rates of shear of 125 sec^{-1} or 210 sec^{-1} indicated that both the static (lower) yield value and the plastic viscosity increased as the % white wax increased (Table 5.4). For both maximum rates of shear the plastic viscosity increased

TABLE 5.4

Influence of white wax content on the values of some rheological parameters for white petrolatum (Kostenbauder and Martin, 1954)

% wax (wt/wt)	maximum rate of shear 125 sec^{-1}		maximum rate of shear 210 sec^{-1}	
	Static yield value (dyne/cm²)	plastic viscosity (poise)	Static yield value (dyne/cm²)	plastic viscosity (poise)
0	9900	38	11 600	26
5	11 600	48	13 600	32
10	14 200	60	14 900	40
15	20 200	94	22 700	56
20	27 400	120	33 900	66

logarithmically with the % white wax. The static yield value, and also the area within the hysteresis loop, increased curvilinearly with % white wax. The addition of liquid petrolatum to white petrolatum reduced both the static yield value and the plastic viscosity, the reduction being particularly marked at liquid petrolatum concentrations in excess of 10% (Table 5.5). Working this mixture by hand produced little further change. Van Ooteghem (1963, 1968) also observed similar effects for mixtures of liquid petrolatum and white petrolatum. When zinc oxide, a powder commonly used in the preparation of ointments, was added in concentrations of 5–20% (wt/wt) to an ointment base consisting of 30% (wt/wt) liquid petrolatum in 70% (wt/wt) white petrolatum the plastic viscosity increased slowly, but continuously, with increasing zinc oxide content. The static yield value was affected, how-ever, in a more complex manner (Table 5.6). It decreased from initially low vulaes as the zinc oxide concentration rose to 10% (wt/wt), but then it increased substantially as the zinc oxide concentration moved upward to

20% (wt/wt). The area within the hysteresis loops also increased somewhat with increasing zinc oxide concentration. In all three series of tests, i.e. in which either white wax or liquid petrolatum was added to white petrolatum, or zinc oxide was added to mixtures of liquid petrolatum and white petrolatum, the static yield value was influenced to a much greater extent than the plastic viscosity or the area within the hysteresis loop when the concentration of additive was increased.

TABLE 5.5

Influence of liquid petrolatum on the values of some rheological parameters for white petrolatum (Kostenbauder and Martin, 1954)

% liquid petro-latum (wt/wt)	maximum rate of shear 125 sec^{-1}		maximum rate of shear 210 sec^{-1}	
	static yield value (dyne/cm^2)	plastic viscosity (poise)	static yield value (dyne/cm^2)	plastic viscosity (poise)
0	9900	38	11 600	26
5	5000	35	6400	22
10	4000	30	5400	20
20	900	31	2500	18
30	800	23	2000	14

TABLE 5.6

Influence of zinc oxide on the values of some rheological parameters for a 30% (wt/wt) liquid petrolatum in 70% white petrolatum mixture (Kostenbauder and Martin, 1954)

% zinc oxide (wt/wt)	maximum rate of shear 125 sec^{-1}		maximum rate of shear 210 sec^{-1}	
	static yield value (dyne/cm^2)	plastic viscosity (poise)	static yield value (dyne/cm^2)	plastic viscosity (poise)
0	800	23	2000	14
5	200	33	1700	18
10	0	36	1500	21
15	3700	38	6600	20
20	4700	43	6500	24

C. Model Emulsions

There is a definite relationship between the degree of dispersity of a W/O emulsion, i.e. the particle size distribution, and the release of water soluble medicament contained therein (Voigt and Wolf, 1965). When a standard procedure is used to prepare a W/O emulsion based product the degree of dispersity depends upon the water content. It is highest when the water concentration is at its lowest level, and it decreases as the water content increases. Release of the medicament improves as the degree of dispersity decreases. Since the degree of dispersity influences the rheological properties of an emulsion (Chapter 3, Sections 3.A.3 and 3.A.4), the latter should also directly influence the rate of medicament release.

When mild prolonged action is required of antiseptics they are best formulated as W/O emulsions (Rotteglia, 1958). Quicker action is obtained when the antiseptic is incorporated in an O/W emulsion, e.g. in the rectal application of sodium salicylate it is more readily absorbed from an O/W emulsion than from a W/O emulsion.

Ointment bases undergo significant changes in their rheological properties when water is added to them to form W/O emulsions. Following the addition of 2% (wt/wt) of the non-ionic emulsifier Span 80 (sorbitan monooleate) to white petrolatum both the static yield value and plastic viscosity decrease gradually as the water content is increased. Table 5.7 shows data derived by Kostenbauder and Martin (1954) using the same shear programmes as those described in the previous section. It is interesting to note

TABLE 5.7

Influence of water content on the values of some rheological parameters for white petrolatum + 20% (wt/wt) Span 80 (Kostenbauder and Martin, 1954)

gm. water per 100 gm. white petrolatum − Span 80	maximum rate of shear 125 sec^{-1}		maximum rate of shear 210 sec^{-1}	
	static yield value (dyne/cm^2)	plastic viscosity (poise)	static yield value (dyne/cm^2)	plastic viscosity (poise)
0	9400	42	10 300	29
10			9400	25
25	8000	33	9200	25
50	7900	35	9400	26
75	8000	39	9400	30
100	8600	33	9500	26

that although the addition of water to Span 80–white petrolatum bases produced acceptable W/O emulsions, increasing the concentration of dispersed phase appears to have had the reverse effect to that observed when a fluid oil phase is used (Chapter 3, Section 3.A.1). The area within the hysteresis curves also decreased as the water content increased. Hydrophilic petrolatum U.S.P. XIV was affected by the addition of water in a more complex way (Table 5.8). The static yield value decreased with increasing water concentration, but both the plastic viscosity and the area within the hysteresis curves increased.

W/O emulsions prepared with a fluid continuous oil phase and an aqueous immunological agent as the internal phase have been shown to enhance antibody production in both animals and humans when they are injected intramuscularly to provide a depot of slowly released antigen (Freund and McDermott, 1942; Henle and Henle, 1945; Loveless, 1947; Brown, 1959; Woodhour et al., 1961; Rapaport, 1961). The most satisfactory clinical results were obtained using light mineral oil as the continuous phase since it is not metabolised in the body. There is, however, a wide diversity of opinion as to the most suitable form which the emulsion should take in terms of the degree of dispersity, and hence also in so far as the rheological properties are concerned. Fox and Shangraw (1966) developed a standardised procedure for making W/O emulsions so that they could study the factors which influence the rheological properties of W/O emulsions. Mannide monooleate was used as the emulsifying agent because of its general

TABLE 5.8

Influence of water content on the values of some rheological parameters for hydrophilic petrolatum (Kostenbauder and Martin, 1954)

g water per 100 g hydrophilic petrolatum	maximum rate of shear 125 sec^{-1}		maximum rate of shear 210 sec^{-1}	
	static yield value (dyne/cm^2)	plastic viscosity (poise)	static yield value (dyne/cm^2)	plastic viscosity (poise)
0	21 000	47	23 000	34
10	14 000	49	16 000	37
20	11 000	61	15 000	37
30	11 000	63	16 000	37
40	11 000	63	16 000	35

useage in W/O repository emulsions. In agreement with the discussion given in Chapter 3 emulsion viscosity was influenced by the mean particle size and particle size distribution of the freshly prepared emulsions, the rate of particle coalescence, the volume concentration of dispersed phase, the emulsifier concentration, and the homogenising procedure and homogeniser geometry. Similar conclusions were reached by Wurdack (1959), who also studied the influence of emulsifier hydrophile–lipophile balance (HLB) on viscosity at a constant rate of shear. The HLB was varied by blending together different emulsifiers in varying proportions. For a constant HLB the viscosity depended on the chemical composition of the particular emulsifiers used, but when the HLB was altered within the range 4·4 to 7·8 by varying the proportions of two specific emulsifiers the viscosity increased as the HLB decreased.

Apart from the effects exerted by the chemical nature of emulsifier blends, and their HLB values, emulsion viscosity is also influenced by which phase the emulsifiers are dissolved or dispersed in. For example, the HLB value of a mixture of Tween 80 (polyoxyethylene sorbitan monooleate; HLB 15) and Arlacel 80 (sorbitan sesquioleate; HLB 4·3) can be varied by altering the respective concentrations of the two emulsifiers in a mixture. Two series of tests have been made (Lin, 1968) in which the total emulsifier concentration was held steady at 5% (wt/wt) and the HLB value was varied between 6 and 14. In the first series of tests the Tween 80 was introduced into both the water and oil phases, and the Arlacel 80 was placed only in the oil phase, and emulsions containing 30% (wt/wt) mineral oil were prepared. The

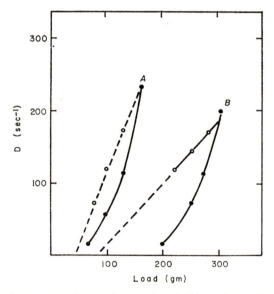

Fig. 5.5. Consistency curve for petrolatum mineral oil bases (Kostenbauder and Martin, 1954). A, petrolatum mineral oil base; B, petrolatum mineral oil base containing 10% zinc oxide. ● upcurve. ○ downcurve.

emulsions were pseudoplastic, and readings in a Brookfield viscometer using a No. 1 spindle at 30 rpm indicated that viscosity was influenced both by the HLB value and the concentration of Tween 80 initially in the aqueous phase (Fig. 5.6). At constant HLB values of 6 or 8, for example, the viscosity increased as the concentration of Tween 80 dissolved in the aqueous phase increased, the viscosity rise being particularly marked at the lower end of the Tween 80 concentration range. On the other hand, when a higher constant HLB value of 10 was used the viscosity decreased as low concentrations of Tween 80 were introduced into the aqueous phase. When 50% of the total Tween 80 concentration was added to the aqueous phase the viscosity began to increase and continued to do so with further additions of Tween 80 to the aqueous phase. At HLB values of 14 and above the location of the Tween 80, i.e. whether it was introduced into the aqueous phase, or the oil phase, or into both phases, had no effect on emulsion viscosity. When the Tween 80 was introduced only into the aqueous phase, and the Arlacel 80 was partitioned between the oil and aqueous phases along the lines previously adopted for the Tween 80, the effect on viscosity was less pronounced than before at HLB values of 8–10 (Fig. 5). Once again it was observed that the location of the Arlacel 80 was not important at the higher HLB values (12–14).

In the second series of tests a much higher concentration (70% wt/wt) of mineral oil was used. At HLB 6 the viscosity increased sharply when 0–10% (wt/wt) of the total Tween 80 concentration used was introduced into the

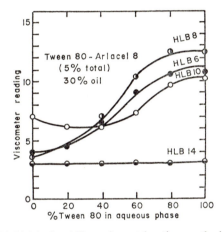

Fig. 5.6. Effect of initial hydrophilic surfactant location on the immediate viscosity of emulsions (Lin, 1968).

aqueous phase (Fig. 5.8). Conductivity measurements on these emulsions indicated that the viscosity changes were due to phase inversion from a W/O to an O/W emulsion. When the HLB was 10 the inversion did not occur until about 70% of the total concentration of Tween 80 used was introduced into the aqueous phase. Normally, inversion from W/O to O/W is accompanied by a sharp decrease in viscosity, an observation which the present data would appear to contradict. However, Lin (1968) found that the mean volume diameter of the globules decreased as the concentration of Tween 80 initially present in the aqueous phase increased, which, on the basis of the discussion in Section 3.A.3, Chapter 3, would favour an increase in viscosity.

Münzel (1947) found that a combination of emulsifiers could provide a high initial degree of dispersion in W/O emulsions, but that in many cases the emulsions were not very stable and the globule sizes changed rapidly. This would be associated with a decrease in the viscosity of each emulsion. The only blend of emulsifiers which was really satisfactory was cholesterol and lanolin.

Many O/W emulsions which are used internally contain acacia as the stabilising agent. It is a complex polysaccharide with a molecular weight which is believed to lie somewhere between 250 000 and 1 000 000. Studies (Shotton and Wibberley, 1959) on the interfacial tension between benzene and 0·1–5·0% (wt/vol) aqueous solutions of acacia, arabic acid, calcium

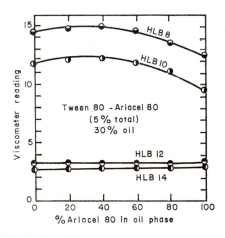

FIG. 5.7. Effect of initial lipophilic surfactant location on the immediate viscosity of emulsions (Lin, 1968).

arabate, and magnesium arabate, indicated that the reduction in interfacial tension due to addition of the emulsifier was time dependent. The reduction continued for several hours before slowing down, the rate of fall increasing with increasing concentration of emulsifier. All the interfacial tension–time curves were similar in form, which indicated that orientation of emulsifier molecules at the oil–water interface was not seriously influenced by the valency of the cation, or by the pH of the aqueous phase. The long time taken for the interfacial tension to fall to a minimum steady value was considered as evidence for adsorption at the oil–water interface giving rise to an interfacial layer which was more than one molecule thick. This belief was further substantiated by forming small oil drops (light liquid paraffin–benzene mixtures) at the tip of a micrometer syringe fitted with a fine metal needle which was immersed in 15% (wt/vol) aqueous solutions of potassium arabate (Shotton and White, 1961, 1963). The drops were withdrawn back into the syringe after they had been in contact with the aqueous solution for various times. When this time exceeded 20 sec microscopic examination of the drop during its withdrawal indicated that the surface was wrinkled, and that folds developed in it. After 15 min contact with the aqueous solution folds appeared in the surface of the drop during withdrawal and it lost its spherical shape and became polygonal.

Information about the thickness of the acacia layer around the oil drops was obtained (Shotton and Wibberley, 1960) by centrifuging the O/W emulsions so as to separate and remove as much of the aqueous phase as possible. The residue was washed several times with distilled water and centrifuged after each washing. The thickness of the interfacial acacia film derived in this fashion was found to be one tenth ($\sim$ 150 Å) of the value calculated from the anomalous viscosity shown by the original O/W emulsion. In the latter case, the (η_∞/η_0)–ϕ relationships for homogenised and non-homogenised emulsions were compared so as to determine the apparent increase in ϕ produced by the homogenisation procedure. This procedure would of course, reduce the mean particle size, but if it is assumed that a thick interfacial film of acacia is also partly responsible then the thickness can be estimated, when the particle size distribution is known, by determining the film thickness which would give the same apparent increase in ϕ. The latter is given by the formula

$$\begin{pmatrix} \text{apparent increase} \\ \text{in } \phi \end{pmatrix} = \begin{pmatrix} \text{increase in} \\ \text{surface area} \end{pmatrix} \times \begin{pmatrix} \text{interfacial film} \\ \text{thickness} \end{pmatrix} \quad (5.7)$$

The influences of particle size and particle size distribution can be minimised by working with dilute emulsions. Emulsions in which $\phi \not> 0.163$ and the oil phase was light liquid paraffin or n-heptane gave an apparent film thickness of $0.16\ \mu$. When the oil phase was a mixture of n-heptane and CCl_4 with the same density as the aqueous phase the apparent film thickness was slightly lower, with a value of $0.14\ \mu$.

Similar studies have also been made on O/W emulsions stabilised by gelatin, sodium alginate, and propylene glycol alginate, with $\phi = 0.04$–0.175 (Shotton et al., 1967). Table 5.9 shows the apparent film thicknesses calculated from the viscosity data in accordance with Eq. (5.7) when using different oil phases. Both gelatin and sodium alginate appear to give interfacial films which are thicker than acacia films. When the emulsions stabilised with gelatin were washed repeatedly with distilled water and then centrifuged the residual film thickness was substantially reduced, as with gum acacia stabilised O/W emulsions. These observations indicate that the adsorbed layer of stabiliser of gum acacia or gelatin consists of two parts, one of which is reversibly adsorbed and can be removed by the washing procedure while the other part is irreversibly adsorbed and cannot be removed by washing. For gelatin the irreversibly adsorbed layer is unlikely to exceed 150 Å in thickness. Emulsions stabilised by sodium alginate or propylene glycol alginate proved unstable on repeated washing, thus indicating that the alginate derivatives are only reversibly adsorbed.

TABLE 5.9

Apparent thickness of adsorbed layer in O/W emulsions calculated from viscosity data (Shotton *et al.* 1967)

Emulsifier	Dispersed phase	Apparent film thickness (mean value in μ)
gelatin pH 3·0	liquid paraffin	0·28
	n-heptane	0·23
sodium alginate	liquid paraffin	0·36
	light liquid paraffin	0·30
propylene glycol alginate	light liquid paraffin	0·12

Viscosity studies on more concentrated O/W emulsions ($\phi \not> 0.64$) stabilised with 0·75–12·0% (wt/wt) potassium arabate, using a Ferranti–Shirley cone–plate viscometer with a maximum shear rate of 1692 $\sec^{-1}$, and a sweep time of 600 $\sec^{-1}$, indicated (Shotton and Davis, 1968) that with increasing concentration of emulsifier the more concentrated emulsions gave upcurves and downcurves which did not coincide, i.e. hysteresis loops appeared, and the loop area increased as ϕ and emulsifier concentration increased (Table 5.10). All emulsions, apart from the most dilute ones, showed pseudoplastic flow. When the relative viscosity was calculated from the straight part of each downcurve, i.e. η_{rel} following structure breakdown, it was proportional to ϕ in the manner prescribed by Eq. (3.129) with $K_R = 2·2$. This limiting η_{rel} was not influenced by the concentration of potassium arabate when $\phi < 0·53$. At larger values of ϕ it increased as the potassium arabate concentration increased up to 8%, after which it became independent of emulsifier concentration again.

Apparent relative viscosities were calculated from the p/D ratio at maximum D, and these were found to be greatly influenced by the shapes of the upcurves, thus providing information about the residual aggregation of particles in the emulsions. The value of η_{app} increased as the concentration of potassium arabate increased provided $\phi > 0·33$, so that residual

TABLE 5.10

The influence of ϕ and potassium arabate concentration on the rheological behavior of O/W emulsions (Shotton and Davis, 1968)

Concentration of potassium arabate (%wt/wt)	ϕ	Relative viscosity		Hysteresis loop area (cm²)	Globule size		Flow behavior
		limiting	apparent		D_s (μ)	standard deviation	
0·75	0·11	1·59	—	—	4·66	2·12	N_F
	0·22	2·17	—	—	5·04	2·23	N_F
	0·33	4·20	—	—	3·57	3·76	N_F
	0·43	7·12	7·30	—	8·35	4·85	P/P
	0·53	9·30	10·4	—	21·5	2·01	P/P $Hy?$
	0·63	13·8	18·5	—	28·3	2·13	P/P $Hy?$
2·25	0·11	1·76	—	—	3·84	2·26	N_F
	0·22	2·58	—	—	4·84	2·26	N_F
	0·33	5·03	—	—	6·31	2·19	N_F
	0·43	8·33	9·59	30	13·8	1·54	P/P Hy
	0·53	11·9	15·1	27	14·5	1·62	P/P Hy
	0·63	18·6	24·4	30	17·0	1·75	P/P Hy
4·5	0·11	1·52	—	—	4·02	2·38	N_F
	0·22	2·37	—	—	6·04	2·18	N_F
	0·33	5·83	—	—	5·89	2·19	N_F
	0·43	8·84	11·7	31	13·5	1·60	P/P Hy
	0·53	12·5	18·1	40	12·3	1·63	P/P Hy
	0·63	20·6	26·9	65	13·6	1·70	P/P Hy
8·0	0·12	1·48	—	—	3·78	2·28	N_F
	0·23	2·40	—	—	5·27	2·05	P/P
	0·33	5·31	6·92	—	9·91	2·28	P/P
	0·44	8·60	11·0	54	10·3	1·65	P/P Hy
	0·54	13·4	19·3	115	13·6	1·62	P/P Hy
	0·64	20·3	30·3	155	16·8	1·74	P/P Hy
12·0	0·12	1·67	—	—	4·61	2·24	N_F
	0·23	2·57	2·58	—	4·88	2·10	P/P
	0·34	5·00	5·99	57	9·46	2·01	P/P Hy
	0·44	8·71	10·8	132	11·0	1·72	P/P Hy

N_F = Newtonian flow P/P = pseudoplasticity Hy = hysteresis loop

aggregation became more pronounced as the emulsifier concentration increased. Repeated homogenisation of non-aggregated ($\phi = 0\cdot22$) and aggregated ($\phi = 0\cdot54$) emulsions containing $1\cdot0$–$6\cdot0\%$ (wt/wt) potassium arabate resulted in the viscosity in all cases rising to a maximum and then falling away, but the aggregated emulsions required more homogenisations to reach their maximum viscosity (Table 5.11). The hysteresis loop area of the aggregated emulsions followed a similar pattern. Homogenisation breaks down the aggregates to smaller sizes but it thereby increases the number of aggregates per unit volume of emulsion. Shotton and Davis (1968) explained globule aggregation by potassium arabate along similar lines to those proposed for flocculation of solid particles by gelatin and by other polymer molecules (Smellie and La Mer, 1958; Kragh and Langston, 1962). Different globules link together because potassium arabate molecules are only partially adsorbed around them thus forming interglobular bridges.

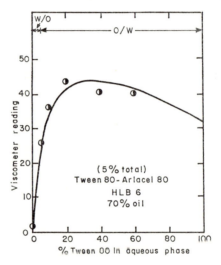

Fig. 5.8. Effect of initial hydrophilic surfactant location on viscosity and type of emulsions (Lin, 1968).

The ability of any molecule to form a bridge between globules will depend on a_m, which is calculated by Eq. (3.137). If a_m is greater than the length of the potassium arabate molecule, as it would be in dilute emulsions, then interglobular bridges cannot be formed. On the other hand, when a_m is small, which can be brought about by increasing ϕ or by substantially decreasing D_s, bridge formation becomes possible. The value of a_m at which aggregation was noticeable suggested that the length of the potassium arabate molecule is $\sim 0\cdot20$ μ. This value accords with the views of Veis and Eggenberger (1954) and Anderson and Rahman (1967) that in solution

TABLE 5.11
The influence of homogenisation on some rheological properties of O/W emulsions stabilised with potassium arabate
(Shotton and Davis, 1968)

Concentration potassium arabate (%wt/wt)	Number of passes through homogeniser	Non-aggregated emulsions ($\phi = 0.22$)		Aggregated emulsions ($\phi = 0.54$)				
		viscosity (cP)	D_s (μ)	Viscosity (cP) limiting	apparent	Loop area (cm²)	D_s (μ)	
1·0	1	6·43	18·3	16·8	20·8	4	15·1	
	2	8·05	13·2	22·6	27·2	18	14·4	
	3	7·77	12·9	23·5	28·6	17	13·8	
	4	7·40	11·8	23·7	28·7	17	13·0	
	5	7·19	9·7	24·3	29·0	16·5	12·8	
	6	6·41	8·6	25·3	30·4	17·5	12·5	
	7	6·11	7·6	24·1	29·1	14	11·7	
3·0	1	8·90	14·4	31·9	35·5	18	13·3	
	2	12·8	13·1	40·5	48·5	26	12·7	
	3	12·0	10·5	45·6	48·8	40	11·9	
	4	11·8	9·4	58·1	63·1	47	10·3	
	5	11·1	7·8	54·0	60·8	45	9·9	
	6	9·8	6·1	51·8	57·0	39	9·7	
	7	8·8	5·5					
6·0	1	14·2	10·4	52·7	63·6	31·5	12·8	
	2	13·7	9·3	78·2	99·9	58·5	12·1	
	3	13·6	6·3	92·3	104·0	63·0	9·9	
	4	13·4	5·0	86·9	114·0	63·0	8·9	
	5	13·4	4·1	65·7	87·4	52·5	8·2	
	6	13·1	3·5					
	7	13·1	3·2					

the acacia molecule is a stiff coil with a minimum length of 0·06 μ at zero charge to 0·24 μ at maximum charge, since the experimental conditions employed by Shotton and Davis (1968) were such that the acacia coil was partly unfolded due to interparticulate repulsion between ionised carboxyl groups on neighboring glucuronic acid residues.

Many O/W emulsion based pharmaceutical products contain fatty alcohols in the oil phase, and the rheological properties of the product can be altered by varying the concentration of fatty alcohol. When 0·2–16·6% (wt/wt of the oil phase) oleyl, lauryl, or cetostearyl alcohol was incorporated in the oil phase of liquid paraffin-in-water emulsions, with weight concentrations of dispersed phase ranging from 33·0% to 37·5% which contained polyoxyethylene sorbitan monolaurate or polyoxyethylene sorbitan monooleate or sodium lauryl sulphate or potassium laurate or cetomacrogol 1000* in the aqueous phase (0·5% wt/wt of aqueous phase), three patterns of behavior were observed (Table 5.12) as the viscosity altered with increasing concentration of oil soluble additive (Talman et al., 1967).

(a) With lauryl alcohol or oleyl alcohol added to the oil phase and cetomacrogol 1000 added to the aqueous phase the emulsions remained fluid, and the apparent viscosity increased slowly and linearly with increasing concentration of fatty alcohol in the oil phase. The cetomacrogol 1000 had little influence on the viscosity. Both lauryl alcohol and oleyl alcohol are miscible with liquid paraffin in all proportions at 25°C.

(b) Addition of low concentrations of cetostearyl alcohol to the oil phase when the aqueous phase contained cetomacrogol 1000 yielded fluid emulsions similar to those obtained in (a). As the concentration of cetostearyl alcohol was increased further a limiting concentration was reached beyond which the viscosity began to rise more rapidly than (a), and a static yield value appeared. This limiting concentration coincided with the concentration of cetostearyl alcohol in the emulsion which was just sufficient to saturate the oil phase. When the cetomacrogol 1000 in the aqueous phase was replaced by any of the ionic additives both the viscosity and static yield value were slightly higher at any given concentration of cetostearyl alcohol in the oil phase. The viscosity could also be altered by using oils other than liquid paraffin or light liquid paraffin as the oil phase, e.g. isopropyl myristate, arachis oil, and castor oil (Table 5.13). With these oils the limiting concentration of cetostearyl alcohol was below the concentration required for saturation of the oil phase. This also

---

*Cetomacrogol is a nonionic surfactant with a polyoxyethylene monohexadecyl ether structure.

TABLE 5.12

Influence of chemical nature of additive to oil phase on the rheological properties of O/W emulsions containing 0·5% (wt/wt) additive to the aqueous phase (Talman *et al.*, 1967)

| Oil soluble component (%wt/wt) | Water soluble component (0·5% wt/wt) | | | | | | | | | |
| | Cetomacrogol 1000 | | Sorbester Q12 | | Cetrimide | | Sodium lauryl sulphate | | Potassium laurate | |
	η_P (poise)	SYV (dyne/cm²)	η_P (poise)	SYV (dyne/cm²)	η_P (poise)	SYV (dyne/cm²)	η_P (poise)	SYV (dyne/cm²)	η_P (poise)	SYV (dyne/cm²)
oleyl alcohol										
1·0	11	0	10	0	13	0	15	0	14	0
2·0	16	0	11	0	11	0	11	0	17	0
4·0	18	0	13	0	13	0	15	0	16	0
6·0	21	0	15	0	14	0	19	0	21	0
8·0	25	0	18	0	20	0	23	0	22	0
10·0	27	0	22	0	21	0	30	0	27	0
lauryl alcohol										
1·0	15	0	15	0	39	151	19	63	13	0
2·0	15	0	15	0	64	352	69	364	17	0
4·0	16	0	19	0	58	477	105	452	25	0
6·0	18	0	20	0	74	503	130	565	85	†
8·0	21	0	23	0	79	691	161	678	165	†
10·0	28	0	25	0	94	754	198	854	210	†
cetostearyl alcohol										
0·25	12	0	12	0	11	0	14	0	10	0
0·75	15	0	13	0	15	0	18	0	11	0
1·5	51	276	29	126	46	251	53	251	28	151
2·5	126	905	76	1131	130	1005	103	477	42	754
4·0	175	2337	176	1407	269	2412	230	1344	218	2638
7·0	468	3895	336	2638	638	6659	679	6533	‡	‡

† exhibit pseudoplasticity   ‡ values too high for measurement with large cone   SYV = static yield value

TABLE 5.13

Influence of chemical nature of oil phase on the rheological properties of O/W emulsions containing 0·5% (wt/wt) Cetomacrogol in the aqueous phase (Talman et al., 1967)

Oil phase

liquid paraffin

% (wt/wt) cetostearyl alcohol	0·25	0·5	0·75	1·0	1·25	1·5	1·75	2·0	3·5
η_P (cP)	14	18	24	29	58	78	105	125	290
SYV (dyne/cm²)	0	0	0	75	276	440	565	704	2060

light liquid paraffin

% (wt/wt) cetostearyl alcohol	0·25	0·5	0·75	1·0	1·25	1·5	1·75	2·0	2·5
η_P (cP)	14	15	18	21	37	56	71	105	261
SYV (dyne/cm²)	0	0	0	0	88	264	339	628	1885

isopropyl myristate

% (wt/wt) cetostearyl alcohol	4·75	6·0	6·5	6·75	7·0	7·25	7·5	7·75	8·0	9·75
η_P (cP)	31	60	76	114	131	155	183	214	237	395
SYV (dyne/cm²)	0	276	415	653	1817	1005	1281	1533	1771	3317

TABLE 5.13 (Continued)

arachis oil

% (wt/wt) cetostearyl alcohol	0·25	1·75	2·0	2·75	3·0	3·25	3·75	5·0	7·5	10·0
η_P (cP)	10	24	28	60	94	94	143	268	464	591
SYV (dyne/cm²)	0	0	75	377	533	533	1030	2588	4259	6433

castor oil (1) constant oil concentration

% (wt/wt) cetostearyl alcohol	2·0	4·0	4·25	4·5	4·76	5·0	5·25	5·5	7·5	10·0
η_P (cP)	19	85	89	107	129	139	151	161	339	416
SYV (dyne/cm²)	0	276	377	565	766	879	980	1156	2299	3593

castor oil (2) constant total dispersed phase concentration

% (wt/wt) castor oil	48·0	47·0	46·0	45·25	44·75	44·0	43·0
% (wt/wt) cetostearyl alcohol	2·0	3·0	4·0	4·75	5·25	6·0	7·0
η_P (cP)	14	27	57	102	133	198	275
SYV (dyne/cm²)	0	0	138	616	854	1206	1709

SYV = static yield value

applied at temperatures of 28°–35°C when liquid paraffin was used as the oil phase.

(c) When the additive to the aqueous phase was other than cetomacrogol 1000, introducing lauryl alcohol to the oil phase gave emulsions whose properties depended on the nature of the former. With non-ionic additives the emulsions were fluid, but ionic additives gave more viscous emulsions which had viscosities and yield values intermediate between those derived under (a) and (b).

The three categories of emulsions showed only small differences in mean globule size, or in dispersed phase concentration, so that an explanation for the observations cannot be based upon these two factors. Similarly, the viscosity of the oil phase was not the determining factor. The explanation appeared to be that when cetostearyl alcohol was added to the oil phase in excess of its saturation concentration the excess moved across into the aqueous phase where it formed a viscous gel. This raised η_0, and hence the emulsion viscosity. For those emulsions where the emulsion viscosity began to rise noticeably before the concentration of cetostearyl alcohol in the oil phase had reached saturation point, it was suggested that the fatty alcohol was nevertheless partitioned between the oil and water phases although its concentration in the water phase might be lower than in the former case. Lauryl alcohol is completely miscible at 25°C with liquid paraffin so that the equilibrium concentration in the oil and aqueous phases would be influenced by those factors which govern a typical partition between two immiscible fluids.

Evidence to support this explanation was two-fold. First, it was possible to prepare viscous gels by interaction of cetostearyl alcohol with the aqueous phases used to prepare the emulsions (Table 5.14). Lauryl alcohol also gave

TABLE 5.14

Rheological characteristics of gels prepared from 10% (wt/wt) cetostearyl alcohol or lauryl alcohol and 0·5% (wt/wt) water soluble additive (Talman *et al.*, 1967)

Additive to aqueous phase	cetostearyl alcohol		lauryl alcohol	
	η_P (poise)	SYV (dyne/cm²)	η_P (poise)	SYV (dyne/cm²)
cetomacrogol 1000	50	1319	20	370
sorbester Q12	34	942	17	339
cetrimide	118	1922	162	2763
sodium-lauryl sulphate	107	1972	58	704
potassium laurate	129	2525	187	3995

SYV = static yield value

viscous gels with some of the aqueous phases. Second, microscopic examination of liquid paraffin-in-water emulsions prepared with cetostearyl alcohol in the oil phase and cetomacrogol 1000 or cetrimide in the aqueous phase showed filamentous structures either around the globules or dispersed in the aqueous phase (Talman and Rowan, 1968). The fatty alcohol which had migrated to the aqueous phase was believed to have combined with the surfactant initially added to the aqueous phase, the complex being in a liquid crystalline state.

A similar liquid crystalline phase develops when cetyl alcohol reacts with an aqueous solution of sodium dodecyl sulphate. The rheological properties of this system have been studied by hysteresis loop type experiments with a Ferranti–Shirley cone-plate viscometer (Barry and Shotton, 1967a), and also by subjecting the system to small dynamic strains in a Weisssenberg rheogoniometer or to small shearing stresses in creep compliance-time studies (Barry and Shotton, 1967b). The creep compliance-time studies were made with a coaxial cylinder viscometer similar to the one described in Section 2.E, Chapter 2, but more recently a more sophisticated version of this viscometer has been used in which the inner cylinder is supported by an air bearing which also supplies the stress (Warburton and Barry, 1968; Davis et al., 1968). As the molar ratio of cetyl alcohol to sodium dodecyl sulphate was increased the product changed from an off-white liquid at one end of the ratio scale to a white soft solid at the other end. The areas enclosed by the hysteresis loop increased from a minimum value of 2·6 cm² (molar ratio = 1) to a maximum value of 167·8 cm² (molar ratio = 10) when the maximum shear rate was 1632 sec⁻¹ and the sweep time was 600 sec. There did not appear to be a static yield value with any of the samples examined, i.e. the flow was never plastic. The upcurve and downcurve in each test did not coincide, and even when samples were allowed to rest for various times coincidence was not obtained, so that the time required for the destroyed structure to be re-established must have been infinitely long. Thus, the samples did not exhibit true thixotropy. When the temperature was raised all samples showed a maximum viscosity around 42·5°C at

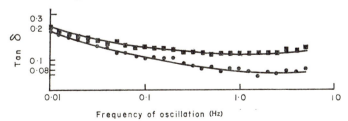

Fig. 5.9. Variation of tan phase angle (ϕ) with frequency of oscillation (Barry and Shotton, 1967b). ■ = 5/1 molar ratio of cetyl alchol to sod. dodceyl sulphate; ● = 10/1 molar ratio of cetyl alcohol to sod. dodecyl sulphate.

816–1632 sec^{-1} after which the viscosity fell away rapidly as the temperature rose still higher, and eventually at about 65°C the viscosity was very low.

Figure 5.9 shows plots of the loss tangent (tan δ) against frequency of oscillation in Weissenberg Rheogoniometer tests for molar ratios of cetyl alcohol to sodium dodecyl sulphate of 5/1 and 10/1. These plots were characterised by low values of tan δ and the slow rate at which tan δ decreased with increasing frequency of oscillation. Both observations suggest that the behavior was mainly elastic under the test conditions applied. Analysis of the shear strain–time curves obtained for the higher molar ratios (e.g. 8/1) of cetyl alcohol to sodium dodecyl alcohol (Fig. 5.10) indicated that the

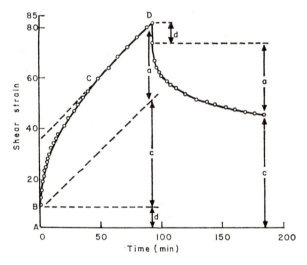

FIG. 5.10. Shear strain against time. For each unit on the shear strain axis, $\epsilon = 17 \cdot 98 \times 10^{-4}$ (Barry and Shotton, 1967b). 8/1 molar ratio of cetyl alcohol to sod. dodecyl sulphate.

creep compliance–time curves were viscoelastic. Analysis of a typical set of data by the method outlined in Section 1.C.1., Chapter 1, gave the creep compliance–time behavior in accordance with Eq. (1.30) as

$$J(t) = J_0 + J_1 \left[1 - \exp\left(-t/\tau_1\right)\right] + J_2 \left[1 - \exp\left(-t/\tau_2\right)\right]$$
$$+ J_3 \left[1 - \exp\left(-t/\tau_3\right)\right] + t/\eta_N \tag{5.8}$$

The magnitudes of the various parameters derived from Eq. (5.8) are given in Table 5.15.

Barry and Shotton (1967a) suggest that for cetyl alcohol and sodium dodecyl sulphate to interact the aqueous solution of sodium dodecyl sulphate must be above a certain critical temperature, which is about 46°C. When cetyl alcohol which has been held at a temperature of around 65°

TABLE 5.15

Viscoelastic parameters for a molar ratio of cetyl alcohol to sodium dodecyl sulphate of 8/1 derived from creep compliance-time study at a mean shear stress of 79.81 dyne/cm^2

(Barry and Shotton, 1967b)

E_0 (dyne/cm^2)	$2 \cdot 22 \times 10^4$	η_N (poise)	$1 \cdot 96 \times 10^7$
E_1 (dyne/cm^2)	$1 \cdot 12 \times 10^4$	η_1 (poise)	$1 \cdot 14 \times 10^7$
E_2 (dyne/cm^2)	$2 \cdot 39 \times 10^4$	η_2 (poise)	$4 \cdot 37 \times 10^6$
E_3 (dyne/cm^2)	$4 \cdot 36 \times 10^4$	η_3 (poise)	$1 \cdot 46 \times 10^6$

maximum shear strain < $4 \cdot 5\%$

is introduced into an aqueous solution of sodium dodecyl sulphate held at the same temperature, the molten cetyl alcohol disintegrates into droplets into which both water and sodium dodecyl sulphate penetrate to form a highly viscous liquid crystal phase. As more water enters the latter phase it becomes less viscous, surface tension forces come into effect, and spherulites form. If the temperature is now lowered below 46°C the cetyl alcohol solidifies so that it can no longer interact with sodium dodecyl sulphate. When the molar ratio of cetyl alcohol to sodium dodecyl sulphate is low a mobile suspension containing solid fatty alcohol and frozen spherulites forms below 46°C. At higher molar ratios two other effects predominate. First, the effective disperse phase concentration rises, due both to the increased fatty alcohol content and to the entrapping of water and a small amount of sodium dodecyl sulphate within the solidified fatty alcohol structural network. Second, sufficient threads of frozen liquid crystal are now available to form a submicroscopic network which entraps both solidified fatty alcohol and frozen spherulites. This results in a gel-like structure exhibiting some of the properties of solids. Microscopic examination of the reaction products of cetyl alcohol and aqueous solutions of sodium dodecyl sulphate provided support for these views.

3. PHARMACEUTICAL AND COSMETIC PRODUCTS

A. Lotions

Some interesting stress relaxation studies at constant shear rates of $3 \cdot 11$, $9 \cdot 67$, $32 \cdot 9$, $84 \cdot 5$, and 196 sec^{-1}, have been made on corticosteroid O/W lotions with an Epprecht Rheomat viscometer (Woodman and Marsden, 1966). The base for this lotion is a thixotropic aqueous dispersion of solid cetostearyl alcohol particles. After each sample had been introduced as carefully as possible into the viscometer, so as to minimise structure break-

N

down, it was allowed to stand undisturbed until the following day. The stress relaxation was found to follow the equation

$$\log (p_1/p_2) = K_e \log (t_1/t_2) \tag{5.9}$$

where p_1 and p_2 are the values of stress at times t_1 and t_2. At shear rates below 200 $\sec^{-1}$ the value of K_e remained approximately constant for each batch of lotion. Values of K_e obtained using the Epprecht Rheomat, and also the Haake "Rotovisko", on two different lotions are given in Table 5.16.

TABLE 5.16

Influence of shear rate on values of K_e derived from Eq (5·9) for two corticosteriod O/W lotions

(Woodman and Marsden, 1966)

Lotion	Viscometer	Apparent rate of shear ($\sec^{-1}$)	K_e
A	Haake " Rotovisko"	571	0·122
		347	0·133
		190	0·155
		127	0·162
		63·4	0·166
		21·2	0·164
		70·5	0·164
B	Epprecht Rheomat	196	0·195
		84·5	0·202
		32·9	0·195
		9·67	0·189
		3·11	0·197

If the shear induced loss of structure (ΔS) can be regarded as a function of the "amount of shear" (Dt) at any given shear rate, then, in general

$$\Delta S = f/(Dt) \tag{5.10}$$

From this it would follow that the shear stress data obtained at shear rate D_2 after a shear time t_1 at shear rate D_1, where $D_2 > D_1$, is identical with shear stress data obtained on a fresh sample provided a time t_2 is added to each value of t since

$$\Delta S (D_1 t_1) = \Delta S (D_2 t_2) \tag{5.11}$$

and

$$t_2 = \frac{D_1 t_1}{D_2}$$

For example, in one test a sample was sheared for 140 min at $3\cdot11$ sec^{-1} before switching over to a higher shear rate of $9\cdot67$ sec^{-1}. According to Eq. (5.11) $t_2 = 45$ min.

Setnikar *et al.* (1968) preferred to minimise the structure alteration occurring when a lotion was poured into a coaxial cylinder viscometer by using a Brookfield viscometer for their rheological tests, even though they recognised the limitations of this instrument in providing quantitative data. They prepared three dermatological lotions at 70°C according to the formulations given in Table 5.17 but omitting the propylene glycol, which was

TABLE 5.17

Composition of lotion bases (Setnikar *et al.*, 1968)

	Additive concentration (g/100ml)		
Additive	Lotion 1	Lotion 2	Lotion 3
Stearic acid	1·00	1·00	—
esters of saturated			
fatty acids	6·00	6·00	—
sorbitan monopalmitate	4·50	6·00	—
polyoxymethylene			
sorbitan monolaurate	2·00	2·00	—
cetostearyl alcohol	—	—	3·00
cetylpolyglycol ether	—	—	0·60
propylene glycol	5·00	5·00	5·00
methyl-p-oxybenzoate	0·15	0·15	0·15
propyl-p-oxybenzoate	0·03	0·03	0·03
deionised water to			
make up to	100 ml	100 ml	100 ml

added subsequently after cooling the mixtures to 30°C. The flow behavior of the lotions over a range of shear rates after carefully introducing the spindle into each sample in turn, indicated that they behaved as Bingham-type non-Newtonian fluids. At any constant low rate of shear the viscosity rose to a maximum value and then decreased somewhat to an approximately steady value as the spindle continued to rotate (Fig. 5.11). The flow behavior also depended on lotion composition, and on the intensity and time of shear, as indicated by typical shear rate–shear stress graphs for the three lotions given in Figs 5.12–5.14 immediately after stirring at 100 rev/min for different times with a two-bladed stirrer. Also included in each of these figures are data for the appropriate lotion after sucking into, and extrusion from, a metered bottling machine with a syringe-type nozzle. It is readily apparent that the structure alteration resulting from the latter process, which involved a high maximum shear rate for a very short time, is similar to that obtained by gently stirring each lotion for some time, which involved a lower maximum shear rate for a much longer time. Another significant observation was that the lotions showed significant changes in their respective viscosities during storage at 24° following preparation. However, whereas samples which had not passed through the filling machine showed a 2·5–3 fold increase in viscosity over several months storage, those samples which had passed through the filling machine were of much lower initial viscosity, and

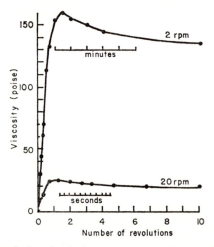

FIG. 5.11. Changes of viscosity during measurement (Setnikar et al., 1968). Spindle No. 2, 24°, at 2 rev/min (approx. 0·6 sec⁻¹ shear rate) and at 20 rev/min (approx. 6 sec⁻¹ shear rate). Viscosity appears related to the number of revolutions of the spindle.

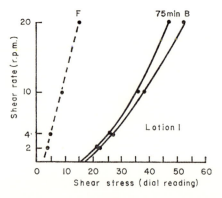

FIG. 5.12. Lotion 1, spindle No. 2, 24°. Rheogram before (B) and after 75 min (75) stirring. The dashline curve (F) is the rheogram after the run through the Farmomac. B is the up curve, the others are down curves (Setnikar et al., 1968).

did not show such viscosity changes on storage. Some typical data for lotion 1 are shown in Table 5.18. Agitation produced a much larger decrease in viscosity in the lotion after it had been stored than when it was in the fresh state. Samples which had passed through the filling machine prior to storage showed little change when agitated with the mechanical mixer, so that irreversible structure breakdown must have occurred in the filling machine.

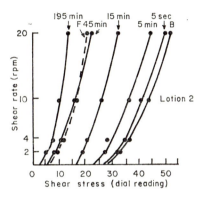

FIG. 5.13. Lotion 2, spindle No. 2, 24°. Rheogram before (B) and after different times of stirring. The total time of stirring to which the sample was subjected is shown. The dash-line curve (F) is the rheogram of the lotion after the run through the Farmomac. B is the up curve, the others are down curves (Setnikar *et al.*, 1968).

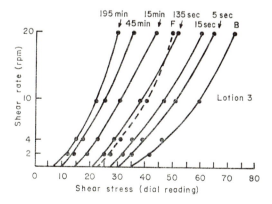

FIG. 5.14. Lotion 3, spindle No. 2, 24°. Rheogram before (B) and after different times of stirring. The total time of stirring to which the sample was subjected is shown. The dash-line curve (F) is the rheogram of the lotion after the run through the Farmomac. B is the up curve, the others are down curves (Setnikar *et al.*, 1968).

l. (1968) stressed the importance of rheological analysis of
formation is required about the following three points

formulation used provides a suitable viscosity for satisfactory
utic usage of the lotion

(b) as a means of quality control during manufacture, and selection of
appropriate bottling equipment

(c) the stability of the lotion during storage after manufacture.

Lotions have more appropriate rheological properties than creams for
use in dermatologic cases since the former require smaller shearing forces
to spread on skin, a point which has to be considered when dealing with
sensitive skin. Acceptable stability during storage, however, requires that the
viscosity is not too low otherwise suspended particles will separate. This is
avoided by formulating the lotion in such a way that it has a yield value. For
the lotions given in Table 5.17 simple calculation indicated that to prevent

TABLE 5.18

Viscosity data for Lotion 1 following preparation, and after storage for 75 days
(Setnikar *et al.*, 1968)

Batch No.	Initial Viscosity at 2 rev/min (poise at 24°)		After storage for 75 days			
			not passed through filling machine before storage		passed through filling machine before storage	
	before agitation	after agitation	before agitation	after agitation	before agitation	after agitation
1	23	22	50	34	2·8	3·3
2	19	17	53	33	4·8	4·5

separation of particles having 4 μ diameter the yield value should not be
less than $6·9 \times 10^{-3}$ dyne/cm². The lotions had yield values which were, in
fact, 100–1000 times larger than this calculated value. Sensory evaluation of
lotions with different viscosities on normal subjects and patients suffering
from skin diseases indicated that the extreme limits of acceptable viscosity,
as measured with a Brookfield No. 2 spindle operating at 20 rpm and 24°,
were 1 and 20 poise. The most desirable viscosity range was 2–5 poise.

B. Penicillin Suspensions

The rheological properties of penicillin suspensions play an important role
in their ease of intramuscular injection from a syringe, and also in their prolonged therapeutic action due to delayed absorption. Ober *et al.* (1958) have
studied in detail the way in which procaine penicillin G powder influences
the flow properties of aqueous suspensions made therefrom. Aqueous media

were prepared from sodium citrate, methyl paraben, propyl paraben, or polysorbate 80, and the penicillin powder was then added in wt/vol concentrations of 40–70%. Viscosity studies with a Hercules Hi–Shear viscometer indicated that they had a consistency similar to toothpaste and a significant static yield value. Figure 5.15 shows typical data for suspensions containing different concentrations of penicillin powder. The maximum rate of shear in these tests was 2200 sec^{-1}. In each case the upcurve and the downcurve did not coincide, so that each sample underwent some structural change when subjected to a shear cycle. Each curve exhibited a point of

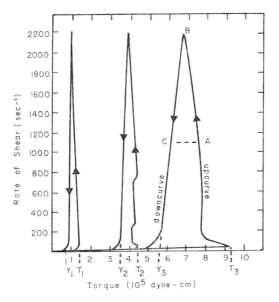

FIG. 5.15. Typical rheograms of thick aqueous procaine penicillin G suspensions (Ober *et al.*, 1958).

maximum torque (T_1, T_2, and T_3) which was supposed ". . . to reflect sharp breakdown points in the suspension structure at very low rates of shear." When a freshly prepared sample was subjected to increasing rate of shear the suspension showed infinitely large viscosity. With continued increase in the shear rate, or with continued application of a constant shear rate, a maximum torque developed and the suspension began to flow. After the shear had been applied for some time an equilibrium was set up with the sample behaving as a plastic system, but it showed slight dilatancy at the highest rates of shear. The rate of structure recovery must have been quite rapid because after a short rest time, amounting to a few minutes, the upcurve of the second hysteresis plot coincided with the first. Both solids content and the specific surface of powders with narrow particle size distributions influenced the value of T (Tables 5.19 and 5.20). According to Ober *et al.* (1958) "It was felt that these rheograms explain why such thick suspensions are inject-able and yet at the same time give prolonged therapeutic blood levels. The

TABLE 5.19

Influence of penicillin powder specific surface upon suspension *T* value
(Ober *et al.*, 1958)

| Sample No. | Particle size distribution (deciles) | | Specific surface (cm²/g) | *T* (dyne-cm) | plug point† (g) |
	first max length (μ)	last min length (μ)			
21	12	105	< 5000	0	< 3
22	8	90	10 800	50 000	3
23	2·5	39	17 000	292 000	5
24	1·9	28	19 100	573 000	5
25	1·9	26	22 000	1 020 000	< 3

†weight of penicillin powder required to plug 2 out of 3 20-gauge hypodermic needles when syringe plunger subjected to 200 lbs/in²

TABLE 5.20

Influence of penicillin powder concentration upon suspension *T* value
(Ober *et al.*, 1958)

Sample No.	% solids	*T* (dyne-cm)	plug point† (g)
29	40	132 000	10
3	50	355 000	10
11	55	578 000	> 15
27	58	780 000	5
20	60	1 000 000	5

†weight of penicillin powder required to plug 2 out of 3 20-gauge hypodermic needles when syringe plunger subjected to 200 lbs/in²

existence of a structural breakdown point, *T*, allows the thick paste to fluidize for passage through the hypodermic needle. On the other hand, the ability of the suspension to recapture quickly its structure probably accounts for the formation of a compact depot after intramuscular injection".

Suspensions with a *T* value approaching 10^6 dyne-cm tended to form plugs rather more easily in 20-gauge hypodermic needles when the syringe plunger was subjected to 200 lb/in^2 pressure (Tables 5.19 and 5.20). Suspensions with the lowest T values behaved in a similar way, but those with intermediate T values (150 000–75 000 dyne-cm) showed much less tendency to form plugs. The tendency to form depots after the penicillin suspension had been injected into rabbits or into 2% gelatin gels was also influenced by the T value. Suspensions with a T value of 100 000 dyne-cm gave depots which were slightly elongated and not as compact as the depots originating from suspensions with T values of 578 000 or 611 000 dyne-cm. Suspensions which had no T value did not form depots. Depot formation is responsible for prolonged therapeutic action since "the compact depot resists the dissolving pull of the tissue fluids by the simple means of presenting a minimum surface to these fluids".

The stability of procaine penicillin G suspensions to storage was found to depend on the temperature of storage. At 5°C hysteresis curves remained unchanged over 24 months when the maximum applied rate of shear was 750 sec^{-1}, but at 26°C or 37°C the yield value and viscosity increased pronouncedly (Boylan and Robison, 1968) with the greater change occurring at the higher temperature. These changes in rheological properties naturally influenced the ejectability of the suspension through a hypodermic needle. After three weeks storage at 37°C the suspension would not eject from a syringe.

C. Vaccines

Immunity to various virus and bacterial infections can be achieved by using emulsions of an aqueous vaccine in mineral oil. An oil-continuous phase retards the diffusion of aqueous phase from the point of injection into the surrounding body fluids. For example, animals have been able to tolerate nearly ten times the lethal dose of botulin toxin when it was emulsified in mineral oil before injection (Freund and Bonauto, 1944). The rheological properties of these W/O emulsions were influenced by all the factors discussed in Section 3, Chapter 3, and Berlin (1960) found an inverse relationship between emulsion viscosity and antibody response. The viscosity should not exceed 100 poise if the vaccine emulsion is to be injected by a conventional hypodermic syringe (Lazarus and Lachman, 1967), and it should be of somewhat lower viscosity if a continuous filling syringe is employed. On the other hand, the viscosity should not be too low or the emulsion will prove unstable on storage, and water drops will separate. In addition, the preparation would behave like an aqueous antigen preparation because there is nothing to prevent its rapid release and diffusion from the site of injection.

D. Creams and Ointments

Many pharmaceutical and cosmetic products which are applied to the skin have an ointment or cream-like basis. In order that they may spread easily on the skin without running they must not be too fluid nor yet too solid. For example, therapeutic creams are often manufactured in the form of W/O emulsions with the therapeutic ingredient contained in the dispersed phase, while the continuous phase, which may be woolwax, vaseline, paraffin etc., contributes to a butter-like consistency (Münzel, 1968). Products of this type would be most suitable for examination by the quantitative methods for semi-solids described in Chapter 3, but as yet few studies of this type have been reported. Qualitative elasticity measurements have been made on water-in-fat ointments using a coaxial cylinder viscometer with the inner cylinder suspended by a thin wire from a torsion head (Münzel and Zwicky, 1956; Zwicky, 1956). By turning the torsion head the wire was made to rotate through a certain angle, which was measured by the usual form of optical system consisting of lamp, small mirror on the wire, and graduated scale. As soon as equilibrium was established in sample deformation the wire began to return to its original position. If s_1 denotes the initial scale reading, presumably corresponding to maximum displacement of the inner cylinder, and s_2 is the difference between s_1 and the scale reading one minute later, i.e. after the cylinder begins to return to its original position, then the ratio s_1/s_2 cannot fall below unity. Water-in-vaseline ointments showed decreasing values of the ratio s_1/s_2 as the water content increased. The main change was in the values of s_2, i.e. the elasticity increased, whereas s_1 was not affected to the same degree (Table 5.21). According to Münzel and

TABLE 5.21

Influence of water content on the elasticity of water-in-vaseline emulsions
(Münzel and Zwicky, 1956; Zwicky, 1956)

%water content	original scale reading	scale reading 1 min after turning torsion head	scale reading 1 min after inner cylinder begins to return	s_1	s_2	s_1/s_2
0	0	6·0	4·5	6·0	1·5	4·0
10	0	5·0	3·5	5·0	1·5	3·3
20	0	6·0	4·0	6·0	2·0	3·0
30	0	5·0	3·0	5·0	2·0	2·5
40	0	6·0	3·5	6·0	2·5	2·4
50	0	6·0	3·0	6·0	3·0	2·0
60	0	6·0	3·0	6·0	3·0	2·0
70	0	6·0	2·5	6·0	3·5	1·7
80	0	6·5	1·5	6·5	5·0	1·3
85	0	7·0	1·0	7·0	6·0	1·2

Zwicky (1956) such elasticity phenomena were not observed with suspension-based ointments, so that the pronounced elastic behavior in the more concentrated emulsions must arise from an interlinked network of deformed water droplets.

The rheological properties of water-in-vaseline products can be modified by the addition of liquid paraffin or hard paraffin to the continuous phase, and thus augmenting the liquid or solid hydrocarbons already present in the vaseline. Van Ooteghem (1968) pointed out that as vaselines are not thixo-

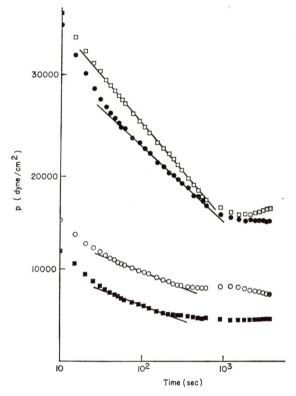

FIG. 5.16. Rheological data for four vaselines (Van Ooteghem, 1968).

tropic it is not possible to define their rate of structure alteration in shear by a thixotropic index derived from hysteresis curves. He observed that when vaselines were subjected to a constant shear rate the shear stress (p) decayed exponentially for some time. When the data was plotted as log p v. time (Fig. 5.16) each curve showed two distinct regions. The first region was approximately linear with a negative gradient whose value depended upon the composition of the vaseline, while the second region was almost flat. The latter region appeared when equilibrium was established between sample destruction and structure re-formation. By selecting the values of shearing stress exhibited at two different times on the first region of each curve it was possible to calculate the velocity (v_s) of structure breakdown from the relationship

$$v_s = \frac{p_1 - p_2}{\log (t_1/t_2)} \tag{5.12}$$

where p_1 and p_2 are the shear stresses at times t_1 and t_2 respectively. Values of v_s for four different natural vaselines shown in Fig. 5.16 are given in Table 5.22, and it is apparent that they are not related to the n–paraffin contents of the vaselines. Thus, the addition of n–paraffins does not retard

TABLE 5.22

Velocity of structure breakdown in 4 different natural vaselines at a constant rate of shear of 250 sec^{-4} as a function of time (Van Ooteghem, 1968)

vaseline	n-paraffin content	cone penetration after 5 sec (mm) (ASTM 20° double cone)	v_s. (dyne cm^{-2} sec^{-1})
1	11·27	164·1 × 10^{-1}	4529
2	8·42	170·2 × 10^{-1}	3539
3	6·14	173·6 × 10^{-1}	1460
4	11·23	164·5 × 10^{-1}	1421

the rate of structure breakdown. When liquid paraffin was added to any vaseline, up to a maximum of 50%, the value of v_s decreased, and it increased when hard paraffin was added in concentrations up to 6%. This suggests that the ratio of oil to wax is of primary importance, since natural vaseline is essentially an interlinked wax network in which is held a super-saturated oil phase (Franks, 1964).

Qualitative rheological data on ointments and vaselines have also been obtained with the cone penetrometer (e.g. Füller and Münzel, 1962; Berneis et al., 1964; Van Ooteghem, 1968). In these studies the yield value was not calculated according to Eq. (2.76), but instead the consistency was quoted as the total penetration after an arbitrarily selected time. Delonca et al. (1967) proposed that ointments be characterised by the velocity of penetration when using a single cylindrical-conical penetrometer. Van Ooteghem (1968) used a double angle ASTM cone (see remarks on this in Section 3.B, Chapter 2) and found that penetrations after 5 sec (Table 5.22) showed smaller differences between the four natural vaselines than did the v_s values. This could be related to the fact that the shear stresses in the two tests were not comparable. Berneis *et al.* (1964) proposed that ointments should be characterised by the total penetration after 20 min using cones with weights between 20 gm and 50 gm, following their observation that there appears to be a linear relationship between the yield value and the reciprocal of the total penetration for mixtures of vaseline and liquid paraffin. However, the analysis is not a simple matter since a different straight line was obtained with each vaseline of different origin. Ointments which gave penetration depths of 150×10^{-1}–200×10^{-1} mm after 5 sec using an ASTM cone penetrometer proved most suitable for dermatological purposes (Füller and Münzel, 1962).

Apart from considerations of how a cream or lotion performs when applied in the desired fashion, it is also important to know whether it will leave a sticky film on the skin. This could result in adhesion of the skin to any protective dressing, or in skin to skin adhesion (Wood and Lapham, 1964). This problem has been studied, along with the increase in stickiness to which some creams and lotions are subject when they dry out, by a modification of the hesion meter described in Section 3.C, Chapter 2. A copper disc of $\frac{3}{4}$ in diameter attached to a 1·5 oz Statham strain gauge was used (Wood *et al.*, 1964). The disc was brought into direct contact with the film of sample on the subject's arm, and the maximum force recorded by the strain gauge as the arm was slowly lowered to the break away point was taken as a measure of stickiness. Prior to the test each sample was rubbed into the skin for 60 sec.

E. Pressurised Foams

In recent years some pharmaceutical and cosmetic products have been developed in the form of pressurised foams, e.g. preparations for contraception or rectal administration of drugs, shaving creams, shampoos etc., but their rheological properties have not been studied in any detail. This may be due to the inherent instability of foams, even in the absence of shear, which results in foam bubble collapse and a rapid change in consistency.

During the second world war the rheological properties of ordinary foams were studied in detail by Blackman and his co-workers. Their findings are summarised most admirably by Matalon (1953), so that they will not be discussed here. Richman and Shangraw (1966) investigated the flow properties of pressurised foams with a Haake "Rotovisko" viscometer, and they found that these properties altered with continued discharge of the contents from a pressurised container. Up to the point where about 80% of the contents had been discharged all the *p–D* curves indicated non-Newtonian flow with decreasing curvature as the container progressively emptied. Figure

5.17 shows some typical data, but they do not provide sufficient detail at low shear stresses to indicate whether a static yield stress is present or not. Furthermore, the rates of shear covered in this study did not extend beyond $\sim 100 \text{ sec}^{-1}$, whereas much higher rates of shear (1000–10 000 sec^{-1}) are probably achieved during extrusion through the container valve (Van Ooteghem, 1967). A capillary viscometer, such as that used by Neu (1960), might, therefore, be more suitable for studying pressurised foams. The changes in flow properties with emptying of the pressurised container were attributed by Richman and Shangraw (1966) to loss of propellant from the fluid product to the increasing vapor space above it, and also to a change in the relative concentrations of the propellants due to fractionation when mixtures of propellants were used.

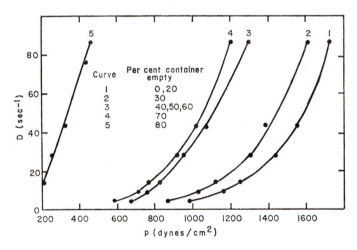

Fig. 5.17. Pressurised foam flow curves–effect of container emptying (Richman and Shangraw, 1966)

Two main factors influence the rheological properties of foams (Richman and Shangraw, 1966), either of which can predominate in a particular situation. These are change in the viscosity of the fluid between the vapor bubbles, which influences the overall foam properties, and the increase in total surface area of vapor bubbles per unit volume due to more or smaller size bubbles, which affects the yield value. In addition, the way in which the bubble walls respond when stress is applied will also be important. When the propellant concentration is increased the number of vapor bubbles per unit volume increases. Consequently, the thickness of the liquid films between the bubbles decreases and the resistance to flow is greater. Foams incorporating dichlorfluoromethane as propellant showed a lower overall consistency than foams prepared with dichlorotetra-fluoroethane or with a 1/1 mixture of the two propellants, because the bubble size was larger in the first case.

Hysteresis curve type experiments have been made with simple foams (Richman and Shangraw, 1966) derived from triethanolamine stearate and coconut fatty acid soap. The shear rate in the "Rotovisko" viscometer was increased to a maximum of $86 \cdot 33 \text{ sec}^{-1}$ and then decreased, the total time for the run being 55 sec with 5 sec intervals between each shear rate which was applied. When the lowest shear rate was reached the shear rate was increased immediately to the maximum value of $86 \cdot 33 \text{ sec}^{-1}$ and kept there for 2 min. Then the shear rate was reduced instantaneously to its minimal level, and the whole procedure was repeated several times. The p–D patterns obtained in this manner resembled to some extent those shown in Figs 5.1– 5.5, but with pressurised foams the downcurves were never linear whereas with the systems depicted in Figs 5.1–5.5 each downcurve was approximately linear over most of its length when the shear rate was reduced quickly. With pressurised foams the structure is probably never wholly broken down under the influence of shear. An equilibrium is ultimately established between foam breakdown by shear and foam development by the agitation provided when the inner cylinder of the viscometer rotates.

BIBLIOGRAPHY

Amman, R. (1956). Doctoral dissertation, Eidgenossischen Technische Hochschule, Zürich.
Anderson, D. M. W. and Rahman, S. (1967). *Carbohydrate Res.* **4**, 298.
Barry, B. W. and Shotton, E. (1967a). *J. Pharm Pharmacol.* **19**, Suppl. 110S.
Barry, B. W. and Shotton, E. (1967b). *J. Pharm. Pharmacol.* **19**, Suppl. 121S.
Barton, B. S. (1960). *J. Immunol.* **85**, 81.
Berneis, K. H. and Münzel, K. (1964). *Pharm-Acta Helv.* **39**, 88.
Berneis, K. H., Münzel, K. and Wader, T. (1964). *Pharm. Act. Helv.* **39**, 604.
Boylan, J. C. (1966). *J. Pharm. Sci.* **55**, 710.
Boylan, J. C. (1967). *J. Pharm. Sci.* **56**, 1164.
Boylan, J. C. and Robison, R. L. (1968). *J. Pharm. Sci.* **57**, 1796.
Brown, E. A. (1959). *Ann. Allergy*, **17**, 34.
Carter, P. (1962). *Am. Perfum. Cosmet.* **77**, 10.
Davis, S. S., Deer, J. J. and Warburton, B. (1968). *J. Sci. Instrum.* Series 2. **1**, 933.
Delonca, H., Dolique, R. and Bardet, L. (1967). *Ann. Pharm. Fr.* **25**, 225.
Fox, C. D. and Shangraw, R. F. (1966). *J. Pharm. Sci.* **55**, 318.
Franks, A. J. (1964). *Soap Perfum. Cosmet.* **37**, 221.
Fredrickson, A. G. (1964). "Principles and Applications of Rheology" Prentice-Hall, New Jersey.
Freund, J. and Bonauto, M. V. (1944). *J. Immunol.* **48**, 325.
Freund, J. and McDermott, K. (1942). *Proc. Soc. Exptl. Biol. Med.* **49**, 548.
Fryklöf, L. E. (1959). *Svensk. Farm. Tidskr.* **63**, 697.

Füller, W. and Münzel, K. (1962). *Pharm. Acta Helv.* **37**, 38.

Green, H. (1949). "Industrial Rheology and Rheological Structures" Wiley, New York.

Hammill, R. D. (1965). Doctoral dissertation, University of Utah.

Henle, W. and Henle, G. (1945). *Proc. Soc. Exptl. Biol. Med.* **59**, 179.

Kabre, S. P., De Kay, H. G. and Banker, G. S. (1964a). *J. Pharm. Sci.* **53**, 492.

Kabre, S. P., De Kay, H. G. and Banker, G. S. (1964b). *J. Pharm. Sci.* **53**, 495.

Kostenbauder, H. B. and Martin, A. N. (1954). *J. Am. Pharm. Assoc.* **43**, 401.

Kragh, A. M. and Langston, W. B. (1962). *J. Colloid Sci.* **6**, 528.

Kragh, G., Berneis, K. H. and Münzel, K. (1966). *Sci. Pharm.* **34**, 42.

Lange, B. and Langenbucher, F. (1966). paper presented at Intern. Soc. Fat Research Meeting, Budapest, October.

Lazarus, J. and Lachman, L. (1967). *Bull. Parenteral Drug Assoc.* **21**, 184.

Lin, J. J. (1968). *J. Soc. Cosmetic Chem.* **19**, 683.

Loveless, M. H. (1947). *Am. J. Med. Sci.* 214, 559.

Matalon, R. (1953). in "Flow Properties of Disperse Systems" ed. J. J. Hermans. p. 322. North Holland Publishing Co., Amsterdam.

Münzel, K. (1947). *Pharm. Acta Helv.* **22**, 247.

Münzel, K. (1968). *J. Soc. Cosmet. Chem.* **19**, 289.

Münzel, K. and Schaub, K. (1961). *Pharm. Acta Helv.* **36**, 647.

Münzel, K. and Zwicky, R. (1956). *Dansk. Tidskr. Farm. Suppl.* **II**, 195.

Neu, G. E. (1960). *J. Soc. Cosmetic Chem.* **11**, 390.

Neuwald, H. (1966). *J. Soc. Cosmet. Chem.* **17**, 213.

Ober, S. S., Vincent, H. C., Simon, D. E., and Frederick, K. J. (1958). *J. Am. Pharm. Assoc.* **47**, 667.

Rapaport, H. G. (1961). *N.Y. State J. Med.* **61**, 2731.

Richman, M. D. and Shangraw, R. (1966). *Aerosol Age* **11**, 32, 39.

Rotteglia, E. (1958). *Boll. chim. farm.* **92**, 72.

Schulte, K. E. and Kassem, M. A. (1963). *Pharm. Acta Helv.* **38**, 358.

Setnikar, I., Gal, C. and Fantelli, S. (1968). *J. Pharm. Sci.* **57**, 671.

Sheth, B. B., McVean, D. E. and Mattocks, A. M. (1962). *J. Pharm. Sci.* **51**, 265.

Shotton, E. and Davis, S. S. (1968). *J. Pharm. Pharmacol.* **20**, 780.

Shotton, E. and White, R. F. (1961). *Boll. Chimico. Farm.* **100**, 802.

Shotton, E. and White, R. F. (1963) in "Emulsion Rheology" (P. Sherman, ed.) p. 59, Pergamon Press, London.

Shotton, E. and Wibberley, K. (1959). *J. Pharm. Pharmacol.* **11**, Suppl. 120T.

Shotton, E. and Wibberley, K. (1960). *J. Pharm. Pharmacol.* **12**, Suppl. 105T.

Shotton, E., Wibberley, K. and Vazin, A. (1967). Proc. 4th Intern. Congr. Surface Activity, **11**, 1211.

Smellie, R. H. and La Mer, V. K. (1958). *J. Colloid Sci.* **13**, 589.

Storz, G. K., De Kay, H. G. and Banker, G. S. (1965). *J. Pharm. Sci.* **54**, 85.

Talman, F. A. J., Davies, P. J. and Rowan, E. M. (1967). *J. Pharm. Pharmacol.* **19**, 417.

Talman, F. A. J. and Rowan, E. M. (1968). *J. Pharm. Pharmacol.* **20**, 810.

Van Ooteghem, M. (1963). Doctoral dissertation, University of Leuven, Belgium.

Van Ooteghem, M. (1967). *J. Pharm. Belg.* **22**, 147.

Van Ooteghem, M. (1968). *Pharm. Acta Helv.* **43**, 764

Veis, A. and Eggenburger, D. N. (1954). *J. Am. Chem. Soc.* **76**, 1560.

Voigt, R. and Wolf, C. (1965). *Pharmazie.* **20**, 509.

Warburton, B. and Barry, B. W. (1968). *J. Pharm. Pharmacol.* **20**, 255.

Wurdack, P. J. (1959). Doctoral dissertation, University of Pittsburgh.
Wood, J. H. (1961). *Amer. Perfumer* **76**, 37.
Wood, J. H., Catacalos, G. and Lieberman, S. V. (1963). *J. Pharm. Sci.* **52**, 354.
Wood, J. H., Giles, W. H. and Catacalos, G. (1964). *J. Soc. Cosmet. Chem.* **15**, 564.
Wood, J. H. and Lapham, E. A. (1964), *J. Pharm. Sci.* **53**, 825.
Woodhour, A., Jensen, K. and Warren, J. (1961). *J. Immunol.* **86**, 681.
Woodman, M. and Marsden, A. (1966). *J. Pharm. Pharmacol.* **18**, Suppl. 198S.
Zwicky, R. (1956). Doctoral dissertation. Eidgenossischen Technischen Hochschule, Zürich.

CHAPTER 6

The Correlation of Rheological and Sensory Assessments of Consistency

1. Introduction 371
2. Consistency Profiling 372
   A. Szczesniak's Classification of Textural Characteristics . . . 372
   B. Amended Classification of Textural Characteristics 377
   C. Basis for Classifying the Consistency Characteristics of Pharmaceutical and Cosmetic Products 378
3. Correlation of Instrumental and Sensory Assessments of Consistency . 380
   A. Steady Shear Stress Tests 380
   B. Scoring Procedures and Sensory Assessment 381
   C. Instrumental Evaluation of Food Texture with a Modified M.I.T. Tenderometer 383
   D. Magnitude of Shear Forces Operating in Sensory Assessment as a Basis for Instrumental Assessment 385
   E. Usage of Statistical Procedures 390
Bibliography 391

1. INTRODUCTION

In Chapter 2 a wide range of rheological instruments was described with particular reference to their construction and principles of operation, and in Chapters 4 and 5 respectively examples were quoted of their application to many foods and pharmaceutical and cosmetic products, over the complete range of shear conditions. As pointed out in Chapter 2, if one desires to obtain information about the rheological properties of any material immediately after its manufacture it should be tested under conditions which produce minimal change during the test, i.e. very small shear stresses and shear strains are used. On the other hand, when one attempts to predict the performance of this same material under practical usage conditions, e.g. chewing a food in the mouth, or smearing a cream on the skin, the test condition requires high shear stress and high shear strain. Unfortunately, very little is known at present about the magnitude of the shear conditions involved in these practical processes, so that the prediction of performance, or consumer response, from rheological data has not reached the degree of sophistication which is desireable if rheological measurements are to be fully utilised. In order to camouflage this gap in our knowledge it is common practice to study rheological properties over as wide limits of shear conditions as a particular instrument permits, even though these limits may not include the shear conditions relevant to practical usage of the sample in question, and to draw conclusions from the derived data.

More detailed knowledge of practical usage shear conditions is essential also if one is to utilise rheological data to predict the consumer's ability to differentiate between two or more samples. Suitably designed rheological equipment is capable of identifying very small differences between samples, which the consumer would not be able to recognise. The true criterion of difference in so far as sensory assessment is concerned is, therefore, one which the test instrument can identify when operating under the same shear conditions as those encountered during product usage. Differences noted at lower shear stresses or shear rates are useful in other respects, but have no relevance to the prediction of sensory response from rheological data.

2. Consistency Profiling

A. Szczesniak's Classification of Textural Characteristics

So far we have used the term texture, or consistency, as if this was a single fundamental property which has to be assessed. In point of fact, this term embraces many properties, or attributes, and it is the way in which these properties are blended together which determines the overall texture or consistency. An analogy can perhaps be drawn with the assessment of colours which result from the blending together of several primary colours. Assessment by an untrained individual is based on his response to the blend and not to the individual constituents, but an experienced observer would be able to identify the contributions from the various primary colours. Szczesniak (1963) defined texture as ". . . the composite of the structural elements . . . and the manner in which it registers with physiological senses . . . The term 'structural elements' as used in this definition refers to the microscopic and molecular structures as well as the macroscopic structure that can be sensed visually". The validity of this extremely useful definition is restricted to the correlation of instrumental measurements with sensory assessment of texture, but it does not extend to the utilisation of instrumental data obtained under very low shear conditions which relate to sample structure only. Furthermore, the definition could perhaps be slightly amended to read " . . . the composite of those properties (attributes) which arise from the structural elements . . . and the manner in which it registers with physiological senses."

In this modified form texture is defined in terms of properties which, as will be shown later, can be assessed instrumentally, whereas in the original version the emphasis is placed on structural elements which can be identified but whose contribution to texture cannot be fully understood without measuring the physical properties arising therefrom. Using the terminology introduced later in this chapter the modified definition of texture emphasises the secondary characteristics of texture instead of its primary characteristics.

Following the methodology of Cairncross and Sjöström (1950), who proposed that the "notes" in a flavor could be identified from the order in which they appear during sensory assessment, Brandt et al. (1963) proposed an analogous procedure for evaluating texture based upon the order in which texture characteristics are identified on the palate. This procedure can be utilised to examine the textural properties of foods, and also of those pharmaceutical products which are taken orally. No similar scheme has been proposed yet for pharmaceutical and cosmetic materials used externally, but some suggestions will be given later as to how this could be done. Identification of textural characteristics during mastication can be subdivided into three phases, viz. the impression obtained on the first bite, mastication, and residual impression (Fig. 6.1). "The first bite, or initial phase, encompasses

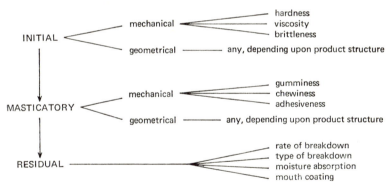

FIG. 6.1. Pattern of textural characteristic identification during mastication (Brandt et al., 1963).

the mechanical characteristics of hardness, brittleness, and viscosity, and any geometrical characteristics observed initially. The second, or masticatory, phase encompasses the mechanical characteristics of gumminess, chewiness, and adhesiveness, and any geometrical characteristics observed during chewing. The third, or residual, phase encompasses changes induced in mechanical and geometrical characteristics throughout mastication."

On the basis of the scheme proposed by Brandt et al. (1963), a system was rawn p for classifying texture characteristics (Szczesniak, 1963) which could be used for correlating instrumental evaluations of texture with sensory evaluations. Textural characteristics were subdivided into the three main classes of mechanical characteristics, geometrical characteristics, and other characteristics, the last named class referring mainly to analytical properties such as moisture content, fat content, etc. Each class was further subdivided into primary parameters and secondary parameters, the latter being parameters "... that could be adequately described by two or more of the primary terms" (Fig. 6.2). The popular terms category refers to terms most commonly used by the consumer to describe his assessment of a product.

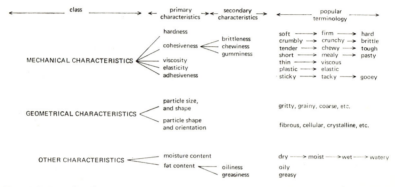

FIG. 6.2. Relation between textural parameters and popular nomenclature (Szczesniak, 1963).

While this texture profile concept is very useful several amendments could be suggested in order to place it on a firmer basis (Sherman, 1969). Some of these amendments, and the reasons for introducing them, will now be discussed.

(1) If one accepts Szczesniak's (1963) definition of a secondary parameter, then the mechanical primary parameters are indeed secondary parameters because they depend upon a blend of primary parameters which fall within the geometrical and "others" categories. This proposal is only a question of semantics, but it is nevertheless rather important. It should be recognised that the terms primary and secondary are not used in the philosophical sense proposed by Locke.

(2) Brittleness is defined "as the force with which the material fractures. It is related to the primary parameters of hardness and cohesiveness. In brittle materials, cohesiveness is low and hardness can vary from low to high." Furthermore, hardness is defined "as the force necessary to attain a given deformation", which is strictly speaking a definition of firmness, and cohesiveness is defined "as the strength of internal bonds making up the body of the product."

In point of fact hard materials have a high modulus of elasticity, which depends on the net potential energy of attraction between the basic structural elements of the material, i.e. on the forces of cohesion. Brittle materials behave as hard materials with a high modulus of elasticity until the applied stress exceeds the value of the yield stress and then cracks become visually apparent in the sample. When the attraction is due to strong primary bonds only, and there are no weak secondary bonds present, a material is brittle (Houwink, 1958). Thus, both hardness and cohesion are related to elasticity.

Biscuits (cookies) are a useful example of a brittle material. They contain very fine cracks due to the internal stresses which develop from uneven contraction during the final stage of baking (Matz, 1962). Therefore, it should be possible to interpret brittle fracture in biscuits in terms of dislocation theory and crack propagation, the latter effect proceeding with a high velocity when the yield stress is exceeded. In plastic materials, on the other hand, the rate of crack propagation depends on the stress developed in the plastic zone at the tip of the crack (Kennedy, 1962). By analogy with two dimensional brittle fracture theory for metals (Kennedy, 1962), and for other materials, the surface energy increases by $4R_d S_e$ when a fracture occurs, where R_d is the radius of the crack, and S_e is the surface energy/unit area. When a crack develops it reduces the strain energy/unit thickness by $\pi R^2_d p/2G$, where p is the applied stress, and G is the elastic shear modulus. If these two factors are the only ones involved in crack propagation, the crack will spread only when the reduction in strain energy associated with an increase in R_d is greater than the increase in surface energy.

At equilibrium

$$\frac{d}{dR_d}\left[4R_d S_e - \left(\frac{\pi R_d^2 p}{2G}\right)\right] = 0 \qquad (6.1)$$

and the criterion for crack propagation is that

$$p = \left(\frac{4GS_e}{\pi R_d}\right)^{\frac{1}{2}} \qquad (6.2)$$

(3) Adhesiveness is defined correctly, viz. "the work necessary to overcome the attractive forces between the surface of the food and the surface of other materials with which the food comes in contact (e.g. tongue, teeth, palate, etc.)." Stickiness is quoted as one of the popular terms related to this, but it depends not only on the adhesion forces but also on the forces of cohesion between the basic structural components. The relative importance of these two types of forces, which are collectively termed "hesion" (Claassens, 1958) depends on their magnitude. By analogy with what was previously stated in

Section 3.C, Chapter 2, one can say that when the adhesion forces are larger than the cohesion forces part of the sample in the mouth will adhere to the teeth etc. as they move upwards following the initial biting motion. Alternatively, when the adhesion forces are smaller than the cohesion forces the food particles will not be retained on the teeth.

The rupture process in solid samples can be treated in terms of the crack propagation model referred to under (2), when R_d is large with respect to the thickness of the crack. In the case of Newtonian fluids one can apply Stefan's law (Bikerman, 1960) for the viscous flow of liquid between two surfaces when one of the surfaces is raised. This leads to

$$\frac{de_d}{dt} = \frac{2Fe_d^3}{3\eta r} \tag{6.3}$$

where e_d is the distance between two surfaces of radius r, η is the viscosity of the liquid, and F is the force required to produce separation. Equation (6.3) indicates that F is directly proportional to the rate at which separation occurs. If the rate of separation is very high, i.e. de_d/dt is very large, there may be a clean break between the surfaces of the teeth etc., and the fluid, or alternatively some of the fluid, may adhere to the teeth etc. When the fluid is non-Newtonian, and exhibits plastic flow, Eq. (6.3) cannot be applied since the viscosity now depends on the operative rate of shear. The non-Newtonian flow of many materials can be defined by a power law type equation [Eq. (3.2)] over several decades of shear rate, and for such materials Eq. (6.3) becomes much more complex,

$$\frac{de_d}{dt} = \frac{A_s F^{n_s} e_d e_{d(0)}^{n_s+1}}{\pi^{n_s} r^{3n_s+1} (\eta_D)^{n_s}} \tag{6.4}$$

where A_S, an infinite convergent series, is a function of n_s and $e_d/e_{d(0)}$, where the limiting thickness $e_{d(0)}$ is equal to $2\pi r^2 p_0/F$. Most non-Newtonian fluids have a value of n_s which is greater than unity, and for these fluids F is less dependent on the rate of separation than are Newtonian fluids.

(4) The definitions of elasticity ("the rate at which a deformed material goes back to its undeformed condition after the deforming force is removed"), chewiness ("energy required to masticate a solid food product to a state ready for swallowing"), and gumminess ("the energy required to disintegrate a semi-solid food product to a state ready for swallowing") have little fundamental significance (Drake, 1966). Strictly speaking, elasticity is "the *property* of a material by virtue of which after deformation and upon removal of stress, it tends to recover part or all of its original size, shape, or both" (Reiner and Scott Blair, 1967). Furthermore, Szczesniak's (1963) definition of elasticity is misleading because, for example in a creep com-

pliance–time study at constant low stress, the sample may not show an instantaneous elastic recovery on removal of the stress which is identical with the instantaneous elastic deformation exhibited when the stress was applied. This phenomenon is known as elastic fatigue (Reiner, 1960).

B. Amended Classification of Textural Characteristics

In the light of these comments on Szczesniak's (1963) texture parameter classification, and also because it has been suggested that texture parameters may be related to more fundamental rheological properties, several modifications have been proposed (Sherman, 1969). Figure 6.3 shows the

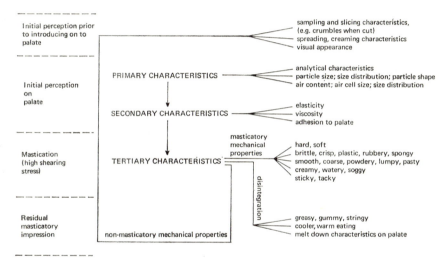

FIG. 6.3. Modified texture profile (Sherman, 1969).

amended version of the classification. No distinction is drawn now between analytical, geometrical, and mechanical attributes. The only criterion for the new classification is whether a characteristic is a fundamental property, or whether it is derived by a combination of two, or more, fundamental properties in unknown proportions. Thus, the properties previously labelled geometric and analytical characteristics are now introduced into the primary category. All other attributes are derived from these. The fundamental rheological properties elasticity, viscosity, and adhesion form the secondary category, and the remaining attributes are established as a tertiary category, since they are a complex mixture of these secondary parameters.

The tertiary category can be sub-divided further according to the mechanical process involved, viz. mastication, disintegration following mastication, or the non-masticatory mechanical treatment of the sample prior to sensory assessment on the palate. The last mentioned mechanical process was not introduced into the schemes shown in Figs 6.1 and 6.2, and yet it contributes to the initial impression of the sample which is formed by the consumer. General terms used to define tertiary masticatory and disintegratory attributes (Fig. 6.4) were selected from housewife panel assessments of, and comments about, various food products. The terms adopted are those which occurred most frequently in the responses of the panel members. It is interesting to note that the same words appeared most frequently during word association tests which were given to a panel of 100 people by Szczesniak and Kleyn (1963) to determine "their degree of texture consciousness and terms used to describe texture." Figure 6.4 indicates how these tertiary terms are distributed between solid, semi-solid, and liquid foods. The subdivisions are inevitably somewhat arbitrary and are employed mainly for ease of tabulation. They do not form distinctive grades of characteristics.

On the basis of the classification proposed in Figs 6.3 and 6.4 attributes which should be assessed in the sensory analysis of cake or ice cream textures are listed in Table 6.1. The questionnaire is divided into three sections comprising initial impression, mastication, and residual impression, thereby enabling each attribute to be assessed independently.

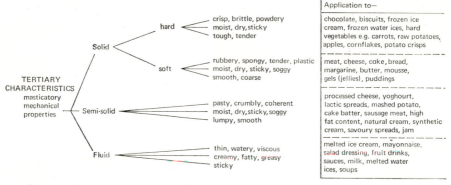

Fig. 6.4. Responses associated with the tertiary characteristics of the modified texture profile (Sherman, 1969).

C. Basis for Classifying the Consistency Characteristics of Pharmaceutical and Cosmetic Products

No similar scheme has been proposed yet for pharmaceutical or cosmetic products which are used externally, but this should not be a difficult task if the above approach is used. For example, Marriot (1961) enumerated the important properties as smoothness, viscosity, thixotropy, spreadability, extrudability (from a tube), appearance of the skin following application, i.e. whether slimy, dull, or matt. To these properties one could also add stickiness or tackiness, and also the residual greasiness of the skin following application of a lotion or cream. These properties could be collectively sub-divided into the three phases of application, viz. initial impression when contact is made between the fingers and the sample, spreading properties on the skin, and the final impression due to the residue on the skin, and whether the sample effectively produces the desired effect from a medical and/or cosmetic viewpoint.

TABLE 6.1

Suggested procedure for sensory assessment of ice cream and cake textures
(Sherman, 1969)

	factors for sensory assessment	ice cream	cake
1. Initial impression	hardness	√	√
(general handling of	elasticity	√	√
of sample)	crumbliness		√
2. Masticatory impression	smoothness	√	
	creaminess	√	
	viscosity	√	
	tackiness	√	√
	Spongy/plastic/rubbery		√
	heavy/light		√
	ease of disintegration (chewiness)		√
	moistness		√
	melt down properties on palate	√	
	residual impression of greasy, or non-greasy coating on palate	√	
3. Any other comments about texture			
4. Overall rating of texture			
5. Factors scored which have greatest influence on the overall assessment of texture			

3. Correlation of Instrumental and Sensory Assessments of Consistency

A. Steady Shear Stress Tests

Scott Blair (1949, 1960) suggested that some of the less specific terms associated with texture are amenable to dimensional analysis. For example, following Nutting (1921), an attribute related to firmness (v_f) can be defined at constant stress in terms of stress, strain, and time. For a viscous fluid

$$v_f = \eta = p\varepsilon^{-1}t = ML^{-1}T^{-1} \tag{6.5}$$

and, for an elastic solid

$$v_f = G = p\varepsilon^{-1}t^0 = ML^{-1}T^{-2} \tag{6.6}$$

so that, in general,

$$v_f = p\varepsilon^{-1}t^u = ML^{-1}T^{(u-2)} \tag{6.7}$$

where u, the "dissipation coefficient", has a value of 1 for viscous fluids and a value of 0 for elastic solids. Materials that fall midway between these two categories show fractional values of u. In this way it is possible to define textural characteristics by a limited number of physical terms.

This concept, which is discussed at greater length by Dr. Scott Blair (1969) in his new book on rheology, has some similarity to the relationship between secondary and tertiary characteristics shown in Fig. 6.3, because the tertiary characteristics describing the mechanical masticatory responses not involving stickiness depend on elasticity and viscosity, which in turn derive from the p–ε–t relationship. Although there is evidence that p varies, for example, with the hardness of a food being masticated (Neil Jenkins, 1966), it apparently remains roughly constant during mastication of a particular food, apart from some small increase towards the end of the chewing period. Presumably, spreading an ointment or a cosmetic on the skin also involves approximately steady shear stress conditions throughout most of the application time. Thus, the criterion which should be studied is the change in ε with t at constant p, as in a creep compliance–time study. Since $d\varepsilon/dt$ does not remain constant but decreases as t increases, and some textural attributes relate to higher values of $d\varepsilon/dt$ than do others, it is possibly more useful to calculate G and η from the ε–t data than to utilise the latter data as they stand.

The secondary characteristic of adhesion must be incorporated now. Each tertiary textural characteristic given in Figs 6.3 and 6.4 is a complex mixture of two or three secondary characteristics, so that in theory the former can be regarded as located in a three dimensional continuum which has the three secondary characteristics as coordinate axes. The position of each tertiary characteristic *at the operative shearing stress* can thus be defined by rectangular coordinates of the form (αG, $\beta \eta$, γA_d), where A_d is the adhesion of a food material to the palate, of a cream to the skin, etc., and α, β and γ define the location of an attribute with respect to the three coordinate axes. Both α and β decrease as one passes from a hard solid to a semi-solid and on to a fluid, while γ varies with the degree of stickiness. All solids are characterised by a high α, although it will be higher for a non-brittle solid than for a brittle one. The magnitude of β also depends on whether the solid is tough or brittle. Table 6.2 illustrates the general levels of α, β, and γ which are associated with some textural characteristics. One can envisage boundary lines running through the three dimensional continuum, as in a Phase Rule diagram, so as to define the transition from possession of a particular characteristic to its absence.

TABLE 6.2

Magnitude of some secondary characteristics associated with some tertiary characteristics (Sherman, 1969)

tertiary textural characteristic	G	η	A_d
hard, tough	very high	high	
hard, brittle (e.g. biscuits)	low	very low	
plastic	low	high	
rubbery (e.g. gels)	low	high	variable
semi-solids (e.g. cake batters, high fat cream, cosmetic creams)	low	medium	variable
fluid (e.g. salad dressing)	nil	low	variable

B. Scoring Procedures and Sensory Assessment

Textural attributes can be evaluated on the basis of the schemes proposed in Figs 6.2–6.4, or after they have been modified to suit a particular product. Each attribute should then be evaluated, i.e. by "a method for securing and recording a judgement concerning the degree to which a stimulus material possesses a specified attribute . . ." (*Inst. Food Technologists*, 1964). The most generally used methods of evaluation are ordinal, interval, and ratio scaling systems (Torgerson, 1965). When using ordinal techniques two or more samples are required which contain a particular attribute to different levels. The samples are then placed in the order which is believed to correspond to an increasing level of the attribute. Using this system of evaluation the actual differences between samples are not specified, and to define these differences one has to use interval or ratio scales. A generally used form of interval scale is based upon a straight line representation, one end of the line representing the absence or minimal presence of an attribute and the other end of the line representing possession of the attribute to a very high level. The samples are each allocated a position on the line corresponding to the level of the attribute which they are considered to possess. Such rating scales are termed hedonic scales. In the ratio scaling procedure one sample is taken as a standard and allocated a number which is supposed to define the attribute level which it possesses. All other samples are then evaluated with respect to this standard, and they are allocated numbers according to the ratio of the attribute levels possessed by each of them with respect to the standard sample. Detailed discussions on rating techniques are given, for example, by Torgerson (1965), Amerine *et al.* (1965) and Ellis (1968).

In Table 6.1 it was suggested that sensory assessment should include an indication of which attributes most influence the overall assessment of texture. This is because when the different attributes are evaluated by similar procedures one cannot deduce from the responses whether all attributes contributed equally to the overall rating for texture, or whether some attributes are considered more important than others.

Detailed sensory analysis of textural attributes is probably too sophisticated and complex a procedure to apply to the lay public in order to predict consumer response. Its main function is in the training of select groups of laboratory personnel who can be used as an analytical tool to establish correlations between instrumentally measured parameters and textural attributes. When such correlations have been established the evaluation procedures can then be extended to include evaluation in a more general manner by a group of lay consumers. Obviously the latter group will be less sensitive than a group of trained personnel, but there may, nevertheless, be some relationship between their respective performances. For example, trained and untrained personnel have been used to evaluate the spreading properties of butter on bread (Prentice, 1954). "The judgements of the panel were compared with those of a larger population of typical consumers on several occasions, and it was found that the panel average showed a bias towards being more critical of the quality of the butter than the public, to the extent of about one step on the rating scale, but that this bias remained consistent throughout the experiments."

Training of laboratory personnel to improve their sensitivity to textural attributes is a long process and can take several months. Brandt *et al.* (1963) selected nine people who had been trained previously in the principles of flavor profiling, and who had been engaged in applying this technique to a variety of food products for about four years. In the initial stages of the texture profile training the personnel were familiarised with attributes such as hardness, brittleness, viscosity, gumminess, chewiness, and adhesiveness, through repeated evaluation of selected samples which possessed these attributes to different degrees. This was followed by evaluation of the mechanical textural attributes of a wide range of food products using the rating procedures developed with the selected samples in the initial phase of training. The third phase of training was used to familiarise panel members with concepts such as powdery, grainy, gritty, coarse, lumpy, flaky, fibrous, crystalline, etc. by use of appropriate food products. Finally, detailed texture profiles were drawn up for several foods "following the prescribed method as to procedures regarding mechanics and order of perception."

C. Instrumental Evaluation of Food Texture with a Modified M.I.T. Tenderometer

Friedman *et al.* (1963) obtained good correlations between attributes evaluated in accordance with their texture profile method and instrumental evaluations by the modified M.I.T. tenderometer described in Section 8, Chapter 4 (Friedman *et al.*, 1963). In the sensory evaluation hardness was evaluated "as the force required to penetrate a substance with molar teeth", brittleness "as the ease or force with which a sample crumbles, cracks, or shatters," chewiness "in terms of the length of time in seconds required to masticate a sample at the rate of one chew per second in order to reduce it to the consistency satisfactory for swallowing," adhesiveness as "the force required to remove the material that adheres to the mouth (generally to the palate) during normal eating," and viscosity "as the force required to draw a liquid from a spoon over the tongue." Force–distance curves (Fig. 6.5)

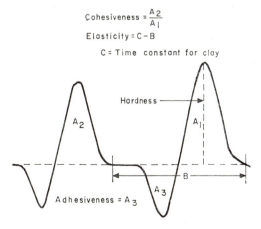

FIG. 6.5. A typical texturometer curve (Friedman *et al.*, 1963).

obtained during instrumental evaluation of texture with $\frac{1}{2}$ in high samples were analysed as follows. Reading from right to left the first peak above the horizontal dotted line traces the first penetration into the sample. The first inverted peak below the horizontal dotted line traces the withdrawal of the plunger. Repeated penetration and withdrawal provided an alternating series of peaks of both types. All textural attributes, with the exception of viscosity, were evaluated from these traces. Viscosity was evaluated using a rotating paddle attachment immersed in a cup containing the fluid sample. The paddle rotated at constant torque.

The height of the first peak in Fig. 6.5 was taken as a measure of hardness. In order to normalise the readings to a one volt input this area was divided by the voltage input. Cohesiveness was evaluated as the ratio of the area (A_2) under the second peak to the area (A_1) under the first peak, the area under each peak representing the work done during each chew. Elasticity was taken to be the distance difference (B) between the commencement of the first chew, when the plunger made contact with the sample, and the commencement of the second chew. The area (A_3) of the first inverted peak, i.e. the work which had to be done to withdraw the plunger from the sample, represented the adhesiveness. Brittle materials gave force–distance curves with many peaks, and brittleness was measured as the height of the first significant break in the peak. Chewiness was calculated as the product of hardness, cohesiveness, and elasticity.

Bourne (1968) pointed out that because the moving parts of the modified M.I.T. texturometer are driven by a lever moving about a fulcrum the plunger moves through an arc of a circle. Consequently, the area of contact between the plunger and the sample increases on the downward stroke of the plunger, and it subsequently decreases as the plunger rises. He suggested that more valid force–distance curves could be obtained by fitting the inverted load cell of an Instron tensile testing machine with a flat horizontal plate and compressing the sample.

In principle it should be possible to predict sensory responses to textural attributes from adhesion tests and mechanical strain–time tests which are made *at shearing stresses corresponding to those which operate during product usage*. Unfortunately, it is not easy to determine these stress values, Furthermore, simulation of the masticatory process, for example, requires knowledge of the way in which a food sample is moistened by the saliva which is produced following the stimuli provided by the presence of food in the mouth, and by the masticatory pressure on the teeth (Neil Jenkins, 1966). The saliva presumably weakens the sample structure, as does enzyme action, so that textural differences between samples following penetration by saliva will be significantly reduced as compared with the textural differences between them before inserting in the mouth. The precise motions of the jaws are also extremely difficult to reproduce (Drake, 1968). At present there is no instrumental method for assessing food texture which can reproduce the complete masticatory process.

D. Magnitude of Shear Forces Operating in Sensory Assessment as a Basis for Instrumental Assessment

Wood (1968) determined the magnitude of the shearing forces operating in the mouth during the assessment of fluid viscosity in a fairly simple way. For this purpose a range of pseudoplastic sauces and a Newtonian glucose syrup were used, the former consisting essentially of a dried sauce base which was mixed with varying proportions of water. A group of volunteers were then asked to taste all the samples and to indicate which of the sauce samples appeared to have a viscosity similar to the standard glucose solutions. Simultaneously the flow properties of all samples were studied using a coaxial cylinder viscometer. Figure 6.6 shows shear rate–shear stress data

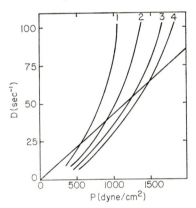

FIG. 6.6. Taste-panel comparison of Newtonian glucose syrup N and non-Newtonian sauces 1, 2, 3, 4 (Wood, 1968). The panel chose sauces 2 and 3 to be closest in consistency to syrup N.

for the samples examined in a typical experiment. The panel responses obtained in this experiment indicated that samples 2 and 3 were believed to show the closest resemblance to the standard glucose solution. The flow curves for both samples intersected the straight line for the standard glucose solution at shear rates in the vicinity of 50 sec^{-1}, viz. 45 and 50 sec^{-1}, but the corresponding shearing stresses were widely different, viz. 506 and 970 dyne/cm^2. Discussion with panel members indicated that samples were assessed orally by isolating small portions on the upper surface of the tongue, and then raising and rolling the tongue about so that the sample flowed through the channel so formed with the roof of the mouth. Thus, the sensory assessment of viscosity could be due to perception of the stress developed at 50 sec^{-1} and/or of the rate of flow through the channel. As a working hypothesis it was assumed that the shear stress developed at 50 sec^{-1} provides the sensory stimulus.

On the basis of this conclusion Wood (1968) compared the viscosities of several cream soups both instrumentally and by analysis of the sensory responses of a panel. In the panel tests the soups were compared randomly in pairs, and the panel members had to state how many times more viscous one sample was considered to be than the other. From the responses the viscosity of each sample was calculated relative to the sample with lowest viscosity. The shear stress (p) developed by each sample at 50 sec^{-1} rate of shear was also determined. When a double logarithmic plot of log p against log (estimated viscosity) was drawn up (Fig. 6.7) a straight line was obtained which had a slope of 1·28. The data was thus in agreement with the formula.

$$\psi = k_s p^{n_b} \tag{6.8}$$

or
$$\log \psi = n_b \log p + \text{constant}$$

where ψ is the sensory response by panel members, and n_b ($= 1·28$, in this case) and k_s are constants. Thus, viscometry data at an appropriate rate of shear can indicate human response to viscosity assessment on the palate.

Stevens (1957) has made detailed studies of the application of Eq. (6.8) and observed that it is valid for a large number of stimuli with n_b lying between $\sim 0·5$ and ~ 2. For example, the formula was found to be valid

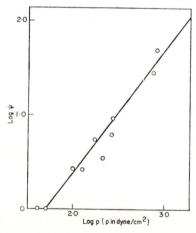

FIG. 6.7. Test of the relationship between the subjectively assessed consistency of cream soup and the measured physical stimulus (Wood, 1968). Plot of the logarithm of the estimated thickness (log ψ) against the logarithm of the shearing stress at 50 sec^{-1} (log p).

when the viscosities of a range of silicone fluids were estimated relative to one another, with $n_b = 0.42–0.46$ (Stevens and Guirao, 1964). At present there does not appear to be any explanation for the variation in n_b with different materials and tests. Treisman (1964) has questioned the general validity of Eq. (6.8), and he has pointed out that in many circumstances, depending on the method of scaling which is adopted, the relationship between stimulus and sensory response may be a logarithmic function rather than a power function, so that

$$\psi = k_s' \log p \tag{6.9}$$

which is of the same form as Fechner's law (Amerine *et al.*, 1965). Equations (6.7) and (6.8) "do not reflect empirical differences between scaling procedures but conventional differences in the assumptions made when interpreting these procedures" (Treisman, 1964). When the lower and upper thresholds are approached Eq. (6.8) may not be valid (Harper and Stevens, 1964) although it applies in the intermediate region.

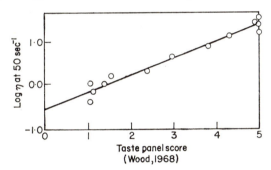

FIG. 6.8. Plot of taste-panel score against log of the viscosity of cream soups (Wood, 1968).

Irrespective of whether Eq. (6.8) or Eq. (6.9) is the true form of the relationship between stimulus and sensory response, or whether both equations are valid within certain limiting conditions, one would anticipate some form of relationship between sensory response and instrumental viscosity measurements if the stimulus is provided by the stress developed at 50 sec^{-1}, since viscosity depends upon the stress/rate of shear ratio. Wood (1968) investigated this point for a range of cream soups. In this study, panel members evaluated the viscosity of the soups by the following 5 point intensity scale.

watery thin—1 point, slightly creamy—2 points, creamy—3 points, very creamy—4 points, thick stodgy—5 points.

The geometric mean of the scores from all panel members for each soup was used for correlation with the viscosity (η) of the same sample as determined at 50 sec^{-1}. A reasonably good straight line was obtained when the values of log η were plotted against mean panel scores (Fig. 6.8), with some scatter at either end of the viscosity scale. Fryklöf (1959) obtained similar linear semi-logarithmic relationships between log η values and sensory assessment of the viscosity of glucose syrups by shaking or by stirring.

The shearing forces operating during sensory assessment of other textural attributes could presumably be determined in a similar way to that employed for viscosity, viz. by comparing the performance of commercial samples with standards. This form of approach would probably yield more reliable data than theoretical calculations of shearing forces since the latter involve many questionable assumptions. For example, Henderson et al. (1961) calculated the shear rate involved during the spreading of ointment on the skin. It was assumed that the ointment was applied at the rate of 4 strokes/sec, that each stroke was 6 cm in length, and that on average a cream layer of 0·2 cm thickness is deposited on the skin. From these data it follows that the velocity of the top layer of ointment is (6 × 4) cm/sec. Assuming that the layer of cream in contact with the skin is stationary, which is highly questionable for such a thin layer of cream, the rate of shear (D) is given by

$$D = \frac{24\text{cm/sec}}{0\cdot2\text{ cm}} = 120 \text{ sec}^{-1} \qquad (6.10)$$

The value of 120 sec^{-1} can be accepted only as indicating the order of magnitude of the operative rate of shear since the precise value of the latter depends on accurate information about all the variables involved. Lange and Langebucher (1966) have calculated the rates of shear corresponding to cream layer thicknesses of 1 cm, 0·1 cm, and 0·01 cm when a constant low velocity is employed, and showed that the corresponding D values are 10^2, 10^3, and 10^4 sec^{-1} respectively. Thus, the only conclusion that can be drawn is that the rate of shear involved in spreading cream on the skin is one which temporarily causes substantial, if not complete, structural breakdown. Kostenbauder and Martin (1954) suggested that the values quoted in Eq. (6.10) have to be slightly amended to 50 cm/sec and 0·25 cm when calculating the rate of shear produced when cream is smeared on skin by a spatula. This would give a rate of shear of 200 sec^{-1}. Van Ooteghem (1967) arrived at a similar figure of 250 sec^{-1}.

When considering the application of cream to skin the temperature of the skin must be allowed for, in addition to the rate of shear. Body temperature should, of course, also be considered when simulating masticatory processes. Skin temperature may vary from 25°–35°C depending on a variety of factors (Rothman, 1954). Boylan (1966) observed that many creams showed similar viscosities, and similar thixotropic behavior, when tested at 35°C because they were all formulated so as to melt at about skin temperature. With stiffer ointments Lange and Langenbucher (1966) found good correlation between sensory assessment of spreadability on the skin and viscosity at a very high shear rate when the latter was measured at 35°C. Viscosity data obtained at the same shear rate but at the lower temperature of 20°C did not correlate satisfactorily with sensory assessment of spreadability.

Kostenbauder and Martin (1954) divided ointments into three classes following discussion with several dermatologists about their relative hardness or stiffness. Group 1 consisted of ophthalmic ointments which are the softest ointments. Group 2 included medicated ointments such as ammoniated mercury ointment or boric acid ointment which were soft but which would, nevertheless, be stiff enough to remain on the area where they were applied. Group 3 included protective ointments such as zinc oxide paste which are sufficiently hard and stiff to remain on the area of application even when it is moist and ulcerated. Table 6.3 shows typical values of the yield stresses and viscosities for Group 1 and Group 2 ointments at rates of shear of 125 and 210 sec^{-1}, i.e. the theoretical rates of shear for smearing on the skin by finger and by spatula respectively. The data appear to show clear demarcations in the viscosity and yield stress limits for the two groups of ointments. Fryklöf (1959), on the other hand, found overlapping in the rheological properties of pharmaceutical pastes, ointments, and creams, when they were plotted as log (yield value) against log (plastic viscosity).

TABLE 6.3

Rheological properties required for effective use of ointments
(Kostenbauder and Martin, 1954)

Rheological parameter	125 sec^{-1}		210 sec^{-1}	
	group 1	group 2	group 1	group 2
yield stress (dynes/cm^2)	800 – 900	4000 – 10 000	2000 – 2500	5000 – 12 000
viscosity (poise)	23 – 31	30 – 38	14 – 18	20 – 26

Additional calculations by Henderson et al. (1961) indicate that when attempting to predict the flow characteristics of a pharmaceutical preparation from a bottle the viscosity should be determined at a somewhat lower rate of shear. Flow was assumed to occur through half of the mouth area of the bottle, and it was treated as flow in one half of a right circular cone. The flow of glycerine from a 16 oz bottle with a neck radius (r) of 0 91 cm was calculated in the following way, assuming a flow rate from the bottle of 21 ml/sec.

half volume of a right circular cone $= \frac{1}{2} (\frac{1}{3}\pi r^2 h_e) = 21$ ml/sec.

$$(6.11)$$

where h_e, the maximum velocity, is 48·5 cm/sec.

Then

$$D = \frac{48 \cdot 5 \text{ cm/sec}}{0 \cdot 91 \text{ cm}} = 53 \cdot 3 \text{ sec}^{-1} \qquad (6.12)$$

Maximum rates of shear of $\sim 90 \text{ sec}^{-1}$ were found when pouring very fluid suspensions, e.g. calamine lotion, from a bottle, with minimum rates of shear $\sim 10 \text{ sec}^{-1}$.

No correlation was found between the extrusion behavior of soft solids from tubes and their flow properties. This was not unexpected since extrusion involves compression in addition to flow. Extrusion of ointment from a pressure pack can be regarded as similar to flow in a pliable tube, and on this basis rates of shear between 1000 and 10 000 sec^{-1} are considered possible (Van Ooteghem, 1969).

E. Usage of Statistical Procedures

The correlation of instrumental and sensory assessments of consistency must be made with care so that conclusions based upon studies within restricted limits are not used as a basis for wider generalisations. When analysing the two sets of data the following points should be considered—the significance of the correlation coefficient, what properties are measured by the instrument(s) employed, the test conditions, homogeneity of the sample(s), physiological and psychological factors influencing panel members in their sensory assessments, and methods of data interpretation. Szczesniak (1968) has recently discussed the pitfalls encountered when attempting correlations between instrumental and panel data, and points out that "The correlation coefficient will continue to be used by many researchers to draw predictive type of inferences from data that, strictly speaking, demonstrate nothing more than the tendency of two factors to vary in a reasonably consistent manner when examined over a range of conditions. This being the case, it is doubly important that the advice of trained statisticians be sought in examining the data for critical factors influencing its significance. It would also be highly desirable if authors reporting correlations between objective and sensory texture measurements would give scatter diagrams from which the range of values covered and the number of samples tested could easily be seen." . . . "A great deal has been said and written about the psychology and physiology of sensory evaluation, but many questions still remain to be answered. There is no universal agreement on the best way of conducting sensory texture measurements and different research groups use different methods. Differences exist in the degree of panel training or sophistication, in the type of scales used, in the physical conditions surrounding the test situation, and whether or not a standard is supplied as a reference. All these factors affect correlations with objective measurements."

BIBLIOGRAPHY

Amerine, M. A., Pangborn, R. M. and Roessler, E. B. (1965). "Principles of Sensory Evaluation of Foods" Chapters 5, 6 and 8. Academic Press, New York.

Bikerman, J. J. (1960) in "Rheology. Theory and Applications" Vol. 3. (ed. F. R. Eirich), p. 479. Academic Press, New York.

Bourne, M. C. (1968). *J. Food Sci.* **33**, 223.

Boylan, J. C. (1966). *J. Pharm. Sci.* **55**, 710.

Brandt, M. A., Skinner, E. Z. and Coleman, J. A. (1963). *J. Food Sci.* **28**, 404.

Cairncross, S. E. and Sjöström, L. B. (1950). *Food Technol.* **4**, 308.

Claassens, J. W. (1959). Doctoral dissertation, University of Reading, England.

Drake, B. (1966). Proc. 2nd. Intern. Congr. Food Sci. Technol. Warsaw, p. 277.

Drake, B. (1968). S.C.I. Monograph No. 27 "Rheology and Texture, of Foodstuffs" p. 29.

Ellis, B. H. (1968). *Food Technol.* **22**, 583.

Friedman, H. H., Whitney, J. E. and Szczesniak, A. S. (1963). *J. Food Sci.* **28**, 390.

Fryklöf, L. E. (1959). *Svensk. Farm. Tidskr.* **63**, 697.

Harper, R. and Stevens, S. S. (1964). *Q. Jl. exp. Psychol.* **16**, 204.

Henderson, N. L., Meer, P. M. and Kostenbauder, H. B. (1961). *J. Pharm. Sci.* **50**, 788.

Houwink, R. (1958). "Elasticity, Plasticity, and Structure of Matter" p. 72. Dover, New York.

Inst. Food Technologists (1964). *Food Technol.* **18**, 1135.

Kennedy, A. J. (1962). "Processes of Creep and Fatigue" p. 353. Oliver and Boyd, London.

Kostenbauder, H. B. and Martin, A. N. (1954). *J. Am. Pharm. Assoc.* **43**, 401.

Lange, B. and Langenbucher, F. (1966). paper presented at meeting of Intern. Soc. Fat Research, Budapest.

Marriot, R. H. (1961). *J. Soc. Cosm. Chemists* **12**, 89.

Matz, S. A. (1962). "Food Texture" p. 145. Avi Publishing Co., Westport, Connecticut.

Neil Jenkins, G. (1966). "The Physiology of the Mouth" 3rd edn. Blackwell, Oxford.

Nutting, P. G. (1921). *J. Franklin Inst.* **191**, 679.

Prentice, J. H. (1954). *Lab. Pract.* **3**, 186.

Reiner, M. (1960). "Deformation, Strain, and Flow" H. K. Lewis, London.

Reiner, M. and Scott Blair, G. W. (1967) in "Rheology. Theory and Applications" Vol. 4. ed. F. R. Eirich, p. 461. Academic Press, New York.

Rothman, S. (1954). "Physiology and Biochemistry of the Skin" Univ. Chicago Press pp. 244-265.

Scott Blair, G. W. (1949). "A Survey of General and Applied Rheology" 2nd edn. p. 188. Pitman, London.

Scott Blair, G. W. (1960). S.C.I. Monograph No. 7. p. 89.

Scott Blair, G. W. (1969). "Elementary Rheology", Academic Press, London.

Sherman, P. (1969). to be published in *J. Food Sci.*

Stevens, S. S. (1957). *Psychol. Rev.* **64**, 157.

Stevens, S. S. and Guirao, M. (1964). *Science* **144**, 1157.

Szczesniak, A. S. (1963). *J. Food. Sci.* **28**, 385.

Szczesniak, A. S. (1968). *Food Technol.* **22**, 981.

Szczesniak, A. S. and Kleyn, D. H. (1963). *Food Technol.* **17**, 74.

Torgerson, W. S. (1965). "Theory and Methods of Scaling" p. 16, Wiley, New York.

Treisman, M. (1964). *Q. Jl. exp. Psychol.* **16**, 11.

Van Ooteghem, M. (1967). *J. Pharm. Belg.* 147.

Wood, F. W. (1968). S.C.I. Monograph No. 27 "Rheology and Texture of Foodstuffs" p. 40.

Yakatan, G. J. and Araujo, O. E. (1968). *J. Pharm. Sci.* **57**, 155.

Nomenclature

A	area, or London–van der Waals' constant
A'	constant in Scott Blair's rationalisation of Casson and Herschel–Bulkley equations
A_b	constant in Arrhenius equation for viscosity
A_B	constant in equation relating η_N and activation energy of flow
A_C	$= (a_x a_3 - 1)$ in Casson equation
A_d	adhesion
A_F	constant for rolling sphere viscometer
A_S	infinite convergent series
A_T	constant in Thomas modification of polynomial form of Einstein equation
A_1	$= I_m (R_2{}^2 - R_1{}^2)/4\pi\, h_c\, R_1\, R_2$
A_2	$= (I_m/32\pi\, h_c) \left[4\ln (R_1/R_2) + \left(\dfrac{R_2}{R_1} \right)^2 - \left(\dfrac{R_1}{R_2} \right)^2 \right]$
A_1', A_2', A_3', A_4'	constants in equation relating viscosity of milk to fat and solids-not-fat
B	constant in Nutting equation
B'	constant in Scott Blair's rationalisation of Casson and Herschel–Bulkley equations
B_A	minor axis dimension for deformed particle
B_C	$= (a_x a_4/D^{\frac{1}{2}})$ in Casson equation
B_F	$=$ rolling sphere viscometer constant
B_g	constant in equation relating viscosity of a gum and its concentration
B_S, B_{SC}	bending stiffness of sample, and of sample plus carrier, respectively in bending vibration experiments
B_T	constant in Thomas modification of polynomial form of Einstein equation
B_1	$= (R_2{}^2 - R_1{}^2)^2/8R_2{}^2$
B_2	$= [(R_2{}^2 - R_1{}^2) (R_2{}^4 - 5R_2{}^2 R_1{}^2 - 2R_1{}^4)$ $+ 12R_2{}^2 R_1{}^4 \ln (R_2/R_1)]/192R_2{}^2$

C	rate of particle coalescence in emulsions
C_b	constant in relationships for E' and E'' by bending vibrations
C_g	concentration
C_v	intercept in $\eta_\infty - a_m$ plot
C_w	measure of curvature in Williamson's plot for pseudo-plasticity
C_1, C_2	constants in generalized form of viscosity equation for nonspherical particles
D	rate of shear
$\mathfrak{D}$	diffusion constant
D_b	diameter of ball used in Brinell hardness test
D_c	diameter of concentric spherical enclosure in Simha's cage model for viscosity
D^e	function in Cross equation
D_f	velocity gradient in interfacial film
D_i	value of D at inflexion point in Ree–Eyring plot of η v. $\log D$
D_m	mean particle diameter
D_p	diameter of pulley in coaxial cylinder viscometer used for creep compliance-time studies
D_s	particle diameter
$D_{s(0)}$	initial value of D_s
$D_{s(t)}$	value of D_s at any time t
$D_\alpha, D_\beta, D_\gamma$	values of D in η v. $\log D$ plot according to Ree–Eyring theory
E	Young's modulus
E^*	complex shear modulus
$|E^*|$	$= \sqrt{[(E')^2 + (E'')^2]} =$ ratio of optimum stress to optimum strain
E'	storage shear modulus
E''	loss shear modulus
E_A	activation energy
E_m	Goodeve's modulus of elasticity
E_0	instantaneous elastic modulus in creep compliance-time study
E_R	mean retarded elastic modulus
E_1, E_2, E_3	components of E_R
F	force
F_A	factor depending on axial ratio
F_c	critical value of F
F_m	fat content of milk

F_r	force of resistance
F_T	total force (Goodeve)
F_t	tension due to axial component of hydrostatic force
F_V	attraction force between adjacent particles in chain
F_η	viscous resistance
G	shear modulus
G^*	complex shear modulus
G', G''	components of G
G_R	modulus of rigidity
G_τ	torque
H	heat generated/sec/unit volume of liquid (cone-plate viscometer)
$H(t)$	relaxation spectrum
H_i	height of inner cylinder of Portable Ferranti viscometer as modified for dynamic viscosity measurements
H_0	distance between flocculated particles
H_s	geometrical shape factor
H_x	correction factor in viscosity measurement by capillary viscometer for layer of continuous phase near wall of capillary
I	probability of particle collisions due to Brownian motion
I_f	impulse (Goodeve theory)
I_H	hydrodynamic interaction factor (Simha's theory)
I_m	moment of inertia
I_p	polar moment of inertia
I_r	rotational moment of inertia
I_1, I_2, I_3	strain invariants
J	axial ratios
$J(t)$	creep compliance at any time t
J^*	complex shear modulus
J'	storage compliance
J''	loss compliance
J_c	probability of particle collision due to shear
J_0	instantaneous creep compliance
J_m	mean retarded elastic compliance
J_R	mean retarded creep compliance
J_1, J_2, J_3	components of J_R
K	compression (bulk) modulus
K_a	constant in Sambuc and Naudet's relationship between penetrometer cone angle and depth of penetration
K_b	constant in equation relating yield value to depth of penetration by penetrometer

K_c $\qquad = \left[\dfrac{\eta_0}{(1-\phi)^{A_c}} \right]^{\frac{1}{2}}$ in Casson equation

K_D $\qquad$ constant relating to dependence of v_f on rate of shear

K_e $\qquad$ constant in stress relaxation-time pattern for O/W lotions

K_g $\qquad$ constant in equation relating viscosity and concentration of gums

K_0 $\qquad = \dfrac{B_c D^{\frac{1}{2}}}{A_c} \left[\left(\dfrac{1}{1-\phi} \right)^{A_c/2} -1 \right] K_h$ in Casson equation

K_R $\qquad$ constant in Richardson's equation for the viscosity of concentrated emulsions

$K_1, K_2, K_3,$
K_4, K_5, K_6 $\qquad$ constants in Gillespie's theory of pseudoplasticity

L $\qquad$ length of capillary in capillary viscometer, or of side of cube

$L(\tau)$ $\qquad$ retardation spectrum

L_a $\qquad$ thickness of sample layer in parallel plane rotational viscometer

L_A $\qquad$ major axis in shear deformed particle

L_b $\qquad$ length of bar

L_n $\qquad$ load on nozzle of relaxometer (stress relaxation of dough)

L_0 $\qquad$ L_v when $D \rightarrow 0$

L_r $\qquad$ length of bar used in free torsional vibration tests on solids

L_s $\qquad$ distance over which particles interact in presence of intervening particle

L_v $\qquad$ average number of links/chain (Cross theory)

L_{11} $\qquad$ load supported by dough at constant extension of 11 cm (Extensograph)

M $\qquad$ moment to which upper disc subjected in torsion testing of hollow cylinder

M_s $\qquad$ stiffness coefficient for coaxial cylinder viscometer in oscillatory shear tests

N $\qquad$ number of particles/cc

N_b $\qquad$ number of links formed between 2 particles (Gillespie)

N_g $\qquad$ average number of links/particle

N_k $\qquad$ number of aggregates containing k particles

N $\qquad$ number of links/cc (Goodeve)

N_0 $\qquad$ original number of particles/cc dispersion

N_t $\qquad$ number of particles/cc at time t

N_1 $\qquad$ number of unassociated particles/cc dispersion

N_2 $\qquad$ number of aggregates containing 2 particles

M	constant in Nutting equation
M_F	reading with Portable Ferranti viscometer
M_0	restoring couple/angular displacement
P	average pressure
P^a	applied pressure (capillary viscometer, etc.)
P_d	pressure distribution in semi-infinite body (Boussinesq's theory)
P_f	proportionality factor (van den Tempel's theory)
P_H	homogenization pressure
P_0	pressure corresponding to unit particle size
P_s	algebraic sum of applied pressure and the initial hydrostatic pressure (capillary viscometer)
Q	volume of liquid flowing/sec through capillary
R	gas constant
(Re)	Reynold's constant
R_c	capillary radius
R_d	radius of a crack
R_{ij}	radius of collision for particles i and j
R_p	radius of cone in cone-plate viscometer
R_s	radius of tube in rolling sphere viscometer
R_t	radius of wide arm of Saunders and Ward's tube for modulus of rigidity of gels
R_1, R_2	radii of cylinders in coaxial cylinder viscometer
S_c	slippage coefficient
S_e	surface energy/unit area
S_i	spreadibility index for fats
S_n	solids-not-fat content of milk
T	temperature on absolute scale
T_e	tension
T_r	temperature coefficient of rigidity
V	potential energy, or volume
V_A	attraction potential
V_{max}	maximum value of V in V-H_0 curve
V_0	original volume of sample
V_R	repulsion potential
V_S	strained volume of sample
W	probability of collisions between particles
W_d	decrease in fat hardness on being worked
W_e	work done in extending dough sample in Extensometer
W_f	work done during shortening of muscle fibre
W_1, X_1	instrumental constants for coaxial cylinder viscometer in oscillatory shear tests

X	displacement in bending vibration tests on hard solids
X_a	apparent increase in volume fraction of dispersed phase
X_s	intercept in plot of log (η_∞/η_0) v. $1/a_m$
Y_R	constant in Broughton and Squires' amended version of Richardson's viscosity equation for concentrated emulsions
Y'	damping coefficient for coaxial cylinder viscometer in oscillatory shear tests
Z	$= [1 + (\alpha_k \rho R^4/16 L^2 \eta^2) (P_0 - C)]^{\frac{1}{2}}$
Z'	instrumental constant for coaxial cylinder viscometer in oscillatory shear tests
Z^0	$= [1 + (\alpha_k \rho R^4/16 L^2 \eta^2) P_0]^{\frac{1}{2}}$
Z_d	from relationship $Z_d \tan Z_d = (m_t/m_c)$
Z_i	valency of ion
Z_t	torque on inner cylinder of coaxial cylinder viscometer in oscillatory shear tests
Z_1	constant in equation for E' using cone-plate viscometer in oscillatory shear tests
a	constant in Einstein's equation for viscosity of very dilute dispersions
a_b	constant in function expansion of $c_2 f(p_a)$ in generalised equation for non-spherical particles
a_c	width of sample used in free torsional vibration studies on solids
a_d	distance from centre of area over which cylindrical die acting in compression testing of fruit and vegetables (Boussinesq's treatment)
a_e	defines influence of elastic after effect
a_h	constant in Fincke–Heinz equation for influence of temperature on viscosity of molten chocolate
a_i	constant in equation relating depth of indentation to application time of stress in cheese studies
a_m	mean distance between deflocculated particles
a_N	constant in Nutting equation
a_n	$= \dfrac{\lambda^e \lambda_2 \lambda_3}{2kT}$ (Ree-Eyring)
a_p	half-angle of cone of cone penetrometer
a_r	amplitude ratio
a_v	constant in equation for increase in viscosity of Veegum suspensions with aging time

a_x	$= \dfrac{(\tan \theta_c \cos \theta_p)^2}{(\tan \theta_c \cos \theta_p)^2 + 1}$ (Casson equation)
a_3, a_4	constants in Casson equation
b	constant in polynomial expansion of ϕ
b_a	constant in equation for t_a
b_b	constant in expansion of $c_2 f(p_a)$ in generalised equation for non-spherical particles
b_j	geometric factor in equations for bending vibration testing of solids
b_G	residual viscosity
b_h	constant in Fincke–Heinz equation for influence of temperature on molten chocolate viscosity
b_i	constant in equation relating time of stress application and depth of indentation (cheese tests)
b_r	constant associated with relaxation time (Eq. 4.20)
b_s	constant in equation for viscosity increase of Veegum suspensions with aging
b_v	coefficient of viscoelastic resistance (Williamson equation)
b_1	constant of proportionality (Williamson equation)
c	constant in polynomial expansion of ϕ
c_a	constant in equation for kinetic energy factor in capillary viscometry
c_d	constant relating shear stress and rate of shear
c_f	constant in Fincke–Heinz equation for influence of temperature on molten chocolate viscosity
c_h	concentration of suspending agent
c_N	constant in Nutting equation relating applied force and deformation
c_s	constant in equation for η_∞ using hesion test
c_1, c_2	constants in generalized equation for non-spherical particles
d	additional constant in polynomial expansion of ϕ
d_f	diffusivity of cone and plate in cone-plate viscometer
d_i	diameter of impression in Brinell hardness test
d_l	width of lateral faces of vibrating metal blade in vibrating reed viscometer
d_n	non-recoverable portion of deformation in compression test
d_p	depth of penetration (penetrometer)
d_s	distance from sphere to base of tube in rolling sphere viscometer
e	power of D in Cross equation

e_c	elementary charge
e_d	distance between two surfaces of radius r
e_x	extension
e_1	length of dough sample when contracted
$(e_1 + e_2)$	length of dough sample after elongation
f	frequency of redistribution of energy between link and surroundings (Goodeve)
$f(g)$	function of geometric parameters of two types of particles in mixed suspensions of spheres and rods
f_i, f_{Ω_0}	amplitude of current i and of angular deflection Ω_0
f_N	natural undamped oscillatory frequency of inner cylinder of coaxial cylinder viscometer in dynamic tests
f_0	applied frequency of oscillation of outer cylinder of coaxial cylinder viscometer in dynamic tests
f_r	amplitude of phase difference between current and pendulum deflection for coaxial cylinder viscometer in oscillatory shear tests
f_s	swelling factor
f_v	$= (8\phi_{max})^{\frac{1}{3}}$
h	Planck constant
h	sample height
h_a, h_b	heights of sample in compression test (creep compliance-time) before and after compression
h_c	depth of immersion of inner cylinder of coaxial cylinder viscometer in sample
h_d	thickness of sample used in parallel plate viscoelastometer creep test
h_e	height of right angle cone, or hook descent in Extensograph tests on dough
$h_{e(2)}$	$= 2 \cdot 2\,(e_1 + e_2) - 1 \cdot 74 \times 10^{-3}$ (Extensograph units)
$h_{e(3)}$	$= 2 \cdot 2\,e_1$ (Extensograph units)
h_f	height of bubble during dough extension in Chopin Extensograph
h_g	displacement of mercury in Saunders and Ward tube for gel modulus of rigidity
h_l	limiting height
h_m	distance through which load lifted by muscle fibre
h_s	factor by which ϕ increased due to hydration of particle emulsifier layer
h_v	constant in equation showing simultaneous influence of time and temperature on the viscosity increase of Veegum suspensions with age

h_x	height of dough sample in open tin during fermentation
i	current
i_d	amplitude of driving current in oscillatory testing by cone-plate viscometer
i_s	intercept on mg axis in $mg - 1/t$ plot (hesion test)
j	phase lag between pendulum deflection and current for coaxial cylinder viscometer in oscillatory shear tests
j_v	constant in equation showing the simultaneous influence of temperature and time on the thickening of Veegum suspensions when aged
k	Boltzmann constant
k'	rate constant for flow of unit in Ree–Eyring theory
k_A	rate constant in Avrami equation
k_c	constant in van Wazer et al.'s re-arranged power law
k_d	rate constant for increase in dough volume during fermentation
k_e	rate constant in plot of log (load at constant extension of 11 cm) v. rest period (Brabender extensograph test)
k_s	constant in equation relating stimulus and sensory response
k_t	torsional constant
k_1, k_2	constants in rate of shear and stress relationships developed for Portable Ferranti viscometer in torque decay experiments
k_{11}, k_{22}	elastic moduli for anisotropic homogeneous solid
l	length of sample in oscillatory testing of solids
l_a, l_b	initial and extended lengths of sample in creep compliance-time studies by extensometer
l_c	distance through which F is transmitted
$\bar{l}_c$	mean value of l_c
l_e	half strand length of dough sample in extension (Extensograph)
l_v	critical distance of separation between particles when link ruptures
m	mass
m_c	mass of affixed body in torsional vibration tests on thread-like materials
m_f	factor depending on mode of vibration in vibrational testing of foodstuffs
m_t	mass of thread in torsional vibration tests on thread-like materials
n	number of linkages in flocculated structure
n_A	constant in modified Avrami equation

n_b	constant in equation relating stimulus and sensory response
n_c	constant in correction for "entrance effect" in capillary flow
n_i	ion concentration
n_j	constant defining power of d_p in equation relating yield value and d_p in penetrometer tests
n_r	relative frequency
n_s	power of stress in power law
p	stress
p^*	complex stress
p_a	axial ratio of ellipsoid
p_0	yield stress, or stress applied at $t = 0$
p_r	residual stress
q	factor characterising efficiency of homogenizer
q_v	mean distance of particle from centre of hypothetical sphere
r	radius
r_a	radius of spherical indenter used for compression tests on fruit and vegetables
r_b	radius of dough bubble in Chopin Extensograph
r_d	radius of cylindrical die in compression tests on fruit and vegetables
r_{eff}	effective radius of contact between particles
r_h	radius of hypothetical sphere $= 0.905\, D_s/\phi^{\frac{1}{3}}$
r_i, r_u	inner and outer radii of hollow cylinder
r_k	kymograph reading (Brabender Extensograph)
r_s	radius of sphere used in rolling sphere viscometer
r_1, r_2	radius of curvature of first and second convex bodies in Hertz treatment
s	$= H_0/r$
s'	deformation
s_a	stress/unit area
s_1	initial scale reading of coaxial cylinder viscometer inner cylinder
s_2	reading after 1 min of coaxial cylinder viscometer inner cylinder
t	time
t_a	critical time for displacement readings in stress relaxation studies
t_b	thickness of bar used in bending vibration studies
t_c	average life of link (Goodeve)

t_f	time for number of unassociated particles/cc to decrease to half the original number by flocculation
t_N	time for which shear pulls on a link (Goodeve)
t_s	sample thickness
t_x	$= 1/k_A$
t_μ	time at which stress applied (Boltzmann superposition theory)
u	dissipation coefficient in equation for "firmness"
v	velocity
v_c	corrected value of v_x
v_d	fraction of voids in dispersed system when closely packed
v_f	volume of fluid held in voids between particles
v_g	volume of dough
v_s	velocity of structure breakdown
v_x	speed (cm/sec) of sphere descent in rolling sphere viscometer
w	width of bar in bending vibration testing
x	extension, or distance between uppermost and lowest layers in model for Newtonian flow
$x_1, x_2, \ldots, x_n$	localised areas of shear unit (Ree–Eyring theory)
x_p	proportion of total number of molecules in emulsifier film in state $(i-1)$
x_s	gradient in $\eta_\infty - a_m$ plot for O/W emulsions
w	coefficient of thixotropy
w_m	weight of muscle strip
α	angle of shear
α_c	factor in stress relaxation equation
α_e	Lamé constant
α_f	angle between line joining centres of two particles and direction of shear
α_k	kinetic energy factor
α_m	constant in modified power law
α_p	phase angle difference
α_v	"constant" indicating proportionality between η_∞/η_0 and a_m
$\alpha_1, \alpha_2, \alpha_3$	relate to $\xi, \delta,$ and ζ displacements
β	damping coefficient or factor
β_c	factor in stress relaxation equation
β_d	displacement in creep compliance–time study in parallel plate viscoelastometer
β_e	instrumental damping factor for coaxial cylinder viscometer in dynamic testing

β_m	constant in modified power law
β_n	mean relaxation time
γ	displacement (creep compliance-time study)
γ_T	interfacial tension
Δ	thickness of wall of dough bubble, or thickness of hypothetical layer of fluid near wall of capillary viscometer
Δ_0	thickness of original sheet of dough
δ	phase lag between stress and strain amplitudes, or displacement in y plane
δ_a	resonance frequency
$\tan \delta$	loss tangent
ΔS	shear induced loss of structure
ΔF_n	free energy of activation for flow of nth unit (Ree–Eyring theory)
ε	strain
ε^*	complex strain
ε_d	dielectric constant
ε_o	strain at time $t = 0$ due to stress p_0
ζ	zeta potential, or displacement in z plane
η	viscosity
η'	dynamic viscosity
η''	ratio of stress out of phase with rate of strain/strain
η^*	complex viscosity
η_{app}	apparent viscosity
η_f	steady state viscosity
η_F	Faxen corrected viscosity (rolling sphere viscometer)
η_i	viscosity of dispersed phase in an emulsion
η_{int}	intrinsic viscosity ($[\eta]$)
η_l	initial viscosity
η_N	Newtonian viscosity (creep compliance-time study)
η_0	viscosity of continuous phase
η_p	plastic viscosity
η_r	reference viscosity
η_{rel}	relative viscosity
$\eta_{rel(a)}$	apparent relative viscosity
η_s	shear viscosity of emulsifier film around emulsion particles
η_{sp}	specific increase in viscosity
η_t	viscosity at time t
η_T	viscosity according to Stokes' law from rolling sphere viscometer data
η_v	viscosity of dispersion of linear polymer chains
η_z	measured surface viscosity

$\eta_\alpha, \eta_\beta, \eta_\gamma$	values of η in Ree–Eyring plot of η v. log D
η_β	area viscosity of emulsifier film around emulsion particles
η_∞	viscosity at infinitely high rate of shear
θ	free energy of deformation
θ_a	angle of tilt of rolling sphere viscometer
$\cos \theta_e$	$= h_e/l_e$
θ_i, θ_p	orientation constants
θ_r	angular rotation
κ	specific conductivity
κ_c	conductivity on cone and plate of cone-plate viscometer
λ	Lamé constant
λ'	$= \left(\dfrac{1}{1 - \frac{9}{8} \cdot (r/d_s)} \right)$
λ_a	constant in equation for dilute suspensions of non-spherical particles
λ_e	distance a unit moves between equilibrium positions (Ree–Eyring Theory)
λ_t	relaxation time
λ'_t	relaxation time of greatest frequency
λ_w	wavelength of intrinsic electronic oscillations of atoms
λ_1	distance between flow units of single type (Ree–Eyring)
λ_2, λ_3	cross-sectional area of flow unit (Ree–Eyring)
μ	Poisson's ratio
μ_r	mobility
ν	dynamic stress frequency (cycles/sec)
ν_f	firmness
ν_r	resonance frequency
ξ	displacement in x plane
ζ_r	thickness ratio of sample to carrier in bending vibration testing of soft solids
ρ	density
ρ_i	frictional coefficient of ion i
ρ_m	density of metal resonator in vibrating reed viscometer
σ	area of molecules in interfacial film
τ_m	mean reaction time
τ_1, τ_2, τ_3	components of τ_m
ϕ	volume concentration of dispersed phase
ϕ_a	apparent value of ϕ
$\phi_{\max}$	maximum concentration of dispersed phase
χ	reciprocal thickness of electrical double layer
ψ	sensory response to a stimulus

ψ_a	angle between cone and plate in cone-plate viscometer
ψ_p	potential of outermost part of electrical double layer
Ω	deflection of inner cylinder (coaxial cylinder viscometer)
Ω_i, Ω_0	angular displacements of inner and outer cylinders of coaxial cylinder viscometer in dynamic testing
ω	dynamic frequency (radian/sec)
ω_a	angular velocity
ω_0	angular velocity of outer cylinder of coaxial cylinder viscometer

Author Index

Numbers in italics refer to pages on which references are listed at the end of the chapter.

A

Abitz, W. 295, *318*
Activos, A. 135, *181*
Agranat, N. N. 84, *94*
Albers, W. 120, 124, 133, 167, *180*
Alfrey, Turner Jr. 21, *31*
Altrichter, F. 51, *94*
American Society for Testing Materials *94*
Amerine, M. A. 382, 387, *391*
Amman, R. 326, *367*
Anderson, D. M. W. 344, *367*
Anderson, G. W. 272, *317*
Andrews, R. D. 270, 271, *316*
Araujo, O. E. 329, *392*
Arbuckle, W. S. 206, *317*
Asay, J. 118, *181*
Autard, P. 86, 87, *94*
Avrami, M. 188, 222, *316*
Axford, D. W. E. 218, 222, 223, *317*
Aylward, F. 207, 215, *316*

B

Babb, A. T. S. 224, 225, 226, *316*
Backinger, G. 278, *318*
Bacon, L. R. 51, *94*
Bailey, C. H. 222, 226, *316, 320*
Banker, G. S. 325, 327, 328, *368*
Bardet, L. 327, 364, *367*
Barlow, A. J. 66, *94*
Baron, M. 233, *318, 319*
Barry, B. W. 324, 325, 326, 327, 351, 352, 353, *367, 368*
Bartlett, J. W. 126, *184*

Barton, B. S. *367*
Bate-Smith, E. C. 275, *316*
Bauman, A. J. 59, *94*
Becher, P. 161, *180*
Bechtel, W. G. 222, *316*
Beeby, R. 307, *317, 318*
Ben Arie, M. M. 69, *94*
Bendall, J. R. 275, *316, 317*
Benis, A. M. 59, *95*
Bernies, K. H. 326, 327, 364, 365, *367, 368*
Bice, C. W. 218, *317*
Bikerman, J. J. 376, *391*
Billington, E. W. 64, *95*
Bingham, E. 10, *31*, 192, *317*
Birdseye, C. 279, *320*
Bloksma, A. H. 252, 253, 254, 255, 272, 273, 274, *317*
Boardman, G. 53, *95*
Boeder, P. 155, *180*
Boedtker, H. 295, *317*
Bonauto, M. V. 361, *367*
Booth, F. 164, 165, *180*
Bourne, M. C. 384, *391*
Boussinesq, J. 237, *317*
Bousso, D. 260, 261, 262, 263, *320*
Bowles, R. L. 82, *95*
Boylan, J. C. 324, 326, 330, 331, 332, 333, 361, *367*, 388, *391*
Brandt, M. 288, *320*
Brandt, M. A. 373, 382, *391*
Bratzler, L. J. 277, *317*
Briant, A. M. 292, 301, *317*
Brill, R. 200, *318*
Brinkman, H. C. 136, *180*
Brodnyan, J. G. 54, *95*, 156, 168, *180*

Brody, A. L. 283, 285, 286, *319*
Broughton, J. 136, 158, *180*
Brown, E. A. 337, *367*
Buckingham, E. 47, *95*
Bulkley, R. 98, *181*
Bunzl, M. 133, 156, *181*
Burckhardt, G. J. 278, *318*
Burnett, J. 68, 69, *96*, 261, 308, 309, 310, 311, 312, 313, *319*

C

Caffyn, J. E. 232, 233, *317*
Cairncross, S. E. 373, *391*
Campbell, A. M. 292, 301, *317*
Campbell, L. E. 212, *317*
Carslaw, W. S. 53, *95*
Carter, P. 324, *367*
Casson, N. 108, 112, 113, 170, *180*
Catacalos, G. 325, 365, *369*
Chan, F. S. 165, *180*
Chang, T. N. 143, *181*
Cheng, P. Y. 133, 143, *180*
Chong, J. S. 154, *180*
Claassens, J. W. 86, *95*, 375, *391*
Cobbett, W. G. 292, 293, 294, 295, 296, *321*
Coleman, J. A. 373, 382, *391*
Collins, D. J. 143, 153, *180*
Colwell, R. E. 18, *31*, 35, 46, 49, 61, *96*, 99, *184*
Conway, B. E. 164, *180*
Coppen, F. M. V. 232, *319*
Coppock, J. B. M. 263, 264, 265, 266, 267, 269, 270, *318, 319*
Cornford, S. J. 218, 221, 222, 223, *317*
Cox, C. P. 305, *317*
Criddle, D. W. 162, *180*
Cross, M. M. 111, 112, 170, *181*
Cunningham, H. 240, *318*
Cunningham, J. R. 272, *317*
Curby, W. A. 284, *321*

D

Davey, C. L. 275, *317*
Davie, R. P. 82, *95*
Davies, J. T. 123, *181*
Davies, P. J. 324, 346, 347, 348, 350, *368*
Davis, C. E. 226, *317*

Davis, J. G. 229, 230, 231, 232, 233, *317*
Davis, S. S. 327, 342, 343, 344, 345, 346, 351, *367, 368*
Davison, S. 283, Fig. 4.65 (facing p. 284), 285, 286, 287, *319*
de Bruijn, H. 133, *181*
Deer, J. J. 327, 351, *367*
De Fremery, D. 276, *317*
De Kay, H. G. 325, 327, 328, *368*
Dekking, P. 80, *95*
Delonca, H. 327, 364, *367*
De Man, J. M. 189, 191, 192, 193, *317, 320*
Dempster, C. J. *317*
Derjaguin, B. V. 119, 157, *181*, 182
de Vries, A. J. 112, *181*
de Waele, A. *31*, 98, *181*
De Witt, T. W. 48, 61, 62, 64, *95, 96*
Dienes, G. J. 60, 74, *95*
Din, F. J. 48, *95*
Dobry-Duclaux, A. 164, *180*
Dolique, R. 327, 364, *367*
Doty, P. 295, *317*
Dowdle, R. L. 223, *320*
Drake, B. 376, 385, *391*
Drake, B. K. 246, 247, 248, 249, 250, 251 Facing page 252, 288, *317*
Duclaux, J. 157, *181*

E

Eggenburger, D. N. 344, *368*
Eilers, H. 133, 149, *181*, 305, *317*
Einstein, A. 130, *181*, 198, *317*
Eirich, F. R. 133, 156, *181*
Elke, M. 305, *319*
Ellis, B. H. 382, *391*
Ellis, R. B. 48, *96*
Elton, G. A. H. 218, 222, 223, *317*
Emmons, D. B. 235, *317, 320*
Evans, W. W. 66, *96*
Eveson, G. F. 149, 151, 152, 154, *181*
Ewell, R. H. 163, *181*
Eyring, H. 115, 118, 163, *181, 182*, 189, *320*

F

Falk, S. 246, *317*
Fantelli, S. 327, 355, 356, 357, 358, *368*

Farkas, E. 303, 304, *320*
Faxén, H. 50, *95*
Feltham, P. 261, *317*
Ferry, J. D. 24, *31*, 35, 66, 79, 81, *95*, *96*, 295, *317*
Feuge, R. O. 208, 210, 211, *318*
Fidleris, V. 154, *181*
Fincke, A. 213, 215, *317*, *318*
Finney, E. E. 240, *317*
Fischer, E. R. 49, *95*
Foley, J. 85, *95*
Foster, W. 282, *321*
Fox, C. D. 324, 337, *367*
Fransden, J. H. 206, *317*
Frankel, N. A. 135, *181*
Franks, A. J. 364, *367*
Frederick, K. J. 327, 358, 359, 360, *368*
Frederickson, A. G. 13, *31*, 331, *367*
Freeman, S. M. 66, *96*
Freund, J. 337, 361, *367*
Freundlich, H. 12, *31*
Friedmann, H. H. 286, *317*, 383, *391*
Frisch, H. L. 155, *181*
Fröhlich, H. 164, *181*
Fryklöf, L. E. 83, *95*, 326, *367*, 388, 389, *391*
Fukada, E. 186, *320*
Fuks, G. I. 157, *181*
Fukushima, M. 186, *320*
Füller, W. 327, 364, 365, *368*
Fulmer, E. J. 51, *96*
Funt, C. B. 255, *317*, *318*

G

Gabrysh, A. F. 118, *181*
Gal, C. 327, 355, 356, 357, 358, *368*
Gaskins, F. H. 54, *95*
Geckler, R. D. 136, 143, *183*
Geddes, W. F. 218, *317*
Gent, A. N. 60, *95*
Gerngross, O. 295, *318*
Giles, W. H. 365, *369*
Gillespie, T. 103, 106, 111, 112, 115, 119, 131, 136, *181*
Glasstone, S. 115, *181*
Glücklich, J. 255, 256, 257, 258, 259, 260, *317*
Gneist, K. 305, *320*
Gohlich, H. 244, 245, *318*

Goodeve, C. F. 103, 104, 106, 112, 115, 119, 170, *181*
Goodier, J. N. 240, *320*
Goring, D. A. I. 165, *180*
Goulden, J. D. S. 151, *181*
Grant, A. M. 51, *95*
Green, H. 86, *95*, 331, *368*
Greenberg, S. A. 143, *181*
Grieco, A. 78, *96*
Grim, W. 104, *183*
Grogg, B. 261, 262, *318*
Gross, B. *31*
Guice, W. A. 208, 210, 211, *318*
Guirao, M. 387, *392*
Guth, E. 131, 133, *181*, 198, *318*

H

Haesing, J. 197, *318*
Haighton, A. J. 84, Fig. 2.14 (facing p. 84), 85, *95*, 187, 189, 190, *318*
Hall, C. W. 240, *317*
Halton, P. 255, *318*
Hammill, R. D. 325, *368*
Hansen, H. 280, 281, *320*
Harbard, E. H. 213, *318*
Harmsen, G. J. 165, *181*
Harper, J. C. 50, *95*
Harper, R. 233, *318*, 387, *391*
Harper, R. C., Jr. 62, 64, *95*
Harrison, G. 66, *94*
Harvey, H. G. 292, *318*
Hastewell, L. J. 295, 297, *318*
Hatschek, E. 135, *181*
Hauser, E. A. 142. *181*
Hawbecker, J. V. 286, 302, 303, *318*
Heaps, P. W. 270, *318*
Heimann, W. 213, 215, *318*
Heinz, W. 213, 215, *317*, *318*
Henderson, N. L. 388, 389, *391*
Henle, G. 337, *368*
Henle, W. 337, *368*
Herrmann, K. 295, *318*
Herschel, W. H. 98, *181*
Hertz, C. H. 246, *317*
Hertz, H. 240, *318*
Higginbotham, G. H. 46, *95*, 132, *181*
Hill, R. D. 306, *320*
Hintzer, H. M. R. 90, *95*
Hirai, N. 115, 118, *182*
Hlynka, I. 270, 272, *317*, *318*

Hofmann-Bang, N. 270, 271, *316*
Hopper, V. D. 51, *95*
Hosking, Z. D. 305, *317*
Houwink, R. 11, *31*, 99, 123, *181*, 375, *391*
Huebner, V. R. 197, *318*
Humphrey, A. E. 166, *183*
Hurwicz, H. 278, *318*

I

Inokuchi, K. 15, *31*
Inst. Food Technologists 381, *391*
Isherwood, F. A. 248, *318*

J

Jaeger, J. C. 53, *95*
Jarnutowski, R. 143, *181*
Jeffery, G. B. 155, *181*
Jellinek, H. G. 200, *318*
Jensen, K. 337, *369*
Joly, M. 162, 163, *182*
Jones, N. R. 299, 300, 301, *318*
Jopling, D. W. 295, 297, *318*

K

Kabre, S. P. 325, 327, 328, *368*
Kambe, H. 63, *95*
Karasova, V. V. 157, *182*
Kassem, M. A. 333, *368*
Kelly, E. L. 168, *180*
Kennedy, A. J. 375, *391*
Kim, K. Y. 18, *31*, 35, 46, 49, 61, *96*, 99, *184*
Kim, W. K. 115, 117, *182*
Kitchener, J. 121, *183*
Klemm, H. J. 60, 74, *95*
Kleyn, D. H. 378, *392*
Kohler, O. C. 61, 64, *96*
Kolvanovskaya, A. S. 87, *95*
Kooper, W. D. 305, *318*
Kostenbauder, H. B. 325, 327, 330, 331, 334, 335, 336, 337, 338, *368*, 388, 389, *391*
Kozma, A. 240, *318*
Kragh, A. M. 344, *368*
Kragh, G. 326, *368*
Kramer, A. 278, 288, **302**, 303, *318*
Kratz. P. D. 224, *319*
Kreiger, I. M. 136, 149, *182*

Krieger, I. M. 54, *95*
Kuhn, H. 155, 156, *182*
Kuhn, W. 155, 156, *182*
Kumetat, K. 307, *317*, *318*
Kuno, H. 50, *95*
Kynch, G. J. 131, 146, *182*

L

Lachman, L. 361, *368*
Ladenburg, R. 50, 51, *95*
Laidler, K. J. 115, *181*
Lamb, J. 66, *94*
La Mer, V. K. 344, *368*
Landel, R. F. 59, *95*
Lange, B. 326, *368*, 388, 389, *391*
Langenbucher, F. 326, *368*, 388, 389, *391*
Langston, W. B. 344, *368*
Lapham, E. A. 365, *369*
Lawrence, A. S. C. 147, *182*
Lawrie, R. A. 274, *318*
Lazarus, J. 361, *368*
Leadermann, H. 17, 22, *31*
Le Beau, D. S. 142, *181*
Leighton, A. 132, 137, 142, *182*, 305, *318*
Lemin, C. E. 52, *95*
Lerchenthal, C. H. 255, *317*, *318*
Leviton, A. 132, 137, 142, *182*, 305, *318*
Levy-Pascal, A. 149, *182*
Lewis, W. K. 156, *182*
Lexow, T. 94, *95*, *96*
Lieberman, S. V. 325, *369*
Lin, J. J. 339, 340, 344, *368*
Lindsley, C. H. 49, *95*
Linton, M. 138, *182*
Lissant, K. J. 155, 177, *182*
Lodge, A. S. 28, *31*
Lovegren, N. V. 208, 210, 211, *318*
Loveless, M. H. 337, *368*
Lucassen-Reynders, E. H. 316, *318*
Lustig, A. 51, *94*
Lyons, J. W. 18, *31*, 35, 46, 49, 61, *96*, 99, *184*

M

McDermott, K. 337, *367*
MacDonald, D. C. 282, *321*
McGann, T. C. A. 305, 306, *318*, *319*

McKennell, R. 52, 53, *95*
McVean, D. E. 325, *368*
Madow, B. P. 136, 149, *182*
Malecki, G. J. 283, Fig. 4.65 (facing p. 284), 284, 287, *319*
Manley, R. St. J. 132, *182*
Margaretha, H. 133, 156, *181*
Markovitz, H. 62, 64, *95*
Maron, S. H. 117, 118, 132, 136, 149, *182*
Marriot, R. H. 378, *391*
Marsden, A. 326, 353, 354, *369*
Marsh, D. M. 75, *95*
Martin, A. N. 325, 327, 330, 331, 334, 335, 336, 337, 338, *368*, 388, 389, *391*
Mase, G. E. 240, *317*
Mason, S. G. 132, 138, 139, 162, *182*
Matalon, R. 365, *368*
Matsui, T. 156, *182*
Matsuo, R. R. 272, *318*
Mattocks, A. M. 104, *183*, 325, *368*
Matz, S. A. 375, *391*
Maude, A. D. 51, *95*
Meer, P. M. 388, 389, *391*
Melms, D. 261, 262, *318*
Metzner, A. B. 98, *182*
Michajlow, N. W. 58, *96*
Mikhailov, V. V. 87, *95*
Ming Fok, S. 132, 149, *182*
Miyoda, D. S. 279, *318*
Mohr, W. 88, *95*, 197, 305, *318*
Mohsenin, N. N. 236, 237, 240, 241, 242, 243, 244, 245, *318*
Mooney, M. 114, 125, 132, 136, 156, *182*
Moore, W. J. 163, *182*
Morgan, P. G. 52, *95*
Morrison, F. R. 50, *95*
Morrison, T. E. 61, 64, *95*
Morrow, T. 236, 237, 240, 241, 242, 243, 244, *318*
Moser, B. G. 59, *95*
Mottram, F. J. 84, *95*
Muers, M. 313, 314, 315, *319*
Mukerjee, P. 167, *182*
Muller, H. G. 255, 263, 264, 265, 266, 267, 269, 270, *317*, *318*, *319*
Münzel, K. 325, 326, 327, 329, 340, 362, 364, 365, *367*, *368*
Murnaghan, F. D. 29, *31*
Murray, W. T. 279, *320*

N

Naudet, M. 84, 88, *96*, 196, 197, *319*
Nawab, M. A. 132, 138, 139, 162, *182*
Nederveen, C. J. 76, 77, 79, 80, *96*, 171, *182*, 186, *319*
Nedonchelle, Y. 312, *319*
Neil Jenkins, G. 380, 384, *391*
Neu, G. E. 366, *368*
Neuwald, F. 83, *96*
Neuwald, H. 326, *368*
Nissan, A. H. 126, *184*
Nutting, P. G. 30, 31, *31*, 380, *391*

O

Ober, S. S. 327, 358, 359, 360, *368*
Oberst, H. 79, *96*
Ogawa, S. 74, *96*
Oka, S. 74, *96*
Oldenburg, F. 305, *318*
Oldroyd, J. C. 142, 162, 164, *182*
Oliver, D. R. 46, *95*, 132, *181*
Oosthuizen, J. C. 51, *96*
Ostwald, W. 98, *182*
Overbeek, J. Th. G. 119, 120, 124, 125, 126, 165, *180*, *181*, *184*

P

Pangborn, R. M. 382, 387, *391*
Parkes, J. 158, *184*
Payne, A. R. 169, 171, *182*
Peterlin, A. 155, *182*
Philippoff, W. 54, *95*
Phipps, L. W. 151, *181*
Pierce, P. E. 117, *182*
Pilkington, D. H. 280, *319*
Platt, W. 221, 224, *319*
Poiseuille, J. L. M. 47, *96*
Pool, M. F. 276, *317*
Posener, L. W. 305, *317*
Powers, R. 221, *319*
Prentice, J. H. 88, 89, *96*, 99, 100, 179, *183*, 192, 196, 315, 316, *319*, 382, *391*
Price, W. V. 235, *317*
Proctor, B. E. 283, Fig. 4.65 (facing p. 284), 284, 285, 286, 287, *319*
Pyke, G. T. 305, 306, *318*, *319*

R

Rabinowitsch, R. 48, *96*
Rahman, S. 344, *367*
Rajagopal, E. S. 142, *182*
Rapaport, H. G. 337, *368*
Ree, T. 115, 118, *182*
Reed, J. C. 98, *182*
Rehbinder, P. 58, *96*
Rehbinder, P. A. 84, *96*
Reiner, M. 20, 21, *31*, 47, *96*, 99, 179, *182*, 376, 377, *391*
Reynolds, O. 47, *96*
Rich, S. R. 55, *96*
Richardson, E. G. 136, 142, 147, *182*
Richman, M. D. 365, 366, 367, *368*
Richter, J. 66, *94*
Rideal, E. K. 123, *181*
Roberts, J. E. 27, *31*
Robinson, J. 136, 149, *182*
Robison, R. L. 326, 361, *367*
Robson, A. H. 90, 91, 92, 93, *96*, 219, 220, *319*
Roessler, E. B. 382, 387, *391*
Rogers, H. P. 278, *318*
Rorden, H. C. 78, *96*
Roscoe, R. 132, *182*, 292, 295, 297, 298, 299, *318*, *319*
Rosenberg, G. F. von 208, *319*
Roth, W. 55, *96*
Rothman, S. 388, *391*
Rothwell, E. 147, *182*
Rotteglia, E. 336, *368*
Roran, E. M. 324, 346, 347, 348, 350, 351, *368*
Russell Eggitt, P. W. 263, 264, 265, 266, 267, 269, 270, *318*, *319*

S

Saal, R. N. J. 305, *317*
Sachs, D. 157, *181*
Sack, R. 164, *181*
Saito, H. 132, *182*
Sale, A. J. H. 274, 276, 277, 279, *319*
Sambuc, E. 84, 88, *96*, 196, 197, *319*
Samel, R. 313, 314, 315, *319*
Sanberg, V. 305, *320*
Samygin, M. H. 157, *181*
Saunders, F. L. 133, 136, 143, 153, 154, 167, *182*, *183*

Saunders, P. R. 67, 68, *96*, 292, 293, 294, 295, 296, 299, 303, 308, *319*, *320*
Schachman, H. K. 133. 143, *180*
Schaller, E. J. 166, *183*
Schaub, K. 325, 329, *368*
Schenkel, J. H. 121, *183*
Scherr, H. J. 192, 196, *319*
Schofield, R. K. 255, 257, *319*
Schramp, F. W. 66, *96*
Schulte, K. E. 333, *368*
Schutz, R. A. 312, *319*
Scott, J. R. 60, *96*, 98, *183*
Scott Blair, G. W. 31, *31*, 68, 69, 95, *96*, 99, 100, 111, 179, *183*, 232, 233, 255, 257, 261, 308, 309, 310, 311, 312, 313, 315, *318*, *319*, *320*, 376, 380, *391*, *392*
Seguin, H. 66, *94*
Seidler, L. 305, *319*
Selby, J. W. 299, *319*
Semenenko, N. N. 84, *96*
Senna, M. 50, *95*
Setnikar, I. 327, 355, 356, 357, 358, *368*
Shalopalkina, T. G. 316, *320*
Shama, F. 15, 16, *31*, 70, 71, *96*, 185, 186, 187, 199, 200, 201, 202, 203, 215, 216, 217, 220, *319*
Shangraw, R. 104, *183*, 365, 366, 367, *368*
Shangraw, R. F. 324, 337, *367*
Shaw, D. J. 69, *96*, 199, *319*
Shelef, L. 255, 256, 257, 258, 259, 260, 261, 262, 263, *317*, *320*
Sherman, P. 15, 16, *31*, 70, 71, *96*, 121, 122, 123, 132, 140, 141, 143, 144, 145, 146, 147, 149, 150, 154, 158, 159, 160, 161, 162, 167, 169, 171, 172, 173, 174, 175, 176, 178, 179, 180, *183*, 185, 186, 187, 198, 199, 200, 201, 202, 203, 204, 205, 206, 207, 215, 216, 217, 220, 302, 303, 316, *319*, *320*, *321*, 374, 377, 378, 379, 381, *392*
Sheth, B. B. 325, *368*
Shimazu, F. 214, *320*
Shimizu, M. 118, *181*
Shimmin, P. D. 306, *320*
Shotton, E. 138, 168, *183*, 324, 325, 326, 340, 341, 342, 343, 344, 345, 346, 351, 352, 353, *367*, *368*
Shoulberg, R. H. 61, 64, *96*
Sibree, J. O. 135, *183*

Simha, R. 131, 133, 134, 146, 148, 155, 181, 183, 198, 318
Simon, D. E. 327, 358, 359, 360, 368
Sisko, A. W. 119, 182
Sjöström, L. B. 373, 391
Skinner, E. Z. 288, 320, 373, 382, 391
Skovholt, O. 223, 320
Slattery, J. C. 53, 96
Sloman, K. 288, 320
Smellie, R. H. 344, 368
Soltøft, P. 93, 94, 96, 188, 192, 194, 195, 196, 320
Somcynsky, T. 134, 183
Somers, F. 236, 238, 239, 246, 320
Sone, T. 74, 96, 186, 188, 320
Spöttel, W. 305, 320
Squires, L. 136, 156, 158, 180, 182
Srivastava, S. N. 123, 183
Stainsby, G. 296, 320
Steggles, J. S. 59, 60, 96
Steller, W. R. 222, 320
Steiner, E. H. 212, 213, 320
Sterling, C. 214, 320
Stevens, S. S. 386, 387, 391, 392
Storz, G. K. 327, 368
Street, N. 165, 183
Sumner, C. G. 158, 183
Sutherland, K. 138, 182
Sweeney, R. H. 136, 143, 183
Szabo, G. 234, 235, 320
Szczesniak, A. S. 274, 286, 288, 303, 304, 317, 320, 372, 373, 374, 376, 377, 378, 383, 390, 391, 392

T

Takano, M. 63, 64, 95, 96
Talman, F. A. J. 324, 346, 347, 348, 350, 351, 368
Tapernoux, A. 305, 320
Tappell, A. L. 279, 318
Taylor, G. I. 130, 137, 183
Tempel, M. van den 75, 96
Thomas, D. G. 126, 131, 133, 148, 183
Thomsen, L. C. 197, 318
Thompson, W. I. 156, 182
Thureson, L. E. 305, 320
Timoshenko, S. 240, 320
Tischer, R. G. 278, 318

Titijevskaya, A. S. 157, 181
Tobolsky, A. 189, 320
Tobolsky, A. V. 270, 271, 316
Todd, W. D. 82, 95
Toms, B. A. 138, 183
Torgerson, W. S. 381, 382, 392
Torgeson, K. W. 274, 320
Torssel, H. 305, 320
Trapeznikov, A. A. 316, 320
Treisman, M. 387, 392
Tressler, D. K. 279, 320
Tourila, P. 101, 183
Tuszyński, W. 312, 313, 320
Twigg, B. A. 278, 318

U

Umstätter, H. 12, 31

V

Vaeck, S. V. 215, 320
Vand, V. 38, 96, 133, 136, 149, 184
van den Tempel, M. 112, 113, 115, 122, 169, 170, 183, 185, 189, 320
van der Waarden, M. 133, 166, 184, 305 317
van Schooten, J. 165, 181
Van Holde, K. E. 69, 96
Van Ooteghem, M. 324, 325, 326, 327, 334, 363, 364, 366, 368, 388, 390, 392
Van Wazer, J. R. 18, 31, 35, 46, 49, 61, 96, 99, 184
Vasić, I. 191, 192, 193, 320
Vazin, A. 324, 341, 342, 368
Veis, A. 344, 368
Verwey, E. J. W. 119, 120, 125, 126, 184
Vincent, H. C. 327, 358, 359, 360, 368
Virgin, H. I. 246, 317
Voigt, R. 335, 368
Voisey, P. W. 235, 280, 281, 282, 320, 321
Volarovich, M. P. 84, 94
Vold, M. J. 121, 123, 126, 156, 184
Volodkevich, N. N . 277, 321
von Smoluchowski, M. 101, 107, 112, 125, 137, 164, 165, 184
Vuillaume, R. 305, 320

W

Wade, P. 227, 228, *321*
Wader, T. 327, 364, 365, *367*
Walters, L. E. 280, *319*
Warburton, B. 326, 327, 351, *367*, *368*
Ward, A. G. 67, 68, *96*, 292, 293, 294, 295, 296, 299, 303, 308, *319*, *320*, *321*
Ward, S. G. 46, *95*, 132, 154, 157, *181*, *184*
Warner, K. F. 277, *321*
Warren, J. 337, *369*
Waterman, H. A. 77, *96*
Watson, E. L. 288, 289, 290, 291, 292, *321*
Wayland, H. 143, 153, *180*
Weissenberg, K. 61, 66, *96*
Welch, M. 283, Fig. 4.65 (facing p. 284), 284, 286, *319*
Wellman, R. E. 48, *96*
Weymann, A. D. 111, 171, *184*
White, R. F. 138, 168, *183*, 324, 341, *368*
White, R. K. 243, *321*
Whitehead, J. 206, 207, *321*
Whiteman, J. V. 280, *319*
Whitmore, R. L. 53, *95*, 154, 157, *181*, *184*
Whitney, J. E. 286, *317*, 383, *391*
Wibberley, K. 324, 340, 341, 342, *368*
Wiechert, E. 261, *321*
Williams, J. C. 51, *96*
Williams, J. W. 69, *96*

Williams, M. V. 263, 264, 265, 266, 267, 269, 270, *318*, *319*
Williams, P. S. 148, 149, *184*
Williamson, R. V. 101, 102, 103, 115, *184*
Wilson, G. L. 158, *184*
Winkler, C. A. 277, *317*, *321*
Witnauer, L. P. 192, 196, *319*
Wolf, C. 335, *368*
Wood, F. W. 189, *317*, 385, 386, 387, *392*
Wood, J. H. 325, 329, 365, *369*
Woodhour, A. 337, *369*
Woodman, M. 326, 353, 354, *369*
Woods, M. F. 54, *95*
Wuhrmann, J. J. 214, *320*
Wurdack, P. J. 325, 337, *369*

Y

Yakatan, G. J. 329, *392*
Yavorsky, P. M. 62, 64, *95*
Yerazanis, S. 126, *184*
Yurkstas, A. A. 284, *321*

Z

Zapas, L. J. 61, 62, 64, *95*
Zimm, B. H. 66, *96*
Zimmerli, F. H. 61, 64, *96*
Zwicky, R. 326, 327, 362, 363, *368*, *369*

Subject Index

A

Acacia
mechanism of emulsion stabilisation,
340, 341, 342, 343, 344, 345, 346
thickness of absorbed layer, 341
Adhesion, 86, 373, 374, 376, 377, 380,
381, 383
Absorption
small particles on large particles, 154
stabilisers at interfaces, 169, 199, 341
Agar gels, 299, 300, 301
brittleness, 300
creep compliance-time studies, 301
load-deflexion response, 300, 301
Angular frequency, 23, 24, 77
Apples
bulk compression, 243, 244
creep compliance-time studies, 241,
242, 243
dynamic testing, 246, 247, 248
force-penetration, 286, 287
impact testing, 244, 245
shear test, 245
stress relaxation, 241, 242, 243
texture assessment by chewing sounds,
249, 250, 251, 252
viscoelastic properties, 236, 237, 241,
242, 243, 246, 247, 248
Attraction energy, 120, 121

B

Bingham body, 10, 47, 53, 213
Biscuits, 215, 226, 227, 228, 229, 375, 378
texture assessment by chewing sounds,
249, 250, 251, 252
Bonds
primary, 189, 190
secondary, 189, 190
Bread, 215, 216, 218, 221, 222, 223, 224
force-penetration tests, 286
texture assessment by chewing
sounds, 249, 250, 251, 252

Brinell hardness tester
chocolate, 208, 209, 210, 211
Brittleness
biscuits, 226, 227, 375, 378
foods, 373, 374, 375, 377, 378, 383, 384
Buckingham-Reiner correction for
plug flow, 47, 48
Butter
consistency, 196
crystal growth, 188
extrusion, 192
firmness, 196
flow, 194, 195
hardness, 190, 191, 192, 193, 197, 198
spreadability, 190, 196, 197, 198, 382
viscosity, 188
work softening, 190, 191, 192, 193

C

Cake, 215, 216, 217, 218, 219, 220, 221,
222, 224, 225, 226, 378, 379
Capillary flow
true rate of shear at wall, 48
Central limit theorem, 261, 262, 263,
310, 311
Cheese
dynamic testing, 246, 247, 248
force-penetration tests, 286, 287
natural, 229, 230, 231, 232, 233, 234
processed, 234, 235
texture assessment by chewing sounds,
249, 250, 251, 252
Chewiness, 373, 374, 376, 383, 384
Chewing sounds
apple, 249, 250, 251, 252
biscuits, 249, 250, 251, 252
bread, 249, 250, 251, 252
cheese, 249, 250, 251, 252
meat, 288
Chocolate
constitution, 207
hardness, 208, 209, 210, 211

molten, 208, 212, 213, 214, 215
thermal treatment, 210, 211
Circulation of fluid in drops, 137, 138, 158
Coalescence, 98, 172, 173, 175, 177, 179, 337
Cocoa butter
  viscosity, 213, 214, 215
Cohesion, 86, 374, 375, 376, 384
Collision probability, 107
Compliance
  instantaneous elastic, 13, 14, 17
  loss, 24, 25, 26
  Newtonian, 14
  retarded elastic, 14, 15, 16, 17
  storage, 24
Compressimeter, 215, 221, 222, 224, 229, 230, 234
Compressibility
  bread, 218, 221, 222
  cake, 224, 225
  cheese, 229, 230, 232, 233, 234
  coefficient, 7
  potato, 237, 240
Compression
  meat, 282, 283
  modulus, 6
  solids, 72, 73, 74, 186
Consistency profiling
  cosmetics, 378, 379
  foods, 372, 373, 374, 375, 376, 377, 378
  pharmaceuticals, 378, 379
Contact radius, 107
Contraction, lateral, 5, 6
Crack propagation, 375, 376
Cream, flow properties, 315, 316
Creep compliance, 12, 13, 14, 15, 16, 17, 22, 57, 69, 70, 71, 72
  agar gels, 301
  apples, 241, 242, 243
  cake, 216, 218
  dough, 254, 255, 256
  emulsions, 175, 176, 177, 178, 179, 351, 352
  foods, 377
  frozen ice cream, 199, 200
  ice cream mix, 204, 205, 206
  melted ice cream, 204, 205, 206, 207
  milk gels, 309
  potato, 240, 241, 242, 243

Crumb strength, 218, 219, 220
  bread, 221

D

Deflocculation, 108, 123, 124
Deformation
  elastic, 4, 5
  Nutting's law, 30, 31, 380
  plastic, 4
Dilatancy, 53, 359
Dislocation theory, 375
Dispersed systems, 12, 49, 53, 57, 58, 64, 97, 98, 99, 100, 101, 114, 119, 120, 127
Dispersed systems viscosity, 48, 127, 128, 129
  and continuous phase, 128, 157, 158
  and dispersed phase, 128, 130, 131, 132, 133, 134, 135, 136, 137, 138, 139, 140, 141
  and electroviscous effects, 129, 164, 165, 166
  and particle shape, 128, 154, 155, 156, 157
  and particle size, 127, 128, 148, 149, 150, 151, 152, 153, 154, 158, 159
  and stabilising agents, 129, 168, 169
  and surfactants, 129, 158, 159, 160, 161, 162
  rheological properties at low shear stress, 169, 170, 171
Dough
  creep compliance-time response, 254, 255, 256, 258
  elasticity, 257, 258
  extension, 255, 256, 257, 258, 259, 260, 263, 264, 265, 266, 267, 268, 269, 270, 271, 272, 273, 274
  fermenting, 272, 273, 274
  rest time, 270, 271, 272
  stress relaxation, 259, 260, 261, 262, 263, 272
  structure, 271, 272
Dynamic rheological tests, 60, 61, 62, 63, 64, 65, 66, 67, 246, 247, 248, 274, 275, 316, 351, 352

E

Edge effect
  coaxial cylinder viscometer, 39
  cone-plate viscometer, 52, 53

Elasticity, 12, 13, 28
  cakes, 379
  foods, 374, 375, 376, 377, 379, 384
  frozen ice cream, 379
  second order theory, 29, 30
  water-in-vaseline emulsions, 362, 363
Electrical double layer, 119, 120, 123, 164, 165, 166
Elongation, longitudinal, 6
Electrophoretic mobility, 123
Electroviscous effect
  first, 164, 165, 166, 167, 168
  second, 165, 166, 168
Emulsifying agents
  absorbed at interfaces, and emulsion rheology, 129, 137, 159, 162, 163, 164
  chemical constitution, 129
  concentration, 129
  HLB, 338, 339
  hydration of adsorbed layer, 135
  location, 338, 339, 344
  rheology of adsorbed layer, 142, 145
  solubility, 162
Emulsions, 97, 98, 114, 119, 120, 122, 127, 323, 324, 325, 335, 336, 337, 338, 339, 340, 341, 342, 343, 344, 345, 346, 347, 348, 349, 350, 351, 352, 353, 362, 363
  compressibility, 136
  inversion, 161, 162, 339
Emulsion viscosity
  and continuous phase, 128, 157, 158
  and dispersed phase, 128, 130, 131, 132, 133, 134, 135, 136, 137, 138, 139, 140, 141, 337
  and electroviscous effect, 129, 164, 165, 166, 167, 168
  and emulsifying agent, 129, 137, 142, 145, 158, 159, 160, 161, 162, 163, 164, 338, 339, 341, 342, 344
  and fatty alcohols, 346, 347, 348, 349, 350, 351, 352, 353
  and HLB, 337, 338, 339, 340
  and particle shape, 128, 154, 155, 156, 157
  and particle size, 127, 128, 141, 142, 143, 144, 145, 146, 147, 148, 149, 150, 151, 152, 153, 154, 158, 159, 173, 174, 337, 340, 343, 345

and stabilising agents, 129, 168, 169
changes in rheological properties when aged, 172, 173, 174, 175, 176, 177, 178, 179, 180
rheological properties at low shearing stress, 171, 172
End effects
  coaxial cylinder viscometer, 39, 40, 45, 49
  rolling sphere viscometer, 51, 52
Entrance effects
  capillary viscometer, 38, 45
Extension
  dough, 255, 256, 257
  gelatin gels, 293, 294, 295, 296
  meat, 275, 276
  solids, 5, 74
Extensograph
  Brabender, 252, 263, 264, 265, 266, 267, 268, 269, 270, 271, 273, 274
  Chopin, 252, 273, 274
Extruder, B.F.M.I.R.A.—N.I.R.D., 37, 88, 89, 90, 196, 197
Extrusion
  butter, 192
  fats, 192
  from tube, 378
  lotion from nozzle, 355, 356
  margarine, 192
  penicillin suspensions from hypodermic needle, 360, 361

F

Fats
  bonds in, 189
  consistency measurements, 185, 192, 196
  crystallisation, 188, 189
  extrusion, 192, 194
  firmness, 196
  hardening on storage, 189
  spreadability, 196
  thermal treatment, 196
Fatty alcohols
  mode of operation in emulsions, 350, 351, 352, 353
  use in pharmaceutical emulsions, 346, 347, 348, 349, 350, 351, 352, 353
Fechner's law, 387

F.I.R.A. jelly tester
  agar gels, 299, 300
  pectin gels, 292
Firmness
  cake, 224, 225
  foods, 374, 380
Flocculate diameter, 126
Flocculation
  orthokinetic, 101, 102, 106, 107, 112, 113
  perikinetic, 101, 102, 106, 107, 112, 113, 125, 177, 179
Flow
  Bingham, 10, 47, 53, 213
  butter, 194, 195
  cream, 315, 316
  dilatant, 10, 11
  fluid, 13
  free energy of activation, 116
  laminar, 47
  lotions, 355, 356, 357, 358
  margarine, 194, 195
  molten chocolate, 212, 213, 214, 215
  Newtonian, 8, 9, 10, 65
  non-Newtonian, 9, 10, 54, 86, 376
  penicillin suspensions, 358, 359, 360, 361
  plastic, 10, 64, 86, 376
  plug, 45, 47
  pressurised foams, 365, 366
  pseudoplastic, 10, 11, 53, 64, 86, 102, 108, 328
  thixotropic systems, 12
  turbulent, 45, 47, 49
  vegetable oil, hardened, 194, 195
Fluids
  dynamic studies, 62, 63
  immobilisation, 114, 160, 161
  Newtonian, 36
  non-Newtonian, 36
  viscoelastic, 27, 36
Foams, 365, 366
Food texture
  definition, 372, 373
  geometrical characteristics, 373, 374, 377
  primary parameters, 374, 377
  Szczesniak's classification of attributes, 372, 373, 374, 375, 376, 377
  secondary parameters, 374, 377, 380
  tertiary parameters, 377, 378, 380, 381
  texture profile, amended version, 377, 378
  texture profile, original version, 373
Force, tangential, 1

G

Gelatine gels, 292, 293, 294, 295, 296, 297, 298, 299, 314, 315
  bloom jelly strength, 294, 295, 296
  concentric cylinder viscometer for testing rigidity, 297, 298
  extension tests, 293, 294, 295, 296
  influence of salt, 294
  rigidity, 292, 293, 296, 297, 298, 299
  rupture load, 293, 294, 296
Gels, 67, 68, 69, 288, 289, 290, 291, 292, 293, 294, 295, 296, 297, 298, 299, 300, 301, 302, 303, 304, 308, 309, 310, 311, 312, 313
Gumminess, 373, 374, 376, 377, 383
Gums
  viscosity of solutions, 327, 328

H

Hardness
  biscuits, 227, 228, 229
  cake, 379
  cheese, 232, 233, 234
  foods, 373, 374, 375, 383
  frozen ice cream, 379
  margarine, 190, 191, 192, 193, 197, 198
Hardness index, 208, 209, 210, 211
Hesion, 86, 365, 375
Hesion meter, 37, 85, 86, 87, 365
Hooke's law, 30
Homogeniser efficiency, 151
Homogenisation pressure, influence in particle size, 150, 151
Hydrocolloid
  gels, 288, 289, 290, 291, 292, 293, 294, 295, 296, 297, 298, 299, 300, 301, 302, 303, 304
  solutions, 304, 305, 325, 327, 328, 329
Hydrodynamic interaction
  coefficient, 146
  particles, 131, 134, 135, 136, 143, 145, 146, 147, 152, 153, 154, 157
Hydrostatic head effect, capillary flow, 48, 49

Hysteresis loops, 330, 331, 332, 335, 343, 359, 363, 367

I

Ice cream
frozen, 198, 199, 200, 201, 202, 203, 205, 206
melted, 199, 203, 204, 205, 206, 207
mix, 198, 199, 204, 205, 206
sensory assessment, 378, 379
Immunological agents, 337
Impact testing, 244, 245
Indentation tests, 37, 94, 237
Inertia effects, cone-plate viscometer, 53
Instron test machine
foods, 384
lard, 196
potato, 236, 237, 238, 239
Interaction, potential energy, 121

K

Kelvin-Voigt element, 19, 20, 21, 25, 259, 316
Kinetic energy effect, capillary viscometer, 45, 46, 47, 48

L

Lamé's constants, 7, 8
Lard consistency, 196
L.E.E.—Kramer shear press
gels, 303, 304
meat, 278, 279
Link extension, 105
Link formation, 105, 106, 107, 108, 111
Liquids
viscosity of thin films, 157
viscous, 4
London-van der Waals' constant, 120, 123, 124
Loss compliance, 24, 25, 26
Loss tangent, 24, 77, 352
Lotions, 326, 327, 353, 354, 355, 356, 357, 358, 388, 389

M

Margarine
consistency, 196
creep compliance-time response, 186, 187
extrusion, 192

firmness, 196, 197
flow, 194, 195
hardness, 187
spreadability, 190, 196, 197
Mastication, 376, 377, 378
Masticatory forces, 371, 373, 376, 380, 384
Maxwell element, 19, 21, 22, 25, 240, 316
Meat
chewing sounds for texture assessment, 288
compression, 282, 283
denture tenderometer, 283, 284, 285, 286
dynamic rheological tests, 274, 275
extension tests, 275, 276
gels, 302, 303
penetrometer tests, 279, 280
tenderometer, 276, 277, 286, 287, 288
L.E.E.—Kramer shear press, 278, 279
texture meter, 279
Warner-Bratzler tests, 277, 278, 279
Warner-Bratzler, modified, 280, 281, 282
Mechanical analogues
dough, 259, 260
ice cream, frozen, 203
ice cream, melted, 206
ice cream, mix, 206
viscoelasticity, linear, 18, 19, 20
viscoelasticity, non-linear, 20, 21
Micelle formation, 159, 160
Milk
age thickening, 314, 315
casein micelles, 306, 307, 308
colloidal phosphate removal, 305, 306
gels, 308, 309, 310, 311, 312, 313
viscosity, 305, 306, 307, 308, 314, 315
Modulus
complex shear, 23, 24
compression, 6, 8
dynamic, 22, 25, 61
elastic, 7, 8, 14, 105, 175, 178, 179
loss, 23, 81
relaxation, 22
rigidity, 6, 67, 68, 69, 292, 293, 297, 298, 308
shear, 6, 8
storage, 23, 76, 81
Young's, 7, 77, 79, 80, 239, 240

O

Ointment bases, 330, 331, 332, 333, 334, 335, 336
Ointments, 326, 327, 362, 363, 364, 365, 388, 389, 390

P

Packing geometry
  deformed particles, 155, 177, 178
  spheres, 126, 135, 136, 154, 156, 175
Panimeter, 37, 90, 91, 92, 215, 218, 219, 220
Parallel plate plastometer, 59, 60
  Diene and Klemm's theory, 60
  Scott's theory, 60
Parallel plate viscoelastometer, 69, 70 71, 72, 185, 186, 199, 215
Particle deformation, 130, 154, 162
Particle rotation
  Brownian motion, 155
  shear, 155
Particle settling, 154
Particle shape, 128, 154, 155, 156, 157, 374, 377
Particle size and
  dispersion viscosity, 127, 128, 141, 142, 143, 144, 145, 146, 147, 148, 149, 150, 151
  distribution, 128, 151, 152, 153, 154, 374, 377
  mean, 143
Peas, viscoelastic properties, 236, 237, 246, 247, 248
Pectin gels, 289, 290, 291, 292
Penetrometers
  cone, 37, 43, 83, 84, 85, 190, 191, 196, 197, 208, 211, 215, 223, 301, 325, 327, 364, 365
  needle, 43, 83, 197, 279, 280
  rod, 37, 84
  sphere, 37, 83, 85, 232, 233
Penicillin suspensions, 326, 327, 358, 359, 360, 361
Petrolatum
  structure, 333, 334
  white, 331, 332, 333, 334, 335, 336, 337
Pharmaceutical creams, 326, 327, 358, 362, 365, 388, 389, 390
Pharmaceutical pastes, 326, 327
Pharmaceutical semi-solids, 324, 325

Phase lag, 23, 62
Plastic flow
  Casson theory, 108, 109, 110, 111, 212
  Cross theory, 111, 112, 113, 114, 115
  D.L.V.O. theory, 119, 120, 121, 122, 123
  kinetics of particle aggregation, 112, 113, 114, 115
  rate process theory, 115, 116, 117, 118, 119
Poisson's ratio, 5, 8
Potatoes
  bulk compression, 243, 244
  creep compliance, 240, 241, 242
  dynamic testing, 246, 247, 248
  force-penetration, 286, 287
  Instron tests, 236, 237, 238, 239
  stress relaxation, 241, 242, 243
Potential energy
  attraction, 120, 121, 124
  interaction, 121, 122, 123
  primary minimum, 122, 125
  repulsion, 119, 120, 121
  secondary minimum, 122
Power laws, 98, 99, 100, 101, 179, 180, 310, 312, 313, 315, 316, 327
Pressure, hydrostatic, 6, 7
Pressurised foams, 365, 366, 367
Proportionality factor, 114
Pseudoplasticity
  Gillespie theory, 106, 107, 108
  Goodeve theory, 104, 105, 106
  Williamson theory, 102, 103, 104

R

Relaxation time, 21, 22, 116
Repulsion energy, 119, 120, 121
Retardation spectrum, 17, 25, 187, 188
Retardation time, 14, 15, 16, 17, 18, 25
Reynold's number, 47
Rheological measurements, general considerations, 33, 34, 35
Rotary cutter, 37, 93, 94, 215, 218, 219, 220

S

St. Venant body, 259, 260
Saliva, 384
Scoring procedures in sensory assessment, 381, 382

Secondary minimum, 122, 177
Sectilometer, 37, 88, 196, 197, 233, 235
Semi-solids, 36, 37, 85, 87
Sensory evaluation
  biscuit hardness, 229
  correlation with instrumental evaluations, 390
  cosmetics, 378, 379, 381, 382
  foods, 250, 251, 372, 373, 374, 375, 376, 377, 378, 379, 380, 381, 382, 383, 384, 385, 386, 387, 388
  hydrocolloid solutions, 304, 305
  meat tenderness, 279, 280, 288
  natural cheese consistency, 230, 231, 232
  pharmaceuticals, 378, 379, 381, 382, 388, 389, 390
  processed cheese consistency, 234, 235
  shear conditions, 372, 385, 386, 387, 388, 389, 390
  statistical methods, 390, 391
Shape factor, 62
Shear
  modulus, 6, 232
  rate, 8, 45, 48, 59, 103, 110
  simple, 2, 3, 4
Slippage
  coaxial cylinder viscometer, 49, 50
  interfaces, 130, 137, 142
Solids
  bending vibrations, 79, 80, 81
  compression, 72, 73, 74
  creep compliance-time studies, 69, 70, 71, 72
  dynamic studies, 62, 63, 76, 77, 78
  elastic, 8
  extension, 74, 75
  F.I.R.A.—N.I.R.D. extruder, 37, 88, 89, 90
  indentation, 37, 94
  large deformations, 74
  panimeter, 37, 90, 91, 92
  penetrometers, 37, 43, 83, 84, 85
  rotary cutter, 37, 93, 94
  sectilometer, 37, 88
  torsion, 37, 75, 76
  torsional vibrations, 37, 76, 77, 80
Spectra
  relaxation, 22, 25
  retardation, 17, 18, 25, 177

Spreadability, 190, 196, 197, 198, 362, 371, 377, 378, 388
Stabilising agents, 129, 140, 141
Staling
  bread, 222, 223, 224
  cake, 216, 217, 218, 224, 225
Starch gels, 301, 302
Storage compliance, 24
Stickiness
  foods, 374, 375, 381
  lotions, 365, 381
  pharmaceutical creams, 365, 379, 381
Stimulus, and sensory response, 386, 387, 388
Strain
  complex, 23
  components in second order theory of elasticity, 29
  compressional, 74
  creep compliance-time studies, 72
  rigidity tests, 68
  sinusoidal deformation studies, 24
  shear, 6, 13, 57
  sinusoidal variation, 23
  tensor, 3, 5
  torsion tests, 75
Stress
  complex, 23
  components in second order theory of elasticity, 29
  compressional, 73
  creep compliance-time studies, 72
  development at constant shear rate, 58
  hydrostatic, 7
  normal, 2, 7, 27
  relaxation, 21, 22, 66, 67, 238, 240, 241, 242, 259, 260, 261, 262, 263, 272, 312, 353, 354
  shear, 6, 7, 9, 13, 45, 57, 103, 110, 116
  sinusoidal, 23
  tangential, 2
  tensor, 1, 15
  torsion tests, 75
Surface pressure gradients, 138
Suspending agents, 327, 328, 329
Swelling factor, 114, 174, 175

T

Temperature rise, cone-plate viscometer, 53

Tenderometer
  bite, 215, 276, 277
  denture, 283, 284, 285, 286, 383, 384,
  385
Tensile strength
  cake, 224
  gels, 69
Tension, simple, 5
Tensor
  strain, 3, 5
  stress, 1, 5
Texture meter
  biscuits, 227, 228, 229
  meat, 279
Thickness of adsorbed emulsifier films,
  121
Thixotropy, 11, 12, 104, 105, 106, 330,
  331, 332, 378
Torsion, of solids, 75, 76, 185

V

Vaccines, 361, 362, 363, 364, 365, 366,
  367
Vaselines, 324, 325, 363, 364, 365
Veegum suspensions, 328, 329
Velocity gradient, 9
Vibrations
  bending, 186
  free torsional, 76, 77, 186
Viscoelasticity
  apples, 236, 237, 241, 242, 243, 246,
  247, 248
  cake, 216, 217, 218
  coaxial cylinder viscometer for
  studying, 56, 57
  dispersed systems, 127
  emulsions containing fatty alcohols,
  351, 352, 353
  fluids, 66
  frozen ice cream, 199, 200, 201, 202,
  203
  ice cream mix, 204, 205, 206
  melted ice cream, 204, 205, 206, 207
  plant tissues, 246
  potatoes, 240, 241, 242, 243, 246
Viscometers
  capillary, 35, 36, 38, 41, 44, 45, 46, 47,
  48, 49, 54, 329, 359, 366
  coaxial cylinder, 35, 36, 39, 40, 44, 45,

49, 50, 54, 58, 59, 61, 62, 63, 64, 65, 66,
  206, 212, 213, 297, 298, 315, 316, 324,
  325, 326, 327, 338, 353, 354, 355, 362,
  365, 366, 367, 385, 386, 387
  cone-plate, 35, 36, 40, 44, 45, 52, 53,
  54, 55, 60, 63, 64, 66, 67, 212, 252,
  253, 254, 324, 326, 330, 331, 332, 342
  extrusion capillary, 36, 42, 196
  falling sphere, 35, 36, 44, 45
  orifice, 43, 83, 194
  parallel plate, 55
  rolling sphere, 43, 44, 45, 50, 51, 52
  rotating spindle, 81, 82, 83, 338, 355
  variable pressure capillary, 36, 42,
  192, 194
  vibrating plate, 186
  vibrating reed, 35, 36, 43, 55
  Weissenberg rheogoniometer, 27, 40,
  61, 63, 351
Viscosity
  acacia stabilised emulsions, 341, 342,
  343, 344, 345
  apparent, 10, 11, 44, 98, 105
  cheese, 230, 231, 232
  complex, 24
  continuous phase, 124
  cosmetics, 324, 325, 326, 378
  dilatant, 11
  dynamic, 22, 24, 61
  foods, 373, 374, 377, 379, 380, 383, 384
  hydrocolloid solutions, 304, 305
  hydrocolloid stabilised emulsions, 341
  ice cream mix, 198, 199
  lotions, 355, 356, 357, 358, 378
  maximum, 58
  minimum, 58
  Newtonian, 9, 14, 44, 50, 51, 175, 178,
  179, 200
  pharmaceuticals, 324, 325, 326
  plastic, 11, 82, 87, 116, 198
  power laws, 98, 99, 100, 101, 179, 180,
  315, 316
  pressurised foams, 365, 366, 367
  pseudoplastic, 11
  relative, 114, 130
  sensory assessment, and 385, 386, 387
  solids, 14, 74
  specific increase in, 130
  steady state, 12
  vaccines, 361

white petrolatum, 330, 331, 332, 333, 334

Visible speech analysis, and food texture, 251, 252

W

Wall effect
  capillary flow, 38, 45
  cone-plate viscometer, 53
Warner-Bratzler equipment
  meat, 277, 278
  modified version, 280, 281, 282

Waxes, 334
Weissenberg effect, 27, 28

Y

Yield value, 4, 10, 11, 45, 82, 83, 84, 85, 104, 195, 315, 328, 334, 336, 337, 346, 347, 348, 349, 350, 358

Z

Zeta potential, 120, 123